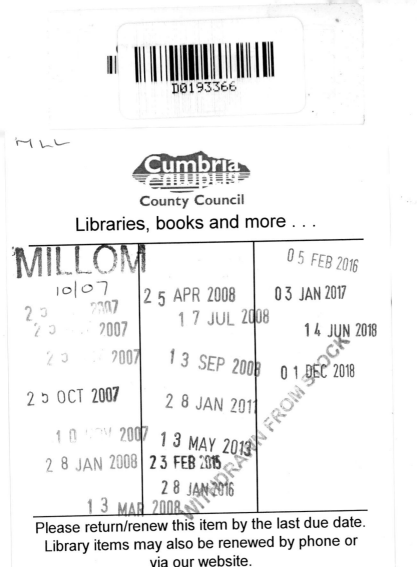

Cumbria
County Council

Libraries, books and more . . .

Please return/renew this item by the last due date.
Library items may also be renewed by phone or
via our website.
www.cumbria.gov.uk/libraries

Cumbria Libraries
CLIC
Interactive Catalogue

Ask for a CLIC password

Collins

Acknowledgements and sources

This book is the result of an unusual collaboration between an author and an illustrator in different countries, united by their common fascination with the sky. International cooperation of another kind is apparent in the photographic illustrations in this book, which are the result of the efforts of professional and amateur astronomers from many nations. Such awesome images rank with the most enduring works of art produced by humankind. It is fitting that the observatories concerned have made their results so readily available. The urge to study the sky transcends national boundaries, and so it should. The skies are open to us all.

The authors would like to thank those whose photographs are used in this book, in particular the staff of the National Optical Astronomy Observatory, Tucson, Arizona. Pictures are credited individually where they appear. The photographic Moon maps are courtesy Mark Rosiek and the U.S. Geological Survey, Flagstaff, Arizona. Dimensions of formations are from the list by Ewen Whitaker in *Norton's Star Atlas*, 19th and 20th editions.

The following publications have been consulted extensively during the preparation of this book: *The Hipparcos and Tycho Catalogues* by M. Perryman *et al.* (European Space Agency, Noordwijk); *The Millennium Star Atlas* by R.W. Sinnott and M. Perryman (European Space Agency; and Sky Publishing Corp., Cambridge, Mass.); *Sky Atlas 2000.0* (2nd edition) by Wil Tirion and R.W. Sinnott (Sky Publishing Corp.; and Cambridge University Press, Cambridge, England); *Sky Catalogue 2000.0*, Vols 1 and 2 (Sky Publishing Corp. and Cambridge University Press); *The Bright Star Catalogue*, 4th edition, by Dorrit Hoffleit (Yale University Observatory); the Washington Double Star Catalogue (online edition); *Burnham's Celestial Handbook* by Robert Burnham, Jr (Dover, New York; Constable, London); *Sky & Telescope* magazine (Cambridge, Mass.). Further details of the origin and mythology of the constellations may be found in *Star Tales* by Ian Ridpath (Lutterworth, Cambridge). For more on the meanings of individual star names see *A Dictionary of Modern Star Names* by Paul Kunitzsch and Tim Smart (Sky Publishing, Cambridge, Mass.).

We thank Peter Gill for commenting on the proofs of this fourth edition, and we remain grateful for the continuing support of our editors at HarperCollins.

I.R., W.T.
http://www.ianridpath.com/
http://www.wil-tirion.com/

CONTENTS

SECTION I

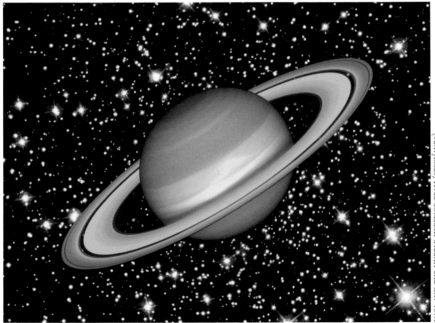

INTRODUCTION

The night sky is one of the most beautiful sights in nature. Yet many people remain lost among the jostling crowd of stars, and are baffled by the progressively changing appearance of the sky from hour to hour and from season to season. The charts and descriptions in this book will guide you to the most splendid celestial sights, many of them within the range of simple optical equipment such as binoculars, and all accessible with an average-sized telescope of the type used by amateur astronomers.

It must be emphasized that you do not need a telescope to take up stargazing. Use the charts in this book to find your way among the stars first with your own eyes, and then with the aid of binoculars, which bring the stars more readily into view. Binoculars are a worthwhile investment, being relatively cheap, easy to carry and useful for many purposes other than stargazing.

Stars and planets

In the night sky, stars appear to the naked eye as spiky, twinkling lights. Those stars near the horizon seem to flash and change colour. The twinkling and flashing effects are due not to the stars themselves but to the Earth's atmosphere: turbulent air currents cause the stars' light to dance around. The steadiness of the atmosphere is referred to as the *seeing*. Steady air means good seeing. The spikiness of star images is due to optical effects in the observer's eye. In reality, stars are spheres of gas similar to our own Sun, emitting their own heat and light.

Stars come in various sizes, from giants to dwarfs, and in a range of colours according to their temperature. At first glance all stars appear white, but more careful inspection reveals that certain ones are somewhat orange, notably Betelgeuse, Antares, Aldebaran and Arcturus, while others such as Rigel, Spica and Vega have a bluish tinge. Binoculars bring out the colours more readily than the naked eye does. Section II of this book, starting on page 267, explains more fully the different types of star that exist.

By contrast, planets are cold bodies that shine by reflecting the Sun's light. They too are described in more detail in Section II, from page 304 onwards. The planets are constantly on the move as they orbit the Sun. Four of them can be easily seen with the naked eye: Venus, Mars, Jupiter and Saturn. Venus, the brightest of all, appears as a dazzling object in the evening or morning sky. Charts showing the positions of Mars, Jupiter and Saturn for a 5-year period can be found on the HarperCollins website: www.collins.co.uk/starsandplanets

About 2000 stars are visible to the naked eye on a clear, dark night, but you will not need to learn them all. Start by identifying the brightest stars and major constellations, and use these as signposts to the fainter, less prominent stars and constellations. Once you know the main features of the night sky, you will never again be lost among the stars.

Constellations

The sky is divided into 88 sections known as constellations which astronomers use as a convenient way of locating and naming celestial objects. Most of the stars in a constellation have no real connection with one another at all; they may lie at vastly differing distances from Earth, and form a pattern simply by chance. Incidentally, when astronomers talk of an object being 'in' a given constellation they mean that it lies in that particular area of sky.

Some constellations are easier to recognize than others, such as the magnificent Orion or the distinctive Cassiopeia and Crux. Others are faint and obscure, such as Lynx

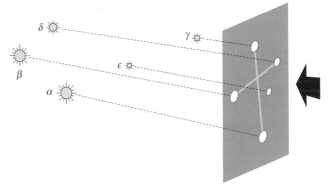

Constellations: Stars in a constellation are usually unrelated to one another. Above, the stars of Crux, the Southern Cross, are shown as they appear from Earth, left, and as they actually lie in space, right. (Wil Tirion)

and Telescopium. Whether large or small, bright or faint, each constellation is given a separate chart and description in this book.

The main constellations were devised at the dawn of history by Middle Eastern peoples who fancied that they could see a likeness to certain fabled creatures and mythological heroes among the stars. Of particular importance were the 12 constellations of the zodiac, through which the Sun passes during its yearly path around the heavens. However, it should be realized that the astrological 'signs' of the zodiac are not the same as the modern astronomical constellations, even though they share the same names.

Our modern system of constellations derives from a catalogue of 48 compiled by the Greek astronomer Ptolemy in AD 150. This list was expanded by navigators and celestial map-makers, notably the Dutchmen Pieter Dirks-zoon Keyser (c. 1540–96) and Frederick de Houtman (1571–1627), the Pole Johannes Hevelius (see page 166) and the Frenchman Nicolas Louis de Lacaille (see page 216).

Keyser and de Houtman introduced 12 new constellations, and Lacaille 14, in parts of the southern sky not visible from Medi-

terranean regions; Hevelius and others invented constellations to fill in the gaps between the figures recognized by the Greeks. The whole process sounds rather arbitrary, and indeed it was. A number of the newly devised patterns fell into disuse, leaving a total of 88 constellations that were officially adopted by the International Astronomical Union, astronomy's governing body, in 1922 (see the table on pages 8–9).

As well as the officially recognized constellations, you can find other patterns among the stars called *asterisms*. An asterism can be composed of stars belonging to one or more constellations. Well-known examples are the Plough or Big Dipper (part of Ursa Major), the Square of Pegasus, the Sickle of Leo and the Teapot of Sagittarius.

Star names

The main stars in each constellation are labelled with a letter of the Greek alphabet, the brightest star usually (but not always!) being termed α (alpha). Notable exceptions in which the stars marked β (beta) are in fact the brightest include the constellations Orion and Gemini. The entire Greek alphabet is given in the table on page 10.

Particularly confusing are the southern constellations Vela and Puppis, which were once joined with Carina to make the extensive figure of Argo Navis, the ship of the

THE 88 CONSTELLATIONS

Name	Genitive	Abbrevn.	Area (square degs.)	Order of size	Origin*
Andromeda	Andromedae	And	722	19	1
Antlia	Antliae	Ant	239	62	6
Apus	Apodis	Aps	206	67	3
Aquarius	Aquarii	Aqr	980	10	1
Aquila	Aquilae	Aql	652	22	1
Ara	Arae	Ara	237	63	1
Aries	Arietis	Ari	441	39	1
Auriga	Aurigae	Aur	657	21	1
Boötes	Boötis	Boo	907	13	1
Caelum	Caeli	Cae	125	81	6
Camelopardalis	Camelopardalis	Cam	757	18	4
Cancer	Cancri	Cnc	506	31	1
Canes Venatici	Canum Venaticorum	CVn	465	38	5
Canis Major	Canis Majoris	CMa	380	43	1
Canis Minor	Canis Minoris	CMi	183	71	1
Capricornus	Capricorni	Cap	414	40	1
Carina	Carinae	Car	494	34	6
Cassiopeia	Cassiopeiae	Cas	598	25	1
Centaurus	Centauri	Cen	1060	9	1
Cepheus	Cephei	Cep	588	27	1
Cetus	Ceti	Cet	1231	4	1
Chamaeleon	Chamaeleontis	Cha	132	79	3
Circinus	Circini	Cir	93	85	6
Columba	Columbae	Col	270	54	4
Coma Berenices	Comae Berenices	Com	386	42	2
Corona Australis	Coronae Australis	CrA	128	80	1
Corona Borealis	Coronae Borealis	CrB	179	73	1
Corvus	Corvi	Crv	184	70	1
Crater	Crateris	Crt	282	53	1
Crux	Crucis	Cru	68	88	4
Cygnus	Cygni	Cyg	804	16	1
Delphinus	Delphini	Del	189	69	1
Dorado	Doradus	Dor	179	72	3
Draco	Draconis	Dra	1083	8	1
Equuleus	Equulei	Equ	72	87	1
Eridanus	Eridani	Eri	1138	6	1
Fornax	Fornacis	For	398	41	6
Gemini	Geminorum	Gem	514	30	1
Grus	Gruis	Gru	366	45	3
Hercules	Herculis	Her	1225	5	1
Horologium	Horologii	Hor	249	58	6
Hydra	Hydrae	Hya	1303	1	1
Hydrus	Hydri	Hyi	243	61	3
Indus	Indi	Ind	294	49	3
Lacerta	Lacertae	Lac	201	68	5
Leo	Leonis	Leo	947	12	1
Leo Minor	Leonis Minoris	LMi	232	64	5
Lepus	Leporis	Lep	290	51	1

Name	Genitive	Abbrevn.	Area (square degs.)	Order of size	Origin*
Libra	Librae	Lib	538	29	1
Lupus	Lupi	Lup	334	46	1
Lynx	Lyncis	Lyn	545	28	5
Lyra	Lyrae	Lyr	286	52	1
Mensa	Mensae	Men	153	75	6
Microscopium	Microscopii	Mic	210	66	6
Monoceros	Monocerotis	Mon	482	35	4
Musca	Muscae	Mus	138	77	3
Norma	Normae	Nor	165	74	6
Octans	Octantis	Oct	291	50	6
Ophiuchus	Ophiuchi	Oph	948	11	1
Orion	Orionis	Ori	594	26	1
Pavo	Pavonis	Pav	378	44	3
Pegasus	Pegasi	Peg	1121	7	1
Perseus	Persei	Per	615	24	1
Phoenix	Phoenicis	Phe	469	37	3
Pictor	Pictoris	Pic	247	59	6
Pisces	Piscium	Psc	889	14	1
Piscis Austrinus	Piscis Austrini	PsA	245	60	1
Puppis	Puppis	Pup	673	20	6
Pyxis	Pyxidis	Pyx	221	65	6
Reticulum	Reticuli	Ret	114	82	6
Sagitta	Sagittae	Sge	80	86	1
Sagittarius	Sagittarii	Sgr	867	15	1
Scorpius	Scorpii	Sco	497	33	1
Sculptor	Sculptoris	Scl	475	36	6
Scutum	Scuti	Sct	109	84	5
Serpens	Serpentis	Ser	637	23	1
Sextans	Sextantis	Sex	314	47	5
Taurus	Tauri	Tau	797	17	1
Telescopium	Telescopii	Tel	252	57	6
Triangulum	Trianguli	Tri	132	78	1
Triangulum Australe	Trianguli Australis	TrA	110	83	3
Tucana	Tucanae	Tuc	295	48	3
Ursa Major	Ursae Majoris	UMa	1280	3	1
Ursa Minor	Ursae Minoris	UMi	256	56	1
Vela	Velorum	Vel	500	32	6
Virgo	Virginis	Vir	1294	2	1
Volans	Volantis	Vol	141	76	3
Vulpecula	Vulpeculae	Vul	268	55	5

* **Origin:**

1   One of the original 48 Greek constellations listed by Ptolemy. The Greek figure of Argo Navis has since been divided into Carina, Puppis and Vela.
2   Considered by the Greeks as part of Leo; made separate by Gerardus Mercator in 1551.
3   The 12 southern constellations of Pieter Dirkszoon Keyser and Frederick de Houtman, c. 1600.
4   Four constellations added by Petrus Plancius.
5   Seven constellations of Johannes Hevelius.
6   The 14 southern constellations of Nicolas Louis de Lacaille, who also divided the Greeks' Argo Navis into Carina, Puppis and Vela.

Argonauts. As a result of Argo's subsequent trisection, neither Vela nor Puppis possesses stars labelled α or β, and there are gaps in the sequence of Greek letters in Carina as well.

The system of labelling stars with Greek letters was introduced by Johann Bayer (see page 246), so these designations are often known as Bayer letters. The genitive (possessive) case of the constellation's name is always used when referring to a star within it; hence Canis Major, for instance, becomes Canis Majoris, and the name α Canis Majoris means 'the star α in Canis Major'. All constellation names have standard three-letter abbreviations; for instance, the abbreviated form of Canis Major is CMa.

In heavily populated constellations, where Greek letters ran out, fainter stars were assigned Roman letters, both lower-case and capital, such as l Carinae, P Cygni and L Puppis. An additional system of identifying stars is that of Flamsteed numbers, originating from their order in a star catalogue drawn up by England's first Astronomer Royal, John Flamsteed (1646–1719). Examples are 61 Cygni and 70 Ophiuchi (see also page 178).

Before 1930, there were no officially recognized constellation boundaries; some constellations overlapped, and some stars were shared between constellations. In that year, the International Astronomical Union published definitive boundaries for all constellations. In the process, certain stars allocated by the Bayer and Flamsteed systems to one constellation found themselves transferred to a neighbour, leading to gaps in the sequence of letters and numbers.

Prominent stars also have proper names by which they are commonly known. For example, α Canis Majoris, the brightest star in the sky, is better known as Sirius. Stars' proper names originate from several sources. Some, such as Sirius, Castor and Pollux, date back to ancient Greek times. Many others, such as Aldebaran, are of Arabic origin. Still others were added more recently by European astronomers who borrowed Arabic words in

THE GREEK ALPHABET			
α	alpha	ξ	xi
β	beta	ο	omicron
γ	gamma	π	pi
δ	delta		
ε	epsilon	ρ	rho
ζ	zeta	σ	sigma
η	eta	τ	tau
θ or ϑ	theta	υ	upsilon
ι	iota		
κ	kappa	φ or ϕ	phi
λ	lambda	χ	chi
μ	mu	ψ	psi
ν	nu	ω	omega

corrupted form; an example is Betelgeuse, which in its current form is meaningless in Arabic. To add to the confusion, spelling of names can vary from list to list, and some stars have more than one proper name. For the maps in this book we have adopted the names used by the *Millennium Star Atlas* (European Space Agency and Sky Publishing Corp., 1997).

Star clusters, nebulae and galaxies have a different system of identification. The most prominent of them are given numbers prefixed by the letter M from a catalogue compiled in the late 18th century by the French astronomer Charles Messier (1730–1817). For example, M1 is the Crab Nebula and M31 the Andromeda Galaxy.

Messier's catalogue contained 103 objects. A few more were added later by other astronomers, bringing the total to 110. A far more comprehensive listing, containing many thousands of objects, is the *New General Catalogue* (NGC) compiled by J.L.E. Dreyer (1852–1926), with two supplements called the *Index Catalogues* (IC).

Messier numbers and NGC/IC numbers remain in use by astronomers, and both systems are used in this book. On the charts, such objects are labelled with their Messier number if they have one, or otherwise by their NGC number (without the 'NGC' prefix) or IC number (prefixed 'I').

Star brightness

Stars appear of different brightnesses in the sky, for two reasons. Firstly, they give out different amounts of light. But also, and just as importantly, they lie at vastly differing distances. Hence, a modest star that is quite close to us can appear brighter than a tremendously powerful star that is a long way away.

Astronomers call a star's brightness its *magnitude*. The magnitude scale was introduced by the Greek astronomer Hipparchus in 129 BC. Hipparchus divided the naked-eye stars into six classes of brightness, from 1st magnitude (the brightest stars) to 6th magnitude (the faintest visible to the naked eye). In his day there was no means of measuring star brightness precisely, so this rough classification sufficed. But with the coming of technology it became possible to measure a star's brightness to a fraction of a magnitude.

In 1856 the English astronomer Norman Pogson (1829–91) put the magnitude scale on a precise mathematical footing by defining a star of magnitude 1 as being exactly 100 times brighter than a star of magnitude 6. Since, on this scale, a difference of five magnitudes corresponds to a brightness difference of 100 times, a step of one magnitude is equal to a brightness difference of just over 2.5 times (the fifth root of 100).

Objects more than 250 times brighter than 6th magnitude are given negative (minus) magnitudes. For example, Sirius, the brightest star in the sky, is of magnitude −1.44. At the other end of the scale, stars fainter than magnitude 6 are given progressively larger positive magnitudes. The faintest objects detected by telescopes on Earth are around magnitude 27.

Any object of magnitude 1.49 or brighter is said to be of first magnitude; objects from 1.50 to 2.49 are termed second magnitude; and so on. The magnitude system may sound confusing at first, but it works well in practice and has the advantage that it can be extended indefinitely in both directions, to the very bright and the very faint.

When used without further qualification, the term 'magnitude' refers to how bright a star appears to us in the sky; strictly, this is the star's *apparent magnitude*. But because the distance of a star affects how bright it appears, the apparent magnitude bears little relation to its actual light output, or *absolute magnitude*.

A star's absolute magnitude is defined as the brightness it would appear to have if it were at a standard distance from us of 10 parsecs (32.6 light years). The origin of the parsec is explained on page 13. The table below shows the apparent and absolute magnitudes of the ten brightest stars visible

THE TEN BRIGHTEST STARS AS SEEN FROM EARTH				
Star name	Constellation	Apparent magnitude	Absolute magnitude	Distance (l.y.)
Sirius	Canis Major	−1.44	+1.45	8.60
Canopus	Carina	−0.62	−5.53	313
Rigil Kentaurus	Centaurus	−0.28	+4.07	4.39
Arcturus	Boötes	−0.05	−0.31	36.7
Vega	Lyra	+0.03	+0.58	25.3
Capella	Auriga	+0.08	−0.48	42.2
Rigel	Orion	+0.18	−6.69	773
Procyon	Canis Minor	+0.40	+2.68	11.4
Achernar	Eridanus	+0.45	−2.77	144
Betelgeuse	Orion	+0.45	−5.14	427

in the night sky. Astronomers calculate the absolute magnitude from knowledge of the star's nature and its distance.

Absolute magnitude is a good way of comparing the intrinsic brightness of stars. For instance, our daytime star the Sun has an apparent magnitude of −26.7, but an absolute magnitude of 4.8 (when no sign is given the magnitude is understood to be positive). Deneb (α Cygni) has an apparent magnitude of 1.3, but an absolute magnitude of −8.7. From comparison of these absolute magnitudes we deduce that Deneb gives out over 250,000 times as much light as the Sun and hence is one of the most luminous stars known, even though there is nothing at first sight to mark it out as extraordinary.

A number of stars actually vary in their light output, for various reasons, and are a favourite subject for study by amateur astronomers. The nature of such so-called variable stars is discussed on pages 284–287.

Star distances

In the Universe, distances are so huge that astronomers have abandoned the puny kilometre (km) and have invented their own units. Most familiar of these is the *light year* (l.y.), the distance that a beam of light travels in one year. Light moves at the fastest known speed in the Universe, 299,792.5 km per second. A light year is equivalent to 9.46 million million km.

On average, stars are several light years apart. For instance, the closest star to the Sun, Proxima Centauri (actually a member of the α Centauri triple system), is 4.2 light years away. Sirius is 8.6 l.y. away and Deneb over 3000 l.y. away.

The distance of the nearest stars can be found directly in the following way. A star's position is measured accurately when the Earth is on one side of the Sun, and then remeasured six months later when the Earth has moved around its orbit to the other side of the Sun. When viewed from two widely differing points in space in this way, a nearby star will appear to have shifted slightly in position with respect to more distant stars (see diagram on the facing page).

This effect is known as parallax, and applies to any object viewed from two vantage points against a fixed background, such as a tree against the horizon. A star's parallax shift is so small that under normal circumstances it is unnoticeable – in the case of Proxima Centauri, which has the greatest parallax shift of any star, the amount is about the same as the width of a small coin seen at a distance of 2 km. Once the star's parallax shift has been measured, a simple calculation reveals how far away it is.

THE TEN CLOSEST STARS TO EARTH

Star name	Constellation	Apparent magnitude	Absolute magnitude	Distance (l.y.)
Proxima Centauri	Centaurus	11.05	15.45	4.22
α Centauri A	Centaurus	−0.01	4.34	4.39
α Centauri B	Centaurus	1.35	5.70	4.39
Barnard's Star	Ophiuchus	9.54	13.24	5.94
Wolf 359	Leo	13.54	16.57	7.8
Lalande 21185	Ursa Major	7.49	10.46	8.31
Sirius A	Canis Major	−1.44	1.45	8.60
Sirius B	Canis Major	8.44	11.33	8.60
UV Ceti A	Cetus	12.52	15.42	8.7
UV Ceti B	Cetus	12.96	15.82	8.7

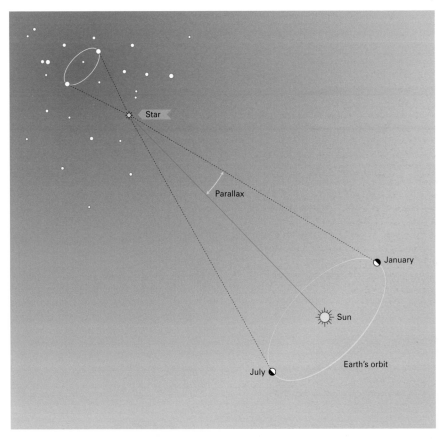

An object close enough to us to show a parallax shift of 1″ (one second of arc) would, in the jargon of astronomers, be said to lie at a distance of one *parsec*, equivalent to 3.26 light years. In practice, no star is this close; the parallax of Proxima Centauri is 0″.77. Astronomers frequently use parsecs in preference to light years because of the ease of converting parallax into distance: a star's distance in parsecs is simply the inverse of its parallax in seconds of arc. For example, a star 2 parsecs away has a parallax of 0″.5, 4 parsecs away its parallax is 0″.25, and so on.

The farther away a star is, the smaller its parallax. Beyond about 50 light years a star's parallax becomes too small to be measured

Parallax: As the Earth moves around its orbit, so a nearby star appears to change in position against the celestial background. The shift in position is known as the star's parallax. The nearer the star is to us, the greater its parallax. Here, the amount of parallax is exaggerated for clarity. (Wil Tirion)

accurately by telescopes on Earth. Before the launch of the European Space Agency's astrometry satellite Hipparcos in 1989, astronomers had been able to establish reliable parallaxes for fewer than 1000 stars from Earth; from space, Hipparcos has increased the number of reliable parallaxes to over 100,000.

For stars that are too distant to have had their parallax measured even by Hipparcos, astronomers first of all estimate the star's absolute magnitude by studying the spectrum of its light. They then compare this estimated absolute magnitude with the observed apparent magnitude to determine the star's distance. The distance obtained in this way is open to considerable error, and the values quoted in various books and catalogues often differ widely as a result.

Distances of stars given in this book are all expressed in light years. Like the star magnitudes, they are taken from *The Hipparcos Catalogue* (European Space Agency, 1997). Most of these distances are accurate to better than 10 per cent, with the uncertainties tending to become greater for the more distant stars.

Star positions

To determine positions of objects in the sky, astronomers use a system of coordinates similar to latitude and longitude on Earth. The celestial equivalent of latitude is called *declination* and the equivalent of longitude is called *right ascension*. Declination is measured in degrees, minutes and seconds (abbreviated °, ′ and ″) of arc from 0° on the celestial equator to 90° at the celestial poles. The celestial poles lie exactly above the Earth's poles, while the celestial equator is the projection onto the sky of the Earth's equator.

Right ascension is measured in hours, minutes and seconds (abbreviated h, m and s), from 0h to 24h. The 0h line of right ascension, the celestial equivalent of the Greenwich meridian, is defined as the point where the Sun crosses the celestial equator on its way into the northern hemisphere each year. Technically, this point is known as the *vernal* (or spring) *equinox*.

Ecliptic: As the Earth moves along its orbit, the Sun is seen in different directions against the star background. The Sun's path against the stars is known as the ecliptic. The constellations that the Sun passes in front of during the year are known as the constellations of the zodiac. Below, the Sun's apparent motion through Gemini, Cancer and Leo is shown. (Wil Tirion)

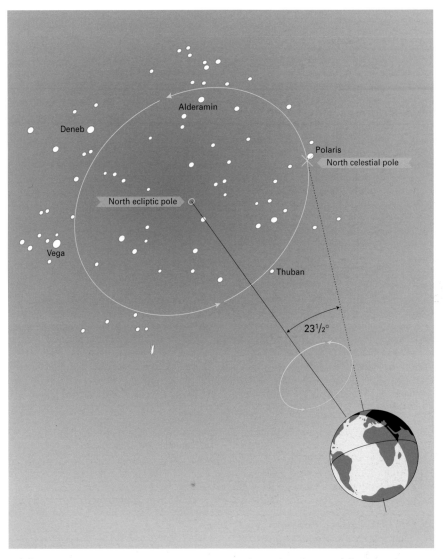

Labels within the image: Alderamin, Deneb, Polaris, North celestial pole, North ecliptic pole, Vega, Thuban, 23½°

Precession: The Earth is very slowly wobbling in space like a tilted spinning top, a movement known as precession. As a consequence, the celestial poles trace out a complete circle on the sky every 26,000 years. Only the path of the north celestial pole is shown here, but the effect applies to the south pole as well. (Wil Tirion)

The Sun's path around the sky each year is known as the *ecliptic*. This path is inclined at 23½° to the celestial equator, because that is the angle at which the Earth's axis is inclined to the vertical. The most northerly and southerly points that the Sun reaches in the sky each year are called the *solstices*, and they lie 23½° north and south of the celestial equator. If the Earth's axis were directly upright with respect to its orbit around the Sun, then the celestial equator and ecliptic would coincide. We would then have no seasons, for the Sun would always remain directly above the Earth's equator.

One additional effect that becomes important over long periods of time is that the Earth is slowly wobbling on its axis, like a spinning top. The axis remains inclined at an angle of 23½°, but the position in the sky to which the north and south poles of the Earth are pointing moves slowly. This wobbling of the Earth in space is termed *precession*.

As a result of precession, the Earth's north and south poles describe a large circle on the sky, taking 26,000 years to return to their starting places (see the diagram on page 15). Hence the positions of the celestial poles are always changing, albeit imperceptibly, as are the two points at which the Sun's path (the ecliptic) cuts the celestial equator.

As an example of the effects caused by precession, Polaris will not always be the Pole Star. Although Polaris currently lies less than 1° from the celestial pole, that is just a matter of chance. In 11,000 years' time the north celestial pole will lie near Vega in the constellation Lyra, having moved through Cepheus and Cygnus in the interim. Similarly, the vernal equinox, which lay in Aries between 1865 BC and 67 BC, now lies in Pisces and in AD 2597 will have reached Aquarius.

The effect of precession means that the coordinates of all celestial objects – the catalogued positions of stars, galaxies, and even constellation boundaries – are continually drifting. Astronomers draw up catalogues and star charts for a standard reference date, or *epoch*, commonly chosen to be the start or middle of a century. The epoch of the star positions in this book is the year 2000. For most general purposes, precession does not introduce a noticeable error until after about 50 years, so the charts in this book will be usable without amendment until halfway through the 21st century.

Proper motions

All the stars visible in the sky are members of a vast wheeling mass of stars called the Galaxy. Those stars visible to the naked eye are among the nearest to us in the Galaxy. More distant stars in the Galaxy crowd together in a hazy band called the Milky Way, which can be seen arching across the sky on dark nights.

The Sun and the other stars are all orbiting the centre of the Galaxy; the Sun takes about 250 million years to complete one orbit. Other stars move at different speeds, like cars in different lanes on a highway. As a result, stars are all very slowly changing their positions relative to one another.

Such stellar movement, termed *proper motion*, is so slight that it is undetectable to the naked eye even over a human lifetime, but it can be measured through telescopes. As with many other aspects of stellar positions, our knowledge of proper motions has been radically improved by the Hipparcos satellite.

If ancient Greek astronomers could be transported 2000 years forward in time to the present day, they would notice little difference in the sky, with the exception of Arcturus, a fast-moving bright star, which has drifted more than two Moon diameters from its position then. Over very long periods of time the proper motions of stars considerably distort the shapes of all constellations. The diagrams on the facing page show some examples of how proper motions will alter some familiar patterns.

An additional long-term effect of stellar motions is to change the apparent magnitudes of stars as they move towards us or away. For example, Sirius will brighten by 20 per cent over the next 60,000 years as its

Proper motions: Three familiar star patterns as they appear today, at left of diagram, and as they will appear in 100,000 years' time, right. The changes in appearance are due to the proper motions of the stars. (Wil Tirion)

URSA MAJOR (The Plough)

In 100,000 years

CYGNUS

In 100,000 years

LEO

In 100,000 years

distance shrinks by 0.8 light years. Then, as it moves away again, it will be superseded as the brightest star in the sky by Vega, which will peak at magnitude −0.8 nearly 300,000 years from now.

Appearance of the sky

Three factors affect the appearance of the sky: the time of night, the time of year and your latitude on Earth. Firstly, let's consider the effect of latitude.

At one of the Earth's poles, latitude 90°, an observer would see the celestial pole directly overhead (at the *zenith*), and as the Earth turned all stars would circle around the celestial pole without rising or setting (see the top diagram on the facing page).

At the other extreme, an observer stationed exactly on the Earth's equator, latitude 0°, would see the celestial equator directly over-head, as shown in the middle diagram opposite. The north and south celestial poles would lie on the north and south horizons respectively, and every part of the sky would be visible at one time or another. All stars would rise in the east and set in the west as the Earth rotated.

For most observers, the real sky appears somewhere between these two extremes: the celestial pole is at some intermediate altitude between horizon and zenith, and the stars closest to it circle around it without setting (they are said to be *circumpolar*) while the rest of the stars rise and set.

The exact angle of the celestial pole above the horizon depends on the observer's latitude. For someone at latitude 50° north, for instance, the north celestial pole is 50° above the northern horizon (bottom diagram opposite). As another example, if you were at latitude 30° south, the south celestial pole would be 30° above the southern horizon. In other words, the altitude of the celestial pole above the horizon is exactly equal to your latitude, a fact long recognized by navigators.

As the Earth turns, completing one 360° rotation every 24 hours, the stars march across the heavens at the rate of 15° per hour. Therefore the appearance of the sky changes with the time of night. An added complication is that the Earth is also orbiting the Sun each year, so the constellations on show change with the seasons.

For example, a constellation such as Orion, splendidly seen in December and January, will be in the daytime sky six months later and hence will then be invisible. The maps on pages 26–73 will help you find out which stars are on view each month of the year, wherever you are on Earth.

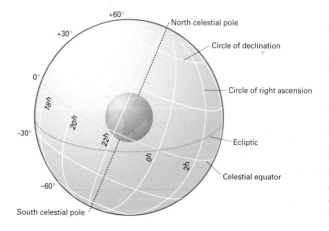

Celestial sphere: To understand celestial coordinates and motions, it helps to think of objects in the sky as lying on a transparent sphere surrounding the Earth, as shown in the diagram on the left. Right ascension and declination can be visualized as circles on this sphere, along with the celestial equator and the ecliptic. (Wil Tirion)

At the north pole

The changing appearance of the sky as seen from different latitudes on Earth.

Left: For an observer at the Earth's pole, only one half of the sky is ever visible, the other half being permanently below the horizon.

Right: At the equator, by contrast, all the sky is visible; as the Earth rotates, stars appear to rise in the east and set in the west.

At the equator

Left: At intermediate latitudes, the situation is between the two extremes. Part of the sky is always above the horizon (the part marked 'Circumpolar'), but an equal part is always below the horizon and hence is invisible. Stars between these two regions rise and set during the night. (Wil Tirion)

At 50° north

THE STAR CHARTS

The hemisphere charts

On the following four pages are charts show-ing the complete northern and southern hemispheres of the sky. As well as the main stars of each hemisphere, the charts depict the hazy band of the Milky Way; the dashed red line is the ecliptic, the Sun's path in the heavens. When planets are visible, they will be found near the ecliptic.

Around the rim of each chart are listed the months of the year, to help you find which constellations are best placed at about 10 p.m. local time each month, or 11 p.m. when daylight-saving time (DST, or summer time) is in operation.

Observers in mid-northern latitudes should take the northern hemisphere chart and turn it so that the month of observation is at the bottom. The chart will show the sky that is visible when you face due south that evening. Rotate the chart 15° anticlockwise for each hour after 10 p.m., and turn it clockwise for each hour before 10 p.m.

Observers in mid-southern latitudes should take the southern hemisphere chart and turn it so that the month of observation is at the bottom. The chart will then show the stars as they appear when you are facing due north. Turn the chart 15° clockwise for each hour after 10 p.m., and 15° anticlock-wise for each hour before.

In all cases above, for 10 p.m. read 11 p.m. when daylight-saving time, DST, is in oper-ation. Up-to-date information on the appli-cation of DST worldwide can be found on the following website:

http://webexhibits.org/daylightsaving/

Spinning Earth: Star trails over the dome of the William Herschel Telescope on La Palma in the Canary Islands, created as the Earth rotated during a time exposure. The bright star near the centre is Polaris. (Nik Szymanek)

The monthly charts

Next comes a series of maps showing the sky as it appears when facing north or south at 10 p.m. (11 p.m. daylight-saving time) in mid-month from various latitudes. The first set of maps is for use in the northern hemi-sphere from latitudes 60° to 10° north; the second set is for use from the equator to 50° south. (They will also be usable for about 10° either side of this range without sig-nificant error.) Curved lines on each map depict the horizon for each latitude. Taking these monthly charts in conjunction with those of the complete celestial hemispheres, you should be able to identify the stars in the sky no matter where you are on Earth.

The constellation charts

The centrepiece of this book, starting on page 74, consists of individual charts of each constellation accompanied by descriptions of the brightest stars and main objects of inter-est. All stars within each constellation down to magnitude 6.0 are shown, with the sizes of the symbols graded in half-magnitude steps, plus the most prominent examples of what are termed *deep-sky objects* (star clusters, nebulae and galaxies). The total number of stars shown is about 5000.

All constellation maps are to the same scale, with the exception of the rambling Hydra, the largest constellation of all, which is drawn to a significantly smaller scale to fit the page. Areas of particular interest in certain constellations, such as the Hyades and Pleiades clusters in Taurus, and the Orion Nebula, are shown to a larger scale in special detail charts.

We hope that the charts and descriptions in this book will serve as trusty companions for many nights of exploration under the stars. Good stargazing!

NORTHERN HEMISPHERE

Overview

NORTHERN HEMISPHERE

Overview

October

November

December

CELESTIAL EQUATOR

ECLIPTIC

CETUS

PISCES

PEGASUS

ARIES

ERIDANUS

LACERTA

ANDROMEDA TRIANGULUM

TAURUS

CASSIOPEIA

PERSEUS

CEPHEUS

AURIGA

CAMELOPARDALIS

ORION

+90°

+80°

URSA MINOR

+70°

LYNX

January

DRACO

+60°

GEMINI

+50°

URSA MAJOR

CANIS
MINOR

+40°

CANES VENATICI

LEO
MINOR

CANCER

MONOCEROS

+30°

February

COMA
BERENICES

+20°

LEO

+10°

HYDRA

VIRGO

Galaxy

Bright nebula

Globular cluster

Open cluster

0°

SEXTANS

March

−10°

April

Magnitudes: >0   0   1   2   3   4   5 var.

SOUTHERN HEMISPHERE

Overview

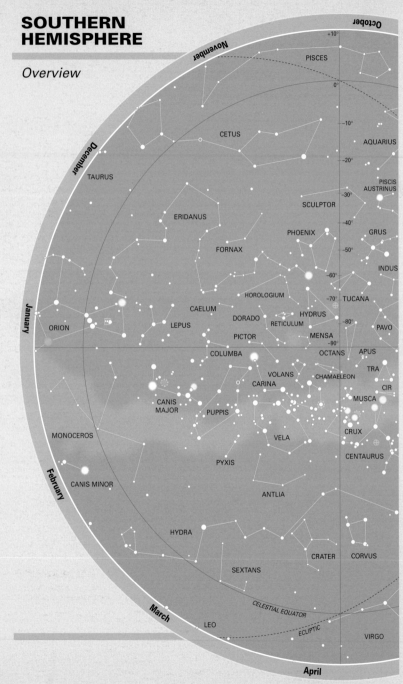

October

November

PISCES

December

CETUS

+10°

0°

AQUARIUS

−10°

−20°

PISCIS AUSTRINUS

TAURUS

ERIDANUS

SCULPTOR

−30°

−40°

GRUS

FORNAX

PHOENIX

−50°

INDUS

January

HOROLOGIUM

−60°

−70°

TUCANA

CAELUM

DORADO

HYDRUS

PAVO

ORION

LEPUS

RETICULUM

−80°

MENSA

−90°

PICTOR

OCTANS

APUS

COLUMBA

VOLANS

CARINA

CHAMAELEON

TRA

CIR

MUSCA

CANIS MAJOR

PUPPIS

VELA

CRUX

MONOCEROS

CENTAURUS

PYXIS

February

CANIS MINOR

ANTLIA

HYDRA

CRATER

CORVUS

SEXTANS

March

LEO

CELESTIAL EQUATOR

ECLIPTIC

VIRGO

April

SOUTHERN HEMISPHERE

Overview

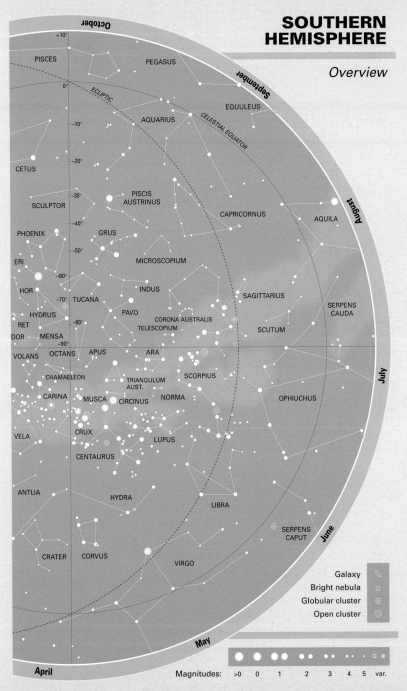

Galaxy
Bright nebula
Globular cluster
Open cluster

Magnitudes: >0 0 1 2 3 4 5 var.

JANUARY
Northern latitudes

Facing North

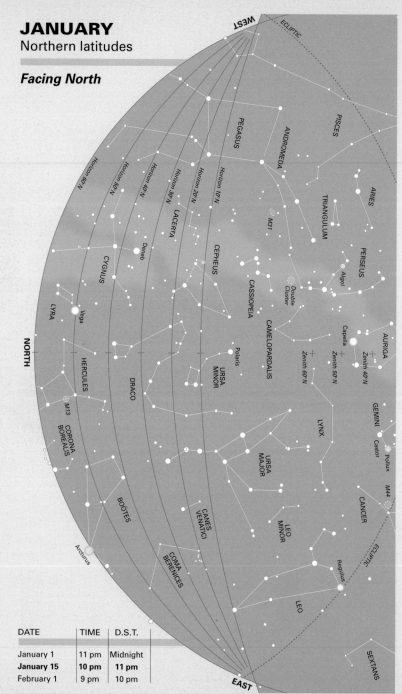

WEST

ECLIPTIC

PISCES

PEGASUS

ANDROMEDA

ARIES

TRIANGULUM

M31

PERSEUS

Algol

Horizon 60° N

Horizon 50° N

Horizon 40° N

Horizon 30° N

Horizon 20° N

Horizon 10° N

LACERTA

CEPHEUS

Double Cluster

Deneb

CYGNUS

CASSIOPEIA

Capella

AURIGA

LYRA

Vega

CAMELOPARDALIS

Zenith 60° N

Zenith 50° N

Zenith 40° N

NORTH

Polaris

URSA MINOR

HERCULES

DRACO

LYNX

GEMINI

Castor

Pollux

M13

CORONA BOREALIS

URSA MAJOR

CANCER

M44

BOÖTES

CANES VENATICI

LEO MINOR

Arcturus

COMA BERENICES

Regulus

ECLIPTIC

LEO

SEXTANS

EAST

DATE	TIME	D.S.T.
January 1	11 pm	Midnight
January 15	**10 pm**	**11 pm**
February 1	9 pm	10 pm

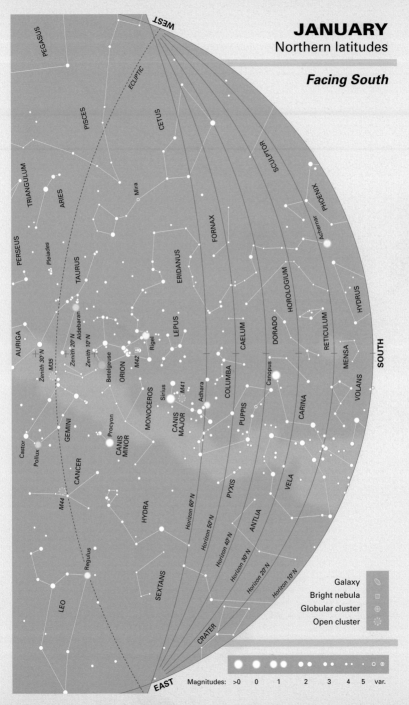

WEST

PEGASUS

ECLIPTIC

PISCES

CETUS

SCULPTOR

TRIANGULUM

ARIES

Mira

PHOENIX

PERSEUS

FORNAX

Achernar

Pleiades

ERIDANUS

HOROLOGIUM

HYDRUS

TAURUS

AURIGA

RETICULUM

Zenith 30° N

Aldebaran

Rigel

CAELUM

DORADO

M35

Zenith 20° N

LEPUS

MENSA

Zenith 10° N

M42

Canopus

VOLANS

Betelgeuse

ORION

COLUMBA

CARINA

Castor

GEMINI

Sirius

M41

Adhara

PUPPIS

VELA

Pollux

Procyon

CANIS
MAJOR

MONOCEROS

CANCER

CANIS
MINOR

PYXIS

M44

Horizon 60° N

ANTLIA

HYDRA

Horizon 50° N

Horizon 40° N

Regulus

Horizon 30° N

Horizon 20° N

SEXTANS

Horizon 10° N

LEO

SOUTH

Galaxy

Bright nebula

Globular cluster

Open cluster

CRATER

EAST

Magnitudes: >0   0   1   2   3   4   5   var.

JANUARY
Southern latitudes

Facing North

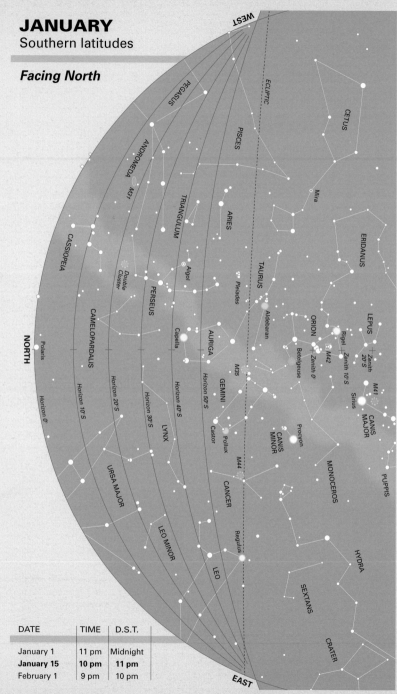

WEST

ECLIPTIC

PEGASUS

ANDROMEDA

M31

CASSIOPEIA

CAMELOPARDALIS

NORTH

Polaris

TRIANGULUM

PISCES

ARIES

Double
Cluster

Algol

PERSEUS

Capella

AURIGA

TAURUS

Pleiades

Aldebaran

CETUS

Mira

ERIDANUS

ORION

Rigel

LEPUS

Zenith
20°S

Betelgeuse

Zenith 0°

M42

Zenith 10°S

Horizon 0°

Horizon 10°S

Horizon 20°S

Horizon 30°S

LYNX

Horizon 40°S

M35

GEMINI

Horizon 50°S

Castor

Pollux

M44

Procyon

CANIS
MINOR

Sirius

M41

CANIS
MAJOR

MONOCEROS

PUPPIS

URSA MAJOR

LEO MINOR

CANCER

Regulus

LEO

SEXTANS

HYDRA

CRATER

EAST

DATE	TIME	D.S.T.
January 1	11 pm	Midnight
January 15	**10 pm**	**11 pm**
February 1	9 pm	10 pm

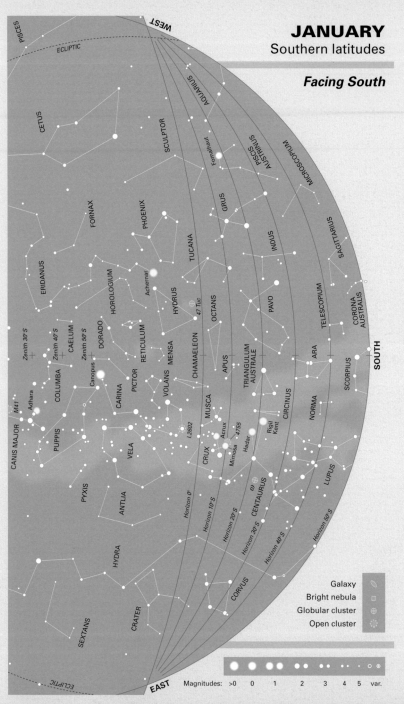

JANUARY
Southern latitudes

Facing South

Galaxy
Bright nebula
Globular cluster
Open cluster

Magnitudes: >0 0 1 2 3 4 5 var.

FEBRUARY
Northern latitudes

Facing North

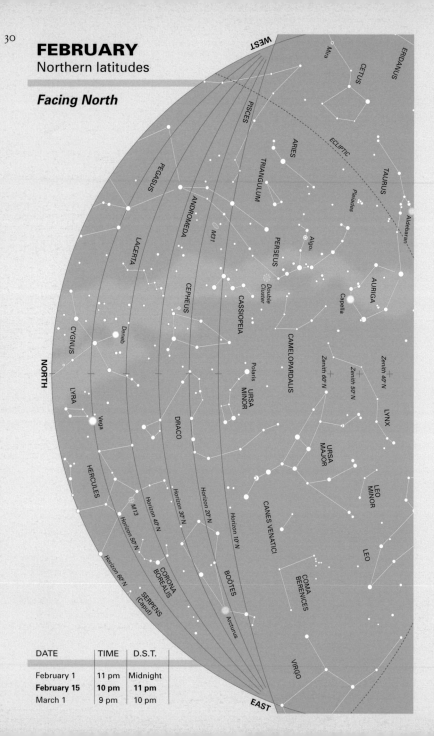

DATE	TIME	D.S.T.
February 1	11 pm	Midnight
February 15	**10 pm**	**11 pm**
March 1	9 pm	10 pm

FEBRUARY
Northern latitudes

Facing South

32

FEBRUARY
Southern latitudes

Facing North

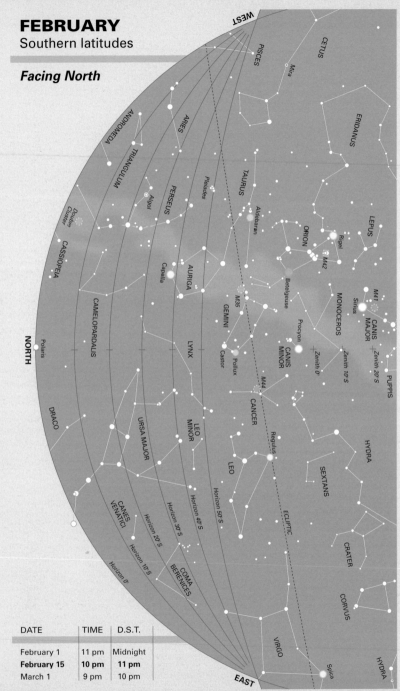

WEST

PISCES

CETUS

Mira

ERIDANUS

ANDROMEDA

ARIES

TRIANGULUM

Algol

PERSEUS

Pleiades

TAURUS

Aldebaran

ORION

Rigel

LEPUS

M42

Double Cluster

CASSIOPEIA

Capella

AURIGA

Betelgeuse

MONOCEROS

CANIS MAJOR

M41

Sirius

PUPPIS

NORTH

Polaris

CAMELOPARDALIS

M35

GEMINI

Castor

Pollux

CANIS MINOR

Procyon

Zenith 0°

Zenith 10° S

Zenith 20° S

LYNX

M44

CANCER

Regulus

HYDRA

DRACO

URSA MAJOR

LEO MINOR

LEO

SEXTANS

CANES VENATICI

Horizon 10° S

Horizon 20° S

Horizon 30° S

Horizon 40° S

Horizon 50° S

COMA BERENICES

ECLIPTIC

CRATER

Horizon 0°

CORVUS

VIRGO

Spica

HYDRA

EAST

DATE	TIME	D.S.T.
February 1	11 pm	Midnight
February 15	**10 pm**	**11 pm**
March 1	9 pm	10 pm

FEBRUARY
Southern latitudes

Facing South

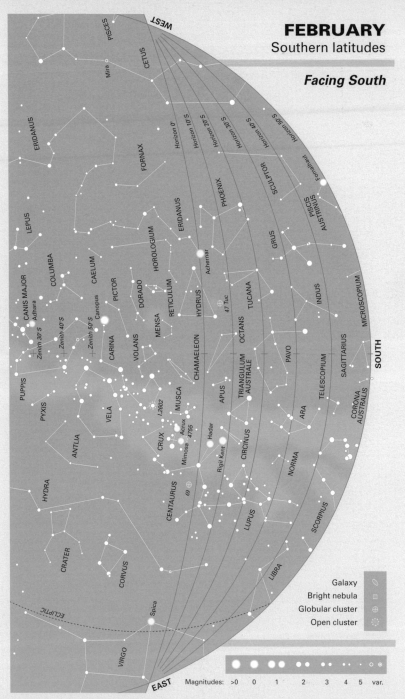

MARCH
Northern latitudes

Facing North

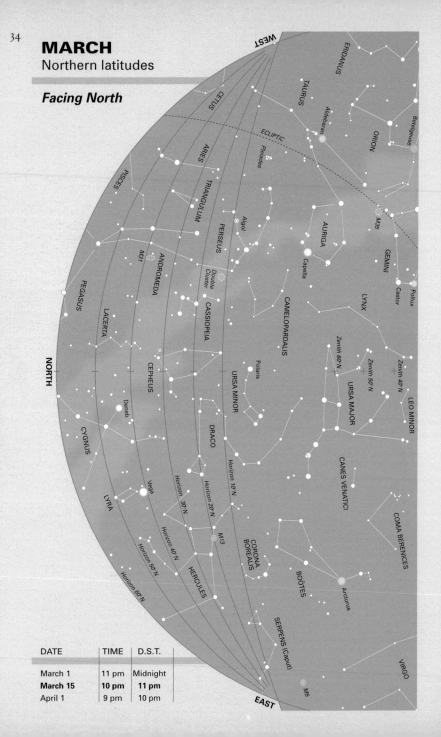

DATE	TIME	D.S.T.
March 1	11 pm	Midnight
March 15	**10 pm**	**11 pm**
April 1	9 pm	10 pm

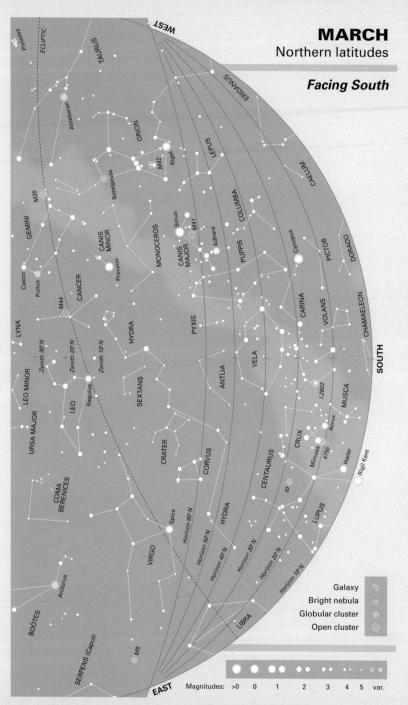

MARCH
Northern latitudes

Facing South

Galaxy
Bright nebula
Globular cluster
Open cluster

Magnitudes: >0 0 1 2 3 4 5 var.

MARCH
Southern latitudes

Facing North

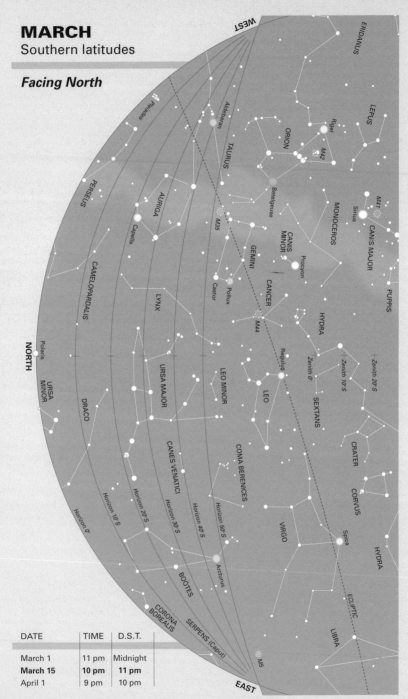

DATE	TIME	D.S.T.
March 1	11 pm	Midnight
March 15	**10 pm**	**11 pm**
April 1	9 pm	10 pm

MARCH
Southern latitudes

Facing South

Galaxy
Bright nebula
Globular cluster
Open cluster

Magnitudes: >0 0 1 2 3 4 5 var.

APRIL
Northern latitudes

Facing North

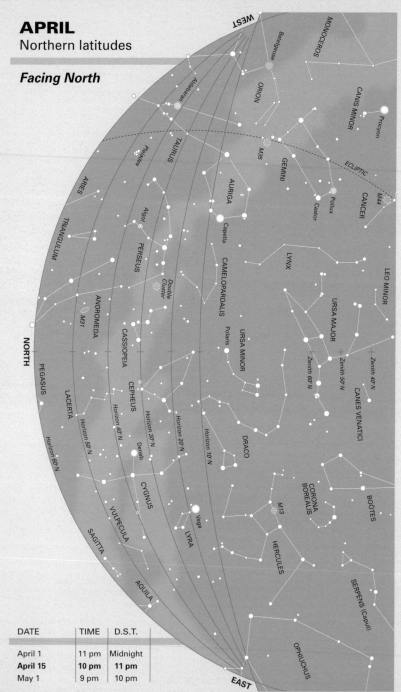

DATE	TIME	D.S.T.
April 1	11 pm	Midnight
April 15	**10 pm**	**11 pm**
May 1	9 pm	10 pm

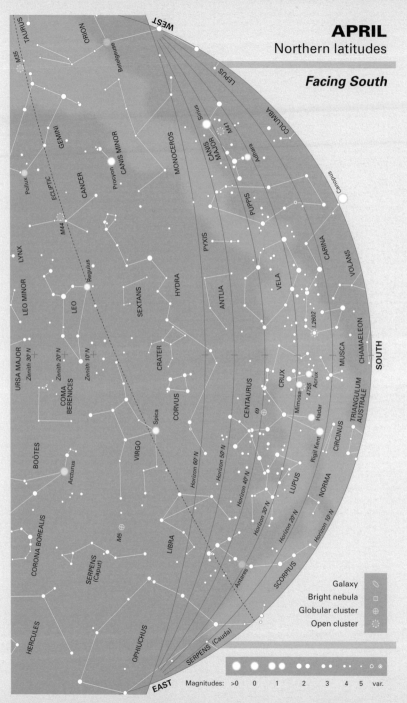

APRIL
Northern latitudes

Facing South

WEST

TAURUS
ORION
Betelgeuse
M35
GEMINI
Pollux
LYNX
CANCER
M44
ECLIPTIC
Procyon
CANIS MINOR
MONOCEROS
LEPUS
Sirius
CANIS MAJOR
M41
Adhara
COLUMBA
PUPPIS
Canopus
CARINA
VOLANS
IC2602
CHAMELEON
MUSCA
VELA
PYXIS
ANTLIA
HYDRA
SEXTANS
LEO
LEO MINOR
Regulus
URSA MAJOR
Zenith 30 N
COMA BERENICES
Zenith 20 N
Zenith 10 N
CRATER
CORVUS
Spica
VIRGO
CENTAURUS
ω
CRUX
Mimosa
4755
Acrux
Hadar
Rigil Kent
TRIANGULUM AUSTRALE
CIRCINUS
NORMA
LUPUS
SCORPIUS
Antares
BOOTES
Arcturus
M5
LIBRA
CORONA BOREALIS
SERPENS (Caput)
HERCULES
OPHIUCHUS
SERPENS (Cauda)

SOUTH

Horizon 60 N
Horizon 50 N
Horizon 40 N
Horizon 30 N
Horizon 20 N
Horizon 10 N

EAST

Galaxy
Bright nebula
Globular cluster
Open cluster

Magnitudes: >0 0 1 2 3 4 5 var.

APRIL
Southern latitudes

Facing North

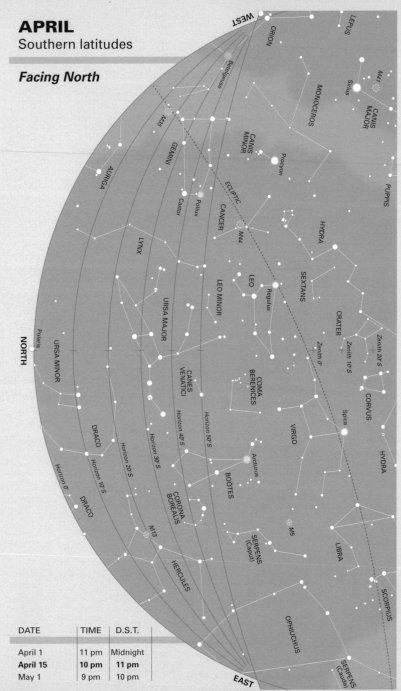

WEST

ORION

LEPUS

CANIS MAJOR

Sirius

M41

Betelgeuse

MONOCEROS

M35

CANIS MINOR

Procyon

AURIGA

GEMINI

Castor

Pollux

ECLIPTIC

CANCER

M44

PUPPIS

HYDRA

LYNX

LEO MINOR

LEO

Regulus

SEXTANS

CRATER

Zenith 20° S

URSA MAJOR

CANES VENATICI

COMA BERENICES

Zenith 0°

Zenith 10° S

Zenith 20° S

CORVUS

NORTH

Polaris

URSA MINOR

Horizon 40° S

Horizon 50° S

VIRGO

Spica

HYDRA

DRACO

Horizon 30° S

Horizon 20° S

Horizon 10° S

BOÖTES

Arcturus

CORONA BOREALIS

M5

DRACO

Horizon 0°

M13

SERPENS (Caput)

LIBRA

HERCULES

SCORPIUS

OPHIUCHUS

EAST

SERPENS (Cauda)

DATE	TIME	D.S.T.
April 1	11 pm	Midnight
April 15	**10 pm**	**11 pm**
May 1	9 pm	10 pm

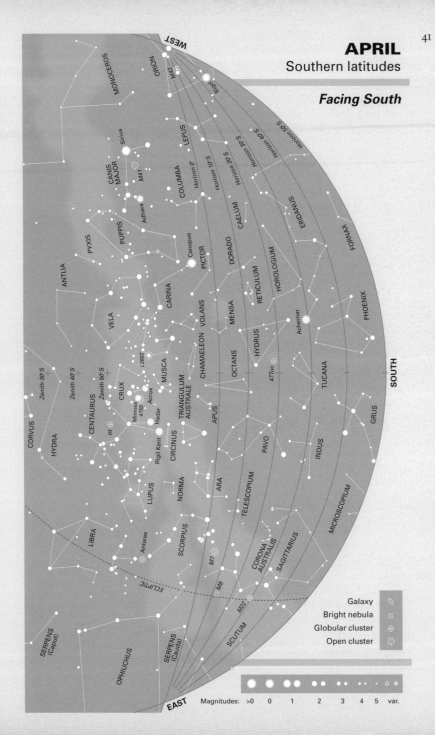

APRIL
Southern latitudes

Facing South

WEST

MONOCEROS
ORION
M42
Rigel
LEPUS
Sirius
CANIS MAJOR
M41
COLUMBA
Adhara
PUPPIS
PYXIS
Canopus
PICTOR
ANTLIA
CAELUM
DORADO
VELA
CARINA
VOLANS
MENSA
RETICULUM
HOROLOGIUM
ERIDANUS
FORNAX
I.2602
MUSCA
CHAMAELEON
HYDRUS
Achernar
PHOENIX
Zenith 30°S
Zenith 40°S
Zenith 50°S
CRUX
TRIANGULUM AUSTRALE
OCTANS
47 Tuc
CORVUS
HYDRA
CENTAURUS
Mimosa
4755
Acrux
ω
APUS
TUCANA
Hadar
Rigil Kent
CIRCINUS
PAVO
INDUS
GRUS
LUPUS
NORMA
ARA
TELESCOPIUM
MICROSCOPIUM
LIBRA
Antares
SCORPIUS
M7
CORONA AUSTRALIS
SAGITTARIUS
SOUTH
ECLIPTIC
M8
M22
SCUTUM
SERPENS (Caput)
SERPENS (Cauda)
OPHIUCHUS

Horizon 50°S
Horizon 40°S
Horizon 30°S
Horizon 20°S
Horizon 10°S
Horizon 0°

EAST

Galaxy
Bright nebula
Globular cluster
Open cluster

Magnitudes: >0   0   1   2   3   4   5   var.

42

MAY
Northern latitudes

Facing North

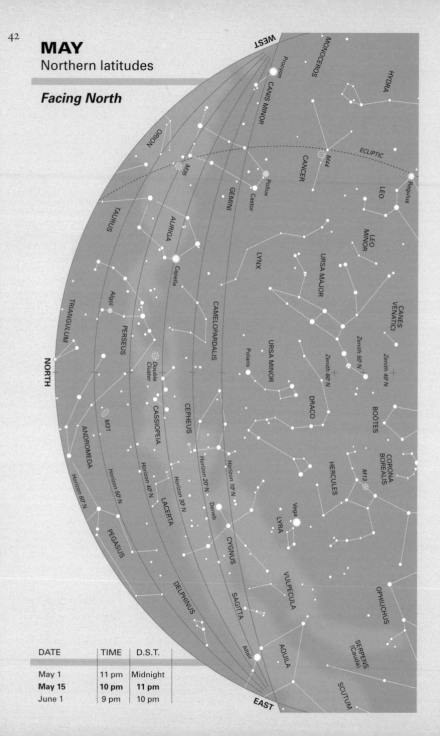

DATE	TIME	D.S.T.
May 1	11 pm	Midnight
May 15	**10 pm**	**11 pm**
June 1	9 pm	10 pm

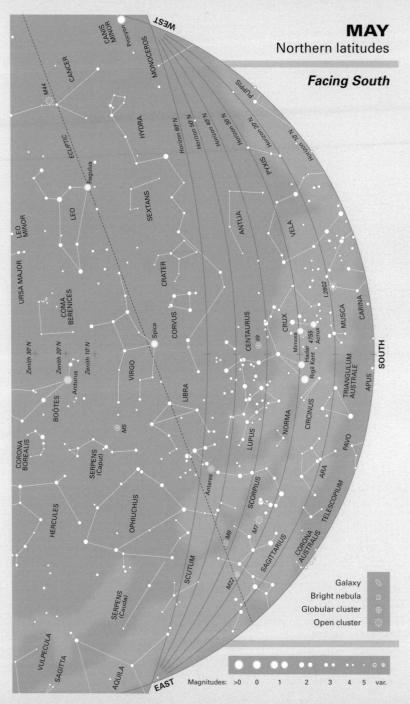

MAY
Northern latitudes

Facing South

WEST

EAST

SOUTH

Galaxy
Bright nebula
Globular cluster
Open cluster

Magnitudes: >0 0 1 2 3 4 5 var.

44

MAY
Southern latitudes

Facing North

DATE	TIME	D.S.T.
May 1	11 pm	Midnight
May 15	**10 pm**	**11 pm**
June 1	9 pm	10 pm

MAY
Southern latitudes

Facing South

WEST

EAST

SOUTH

CANCER

MONOCEROS

HYDRA

CRATER

CORVUS

PYXIS

PUPPIS

ANTLIA

VELA

CARINA

CENTAURUS

CRUX

MUSCA

Sirius

M41

CANIS MAJOR

Adhara

COLUMBA

LEPUS

CAELUM

Canopus

PICTOR

DORADO

HOROLOGIUM

ERIDANUS

VOLANS

MENSA

RETICULUM

CHAMAELEON

OCTANS

HYDRUS

Achernar

PHOENIX

SCULPTOR

I.2602

4755

Mimosa

Acrux

Hadar

ω

Zenith 30°S

Zenith 40°S

Zenith 50°S

Rigil Kent

TRIANGULUM AUSTRALE

APUS

PAVO

47 Tuc

TUCANA

GRUS

PISCIS AUSTRINUS

LIBRA

Antares

SCORPIUS

M7

LUPUS

NORMA

CIRCINUS

ARA

TELESCOPIUM

CORONA AUSTRALIS

INDUS

MICROSCOPIUM

ECLIPTIC

OPHIUCHUS

M8

SAGITTARIUS

M22

SCUTUM

SERPENS (Cauda)

AQUILA

CAPRICORNUS

Horizon 0°S

Horizon 10°S

Horizon 20°S

Horizon 30°S

Horizon 40°S

Horizon 50°S

Galaxy

Bright nebula

Globular cluster

Open cluster

Magnitudes: >0   0   1   2   3   4   5   var.

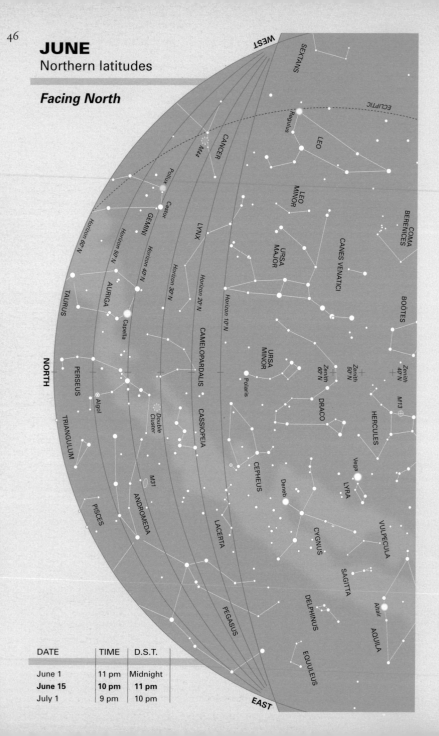

JUNE
Northern latitudes

Facing North

WEST

ECLIPTIC

Regulus

LEO

CANCER

M44

Pollux

Castor

GEMINI

LYNX

LEO MINOR

COMA BERENICES

URSA MAJOR

CANES VENATICI

BOOTES

TAURUS

AURIGA

Capella

Horizon 60°N

Horizon 50°N

Horizon 40°N

Horizon 30°N

Horizon 20°N

Horizon 10°N

CAMELOPARDALIS

URSA MINOR

Polaris

Zenith 60°N

Zenith 50°N

Zenith 40°N

M13

HERCULES

NORTH

PERSEUS

Algol

Double Cluster

CASSIOPEIA

DRACO

Vega

LYRA

TRIANGULUM

M31

ANDROMEDA

CEPHEUS

Deneb

VULPECULA

PISCES

LACERTA

CYGNUS

SAGITTA

DELPHINUS

Altair

PEGASUS

AQUILA

EQUULEUS

EAST

DATE	TIME	D.S.T.
June 1	11 pm	Midnight
June 15	**10 pm**	**11 pm**
July 1	9 pm	10 pm

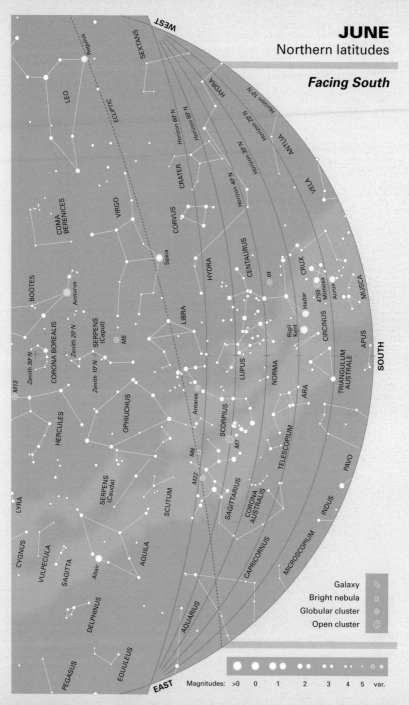

JUNE
Northern latitudes

Facing South

WEST

LEO
Regulus
SEXTANS
ECLIPTIC
HYDRA
Horizon 60° N
Horizon 50° N
Horizon 10° N
ANTLIA
Horizon 20° N
Horizon 30° N
VIRGO
CRATER
Horizon 40° N
COMA BERENICES
CORVUS
VELA
HYDRA
Spica
CENTAURUS
CRUX
BOOTES
Arcturus
ω
4755
Mimosa
MUSCA
Zenith 30° N
SERPENS (Caput)
Hadar
Acrux
Zenith 20° N
M5
LIBRA
Rigil Kent
CIRCINUS
CORONA BOREALIS
Zenith 10° N
LUPUS
NORMA
APUS
M13
HERCULES
Antares
ARA
TRIANGULUM AUSTRALE
OPHIUCHUS
SCORPIUS
TELESCOPIUM
SOUTH
LYRA
M8
M7
PAVO
CYGNUS
SERPENS (Cauda)
M22
SAGITTARIUS
CORONA AUSTRALIS
VULPECULA
SCUTUM
INDUS
SAGITTA
Altair
AQUILA
MICROSCOPIUM
DELPHINUS
CAPRICORNUS
PEGASUS
AQUARIUS
EQUULEUS
EAST

Galaxy
Bright nebula
Globular cluster
Open cluster

Magnitudes: >0 0 1 2 3 4 5 var.

48

JUNE
Southern latitudes

Facing North

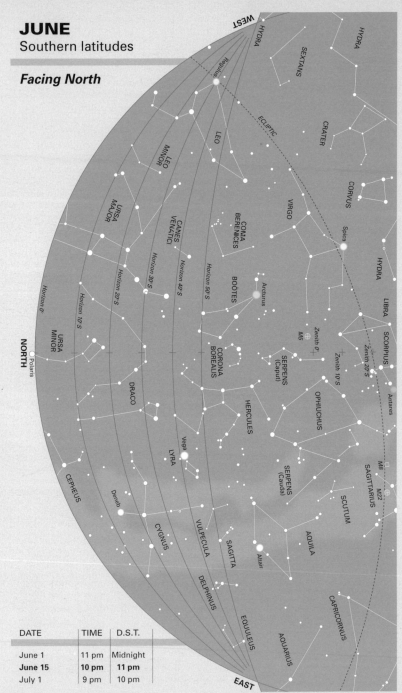

DATE	TIME	D.S.T.
June 1	11 pm	Midnight
June 15	**10 pm**	**11 pm**
July 1	9 pm	10 pm

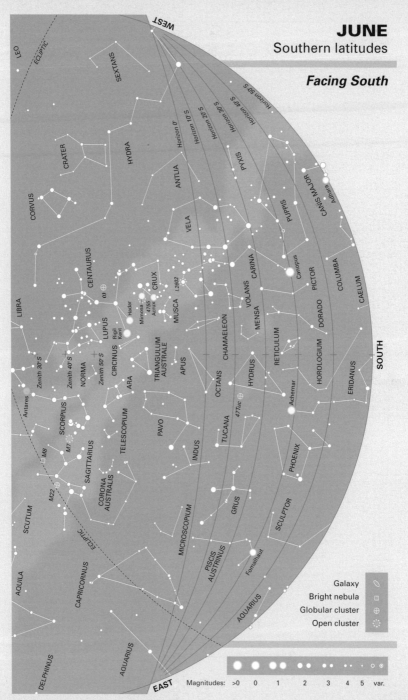

JUNE
Southern latitudes

Facing South

WEST

LEO
ECLIPTIC
SEXTANS
CRATER
HYDRA
CORVUS
ANTLIA
Horizon 0° S
Horizon 10° S
Horizon 20° S
Horizon 30° S
Horizon 40° S
Horizon 50° S
PYXIS
PUPPIS
CANIS MAJOR
Adhara
VELA
CENTAURUS
ω
CRUX
CARINA
VOLANS
Canopus
PICTOR
LIBRA
Hadar
Mimosa
4755
Acrux
MUSCA
I. 2602
COLUMBA
CAELUM
LUPUS
Rigil Kent
CHAMAELEON
MENSA
DORADO
Zenith 30° S
CIRCINUS
TRIANGULUM AUSTRALE
APUS
OCTANS
RETICULUM
HOROLOGIUM
ERIDANUS
Zenith 40° S
NORMA
Zenith 50° S
ARA
HYDRUS
Antares
SCORPIUS
TELESCOPIUM
PAVO
TUCANA
47 Tuc
Achernar
SOUTH
M8
SAGITTARIUS
INDUS
PHOENIX
M7
M22
CORONA AUSTRALIS
MICROSCOPIUM
GRUS
SCULPTOR
SCUTUM
CAPRICORNUS
PISCIS AUSTRINUS
Fomalhaut
AQUILA
ECLIPTIC
AQUARIUS
DELPHINUS
AQUARIUS

EAST

Galaxy
Bright nebula
Globular cluster
Open cluster

Magnitudes: >0 0 1 2 3 4 5 var.

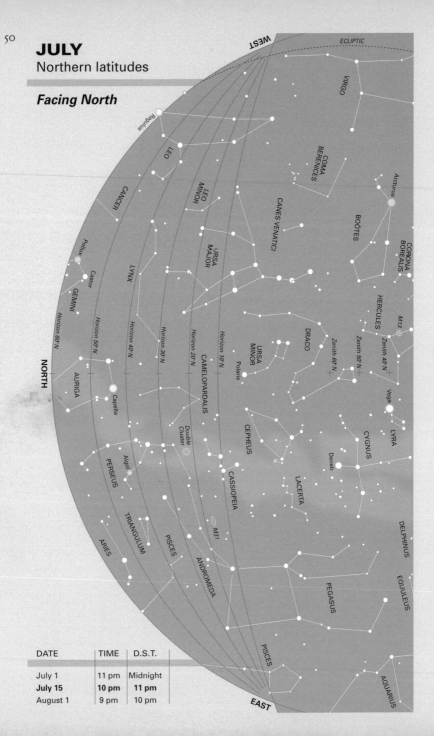

50

JULY
Northern latitudes

Facing North

WEST

ECLIPTIC

VIRGO

Regulus

LEO

COMA BERENICES

Arcturus

CANCER

LEO MINOR

CANES VENATICI

BOÖTES

CORONA BOREALIS

Castor

LYNX

URSA MAJOR

HERCULES

M13

Pollux

GEMINI

Horizon 60° N

Horizon 50° N

Horizon 40° N

Horizon 30° N

Horizon 20° N

Horizon 10° N

DRACO

URSA MINOR

Zenith 60° N

Zenith 50° N

Zenith 40° N

NORTH

AURIGA

Polaris

Vega

LYRA

Capella

CAMELOPARDALIS

CEPHEUS

CYGNUS

Double Cluster

Deneb

Algol

PERSEUS

CASSIOPEIA

LACERTA

TRIANGULUM

M31

PISCES

ANDROMEDA

DELPHINUS

ARIES

PEGASUS

EQUULEUS

PISCES

AQUARIUS

EAST

DATE	TIME	D.S.T.
July 1	11 pm	Midnight
July 15	**10 pm**	**11 pm**
August 1	9 pm	10 pm

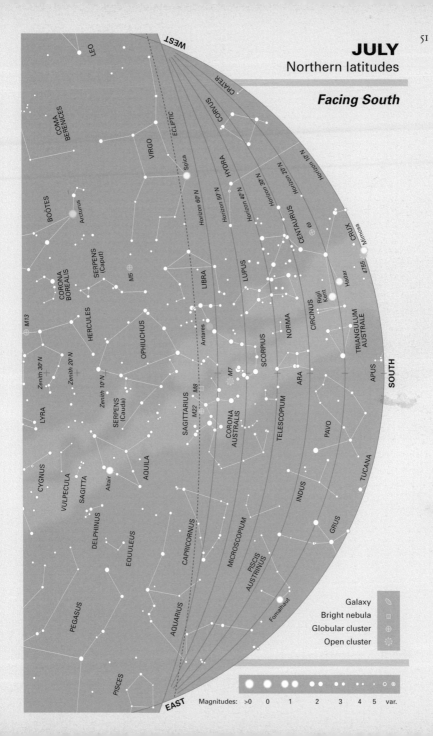

JULY
Northern latitudes

Facing South

WEST

EAST

SOUTH

LEO

COMA BERENICES

BOÖTES

Arcturus

CORONA BOREALIS

SERPENS (Caput)

M5

HERCULES

M13

Zenith 30° N

Zenith 20° N

Zenith 10° N

LYRA

CYGNUS

VULPECULA

SAGITTA

Altair

DELPHINUS

AQUILA

EQUULEUS

PEGASUS

PISCES

AQUARIUS

CAPRICORNUS

SERPENS (Cauda)

OPHIUCHUS

SAGITTARIUS

M22

M8

CORONA AUSTRALIS

MICROSCOPIUM

PISCIS AUSTRINUS

Fomalhaut

VIRGO

ECLIPTIC

Spica

CRATER

CORVUS

HYDRA

Horizon 60° N

Horizon 50° N

Horizon 40° N

Horizon 30° N

Horizon 20° N

Horizon 10° N

LIBRA

Antares

M7

SCORPIUS

LUPUS

NORMA

ARA

TELESCOPIUM

INDUS

GRUS

TUCANA

PAVO

APUS

TRIANGULUM AUSTRALE

CIRCINUS

Rigil Kent

Hadar

CENTAURUS

ω

CRUX

Mimosa

4755

Galaxy

Bright nebula

Globular cluster

Open cluster

Magnitudes: >0   0   1   2   3   4   5   var.

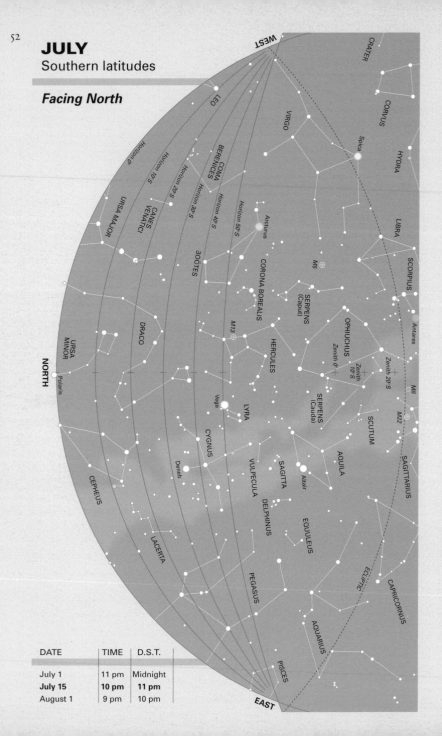

52

JULY
Southern latitudes

Facing North

DATE	TIME	D.S.T.
July 1	11 pm	Midnight
July 15	**10 pm**	**11 pm**
August 1	9 pm	10 pm

JULY
Southern latitudes

Facing South

WEST

VIRGO
ECLIPTIC
CORVUS
CRATER
Spica
HYDRA
Horizon 0°
Horizon 10° S
Horizon 20° S
Horizon 30° S
Horizon 40° S
Horizon 50° S
LIBRA
CENTAURUS
ω
ANTLIA
PYXIS
LUPUS
CRUX
Mimosa
4755
VELA
Acrux
SCORPIUS
Hadar
MUSCA
I.2602
Antares
Rigil
Kent
CIRCINUS
CARINA
PUPPIS
NORMA
TRIANGULUM
Zenith 30° S
AUSTRALE
CHAMAELEON
VOLANS
Zenith
40° S
ARA
APUS
Canopus
M8
Zenith 50° S
MENSA
PICTOR
COLUMBA
M7
CORONA
TELESCOPIUM
PAVO
HYDRUS
DORADO
SOUTH
AUSTRALIS
OCTANS
47Tuc
M22
SAGITTARIUS
RETICULUM
CAELUM
MICROSCOPIUM
Achernar
HOROLOGIUM
INDUS
TUCANA
PHOENIX
ERIDANUS
CAPRICORNUS
GRUS
ERIDANUS
FORNAX
PISCIS
SCULPTOR
AUSTRINUS
Fomalhaut
PEGASUS
AQUARIUS
CETUS
PISCES
EAST

Galaxy
Bright nebula
Globular cluster
Open cluster

Magnitudes: >0   0   1   2   3   4   5   var.

54

AUGUST
Northern latitudes

Facing North

DATE	TIME	D.S.T.
August 1	11 pm	Midnight
August 15	**10 pm**	**11 pm**
September 1	9 pm	10 pm

AUGUST
Northern latitudes

Facing South

WEST

EAST

SOUTH

Arcturus

BOÖTES

VIRGO

CENTAURUS

CORONA BOREALIS

SERPENS (Caput)

M5

LIBRA

LUPUS

CIRCINUS

HERCULES

OPHIUCHUS

SCORPIUS

NORMA

Antares

LYRA

Zenith 30° N

SERPENS (Cauda)

Zenith 20° N

Zenith 10° N

SCUTUM

M22

SAGITTARIUS

M8

CORONA AUSTRALIS

M7

ARA

TELESCOPIUM

PAVO

TRIANGULUM AUSTRALE

CYGNUS

VULPECULA

SAGITTA

Altair

AQUILA

MICROSCOPIUM

INDUS

OCTANS

DELPHINUS

EQUULEUS

AQUARIUS

CAPRICORNUS

PISCIS AUSTRINUS

GRUS

TUCANA

PEGASUS

Fomalhaut

SCULPTOR

PHOENIX

Horizon 60° N

Horizon 50° N

Horizon 40° N

Horizon 30° N

Horizon 20° N

Horizon 10° N

ANDROMEDA

CETUS

ECLIPTIC

PISCES

Galaxy

Bright nebula

Globular cluster

Open cluster

Magnitudes: >0 0 1 2 3 4 5 var.

AUGUST
Southern latitudes

Facing North

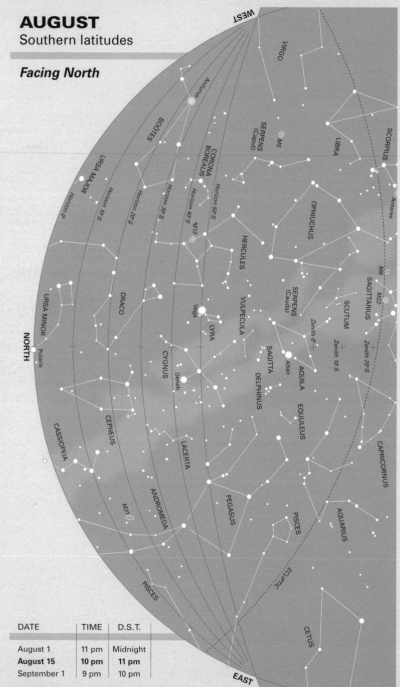

WEST

VIRGO

Arcturus

BOOTES

SERPENS
(Caput) ⊕ M5

LIBRA

SCORPIUS

CORONA BOREALIS

Horizon 0
Horizon 10°S
Horizon 20°S
Horizon 30°S
Horizon 40°S
Horizon 50°S

URSA MAJOR

M13 ⊕

OPHIUCHUS

Antares

HERCULES

SERPENS
(Cauda)

SAGITTARIUS

M8
M22 ⊕

SCUTUM

DRACO

Vega

VULPECULA

Zenith 0°

LYRA

SAGITTA

AQUILA

Zenith 10°S

Zenith 20°S

URSA MINOR

Altair

NORTH

Polaris

CYGNUS

Deneb

DELPHINUS

EQUULEUS

CEPHEUS

LACERTA

CASSIOPEIA

ANDROMEDA

M31

PEGASUS

PISCES

CAPRICORNUS

AQUARIUS

PISCES

ECLIPTIC

CETUS

EAST

DATE	TIME	D.S.T.
August 1	11 pm	Midnight
August 15	**10 pm**	**11 pm**
September 1	9 pm	10 pm

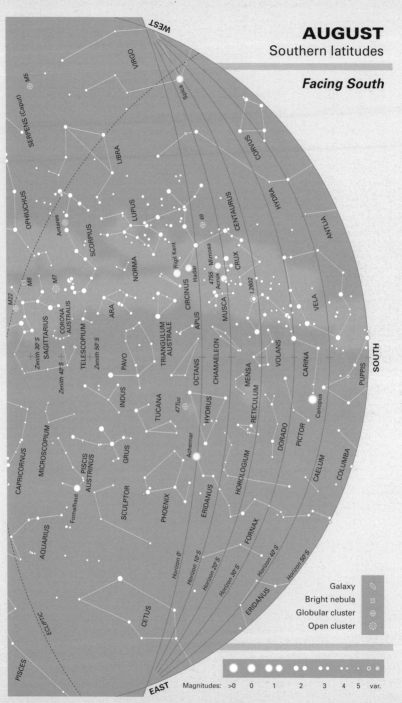

AUGUST
Southern latitudes

Facing South

WEST

VIRGO

SERPENS (Caput)

M5

SCORPIUS

OPHIUCHUS

Antares

LIBRA

LUPUS

NORMA

Rigil Kent

CIRCINUS

Hadar

CENTAURUS

Mimosa

4755

Acrux

CRUX

ω

MUSCA

2602

HYDRA

CORVUS

ANTLIA

VELA

M8

M7

M22

SAGITTARIUS

CORONA
AUSTRALIS

ARA

TELESCOPIUM

Zenith 30°S

Zenith 40°S

Zenith 50°S

TRIANGULUM
AUSTRALE

APUS

PAVO

OCTANS

CHAMAELEON

MENSA

VOLANS

CARINA

PUPPIS

SOUTH

INDUS

TUCANA

47 Tuc

HYDRUS

RETICULUM

DORADO

PICTOR

Canopus

MICROSCOPIUM

GRUS

Achernar

ERIDANUS

HOROLOGIUM

CAELUM

COLUMBA

CAPRICORNUS

PISCIS
AUSTRINUS

Fomalhaut

SCULPTOR

PHOENIX

FORNAX

AQUARIUS

Horizon 0°

Horizon 10°S

Horizon 20°S

Horizon 30°S

Horizon 40°S

Horizon 50°S

CETUS

ERIDANUS

PISCES

ECLIPTIC

EAST

Galaxy
Bright nebula
Globular cluster
Open cluster

Magnitudes: >0 0 1 2 3 4 5 var.

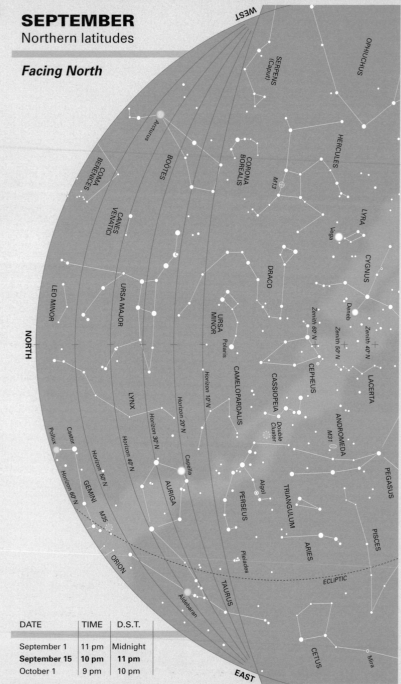

58

Northern latitudes

Facing North

WEST

OPHIUCHUS

SERPENS (Caput)

HERCULES

Arcturus

BOÖTES

CORONA BOREALIS

M13

LYRA

COMA BERENICES

Vega

CANES VENATICI

CYGNUS

DRACO

Deneb

Zenith 60° N

Zenith 50° N

Zenith 40° N

LEO MINOR

URSA MAJOR

URSA MINOR

LACERTA

Polaris

CEPHEUS

CAMELOPARDALIS

CASSIOPEIA

ANDROMEDA

Horizon 10° N

Double Cluster

M31

LYNX

Horizon 20° N

Algol

PERSEUS

TRIANGULUM

PEGASUS

Horizon 30° N

Capella

Pollux

Horizon 40° N

Castor

AURIGA

ARIES

PISCES

Horizon 50° N

GEMINI

Pleiades

NORTH

M35

Horizon 60° N

ORION

TAURUS

ECLIPTIC

Aldebaran

CETUS

Mira

EAST

DATE	TIME	D.S.T.
September 1	11 pm	Midnight
September 15	**10 pm**	**11 pm**
October 1	9 pm	10 pm

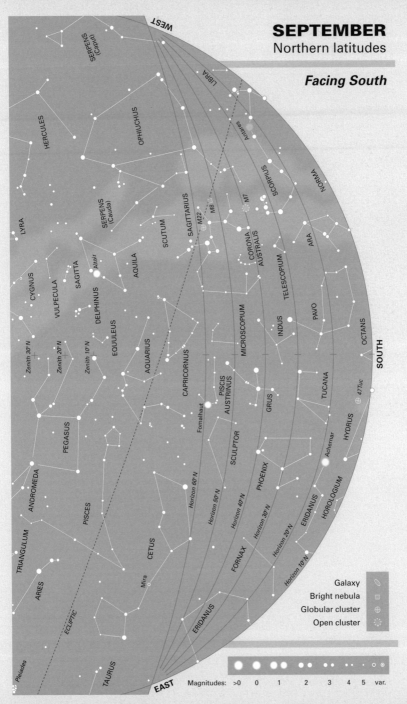

SEPTEMBER
Northern latitudes

Facing South

WEST

SERPENS (Caput)

LIBRA

HERCULES

OPHIUCHUS

Antares

SCORPIUS

NORMA

LYRA

SERPENS (Cauda)

SCUTUM

SAGITTARIUS

M22

M8

M7

CORONA AUSTRALIS

ARA

CYGNUS

VULPECULA

SAGITTA

DELPHINUS

Altair

AQUILA

TELESCOPIUM

PAVO

EQUULEUS

MICROSCOPIUM

INDUS

OCTANS

Zenith 30° N

Zenith 20° N

Zenith 10° N

AQUARIUS

CAPRICORNUS

SOUTH

PISCIS AUSTRINUS

GRUS

TUCANA

47 Tuc

PEGASUS

Fomalhaut

SCULPTOR

PHOENIX

Achernar

HYDRUS

ANDROMEDA

PISCES

Horizon 60° N

Horizon 50° N

Horizon 40° N

Horizon 30° N

Horizon 20° N

ERIDANUS

HOROLOGIUM

TRIANGULUM

CETUS

FORNAX

Horizon 10° N

ARIES

Mira

ECLIPTIC

ERIDANUS

	Galaxy
	Bright nebula
	Globular cluster
	Open cluster

Pleiades

TAURUS

EAST

Magnitudes: >0   0   1   2   3   4   5   var.

SEPTEMBER
Southern latitudes

Facing North

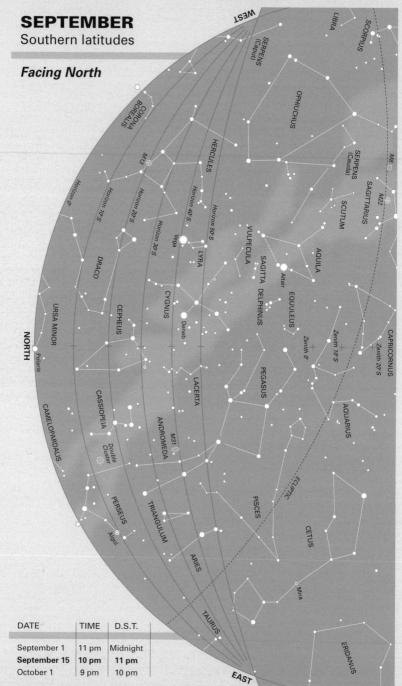

DATE	TIME	D.S.T.
September 1	11 pm	Midnight
September 15	**10 pm**	**11 pm**
October 1	9 pm	10 pm

SEPTEMBER
Southern latitudes

Facing South

WEST

EAST

SOUTH

Galaxy
Bright nebula
Globular cluster
Open cluster

Magnitudes: >0 0 1 2 3 4 5 var.

OPHIUCHUS
SERPENS (Caput)
SERPENS (Cauda)
SCUTUM
SAGITTARIUS
M22
M8
M7
CORONA AUSTRALIS
SCORPIUS
Arcturus
ECLIPTIC
LIBRA
NORMA
LUPUS
HYDRA
CENTAURUS
Rigil Kent
Hadar
ω
4755
Mimosa
Acrux
CRUX
CIRCINUS
APUS
ARA
TELESCOPIUM
PAVO
TRIANGULUM AUSTRALE
I.2602
MENSA
CHAMAELEON
OCTANS
MENSA
VELA
VOLANS
CARINA
Canopus
PUPPIS
PICTOR
DORADO
RETICULUM
HYDRUS
HOROLOGIUM
TUCANA
47 Tuc
Achernar
GRUS
INDUS
MICROSCOPIUM
Zenith 30° S
Zenith 40° S
Zenith 50° S
PISCIS AUSTRINUS
Fomalhaut
CAPRICORNUS
AQUARIUS
SCULPTOR
PHOENIX
FORNAX
CETUS
Mira
PISCES
ERIDANUS
CAELUM
COLUMBA
LEPUS

Horizon 0° S
Horizon 10° S
Horizon 20° S
Horizon 30° S
Horizon 40° S
Horizon 50° S

OCTOBER
Northern latitudes

Facing North

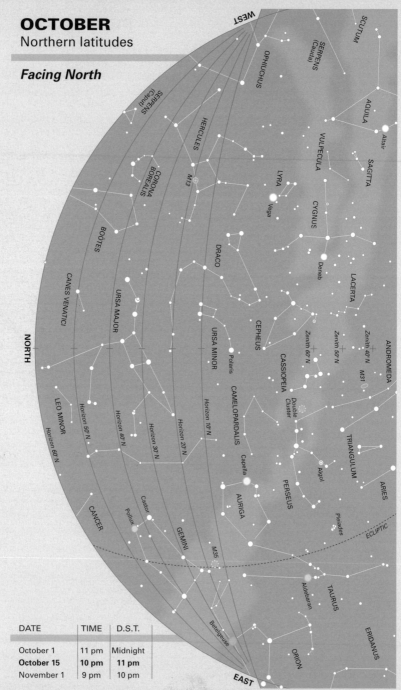

DATE	TIME	D.S.T.
October 1	11 pm	Midnight
October 15	**10 pm**	**11 pm**
November 1	9 pm	10 pm

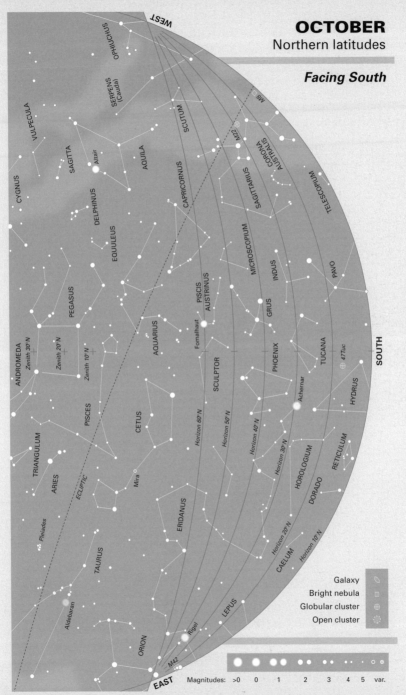

OCTOBER
Northern latitudes

Facing South

WEST

OPHIUCHUS
SERPENS (Cauda)
VULPECULA
SAGITTA
CYGNUS
Altair
DELPHINUS
AQUILA
SCUTUM
CAPRICORNUS
M22
CORONA AUSTRALIS
SAGITTARIUS
MICROSCOPIUM
TELESCOPIUM
EQUULEUS
PEGASUS
PISCIS AUSTRINUS
Fomalhaut
INDUS
GRUS
PAVO
ANDROMEDA
Zenith 30° N
Zenith 20° N
Zenith 10° N
AQUARIUS
SCULPTOR
PHOENIX
TUCANA
47 Tuc
SOUTH
PISCES
CETUS
Achernar
HYDRUS
TRIANGULUM
Mira
Horizon 60° N
Horizon 50° N
Horizon 40° N
Horizon 30° N
RETICULUM
ARIES
ECLIPTIC
ERIDANUS
HOROLOGIUM
DORADO
Pleiades
Horizon 20° N
Horizon 10° N
TAURUS
CAELUM
Aldebaran
LEPUS
Rigel
ORION
M42
EAST

Galaxy
Bright nebula
Globular cluster
Open cluster

Magnitudes: >0   0   1   2   3   4   5   var.

64

OCTOBER
Southern latitudes

Facing North

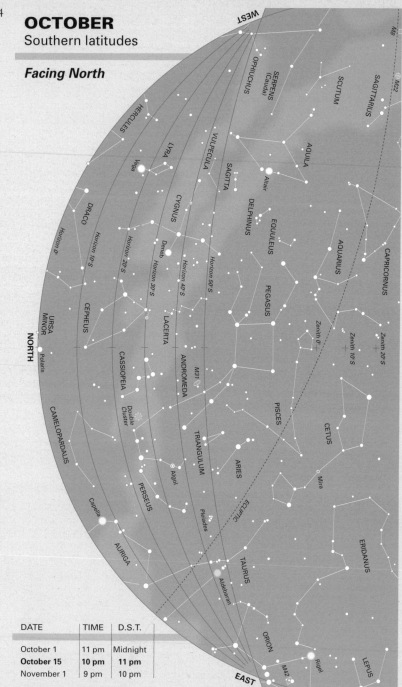

DATE	TIME	D.S.T.
October 1	11 pm	Midnight
October 15	**10 pm**	**11 pm**
November 1	9 pm	10 pm

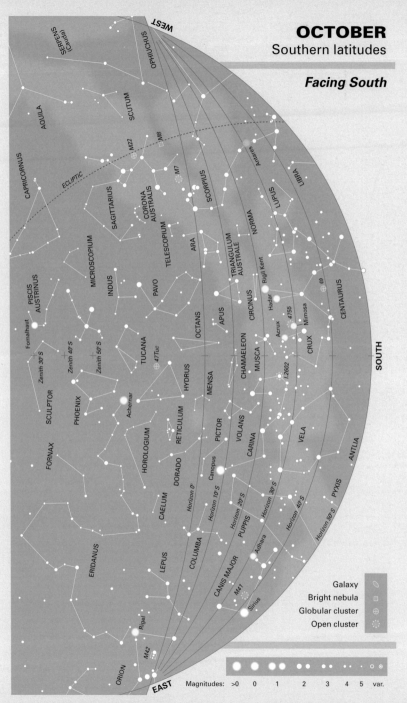

OCTOBER
Southern latitudes

Facing South

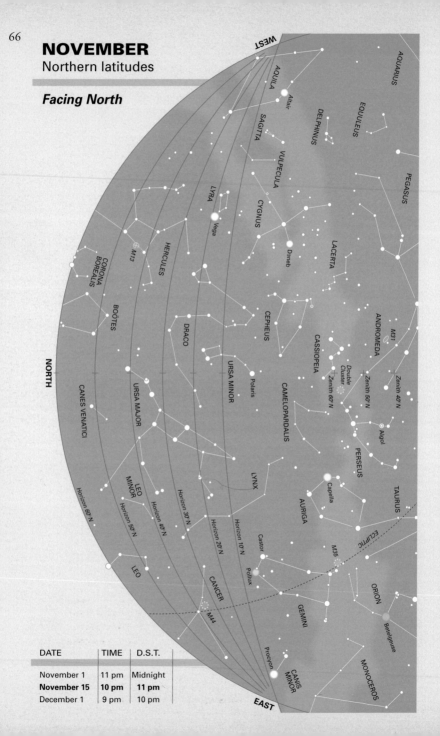

NOVEMBER
Northern latitudes

Facing North

DATE	TIME	D.S.T.
November 1	11 pm	Midnight
November 15	**10 pm**	**11 pm**
December 1	9 pm	10 pm

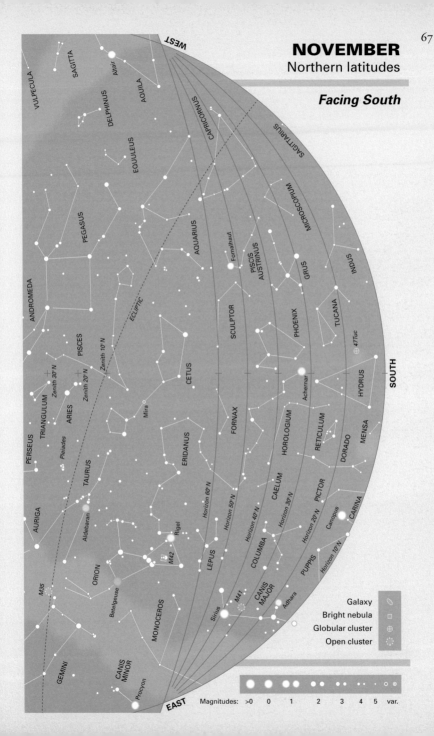

NOVEMBER
Northern latitudes

Facing South

WEST

SOUTH

EAST

Galaxy
Bright nebula
Globular cluster
Open cluster

Magnitudes: >0 0 1 2 3 4 5 var.

VULPECULA
SAGITTA
Altair
AQUILA
DELPHINUS
EQUULEUS
CAPRICORNUS
SAGITTARIUS
PEGASUS
AQUARIUS
Fomalhaut
PISCIS AUSTRINUS
MICROSCOPIUM
GRUS
INDUS
ANDROMEDA
ECLIPTIC
Zenith 10° N
SCULPTOR
PHOENIX
TUCANA
47 Tuc
PISCES
ARIES
Zenith 20° N
Zenith 30° N
TRIANGULUM
CETUS
Achernar
HYDRUS
PERSEUS
Mira
FORNAX
HOROLOGIUM
RETICULUM
MENSA
Pleiades
ERIDANUS
CAELUM
DORADO
PICTOR
AURIGA
Aldebaran
TAURUS
Horizon 60° N
Horizon 50° N
Horizon 40° N
Horizon 30° N
Horizon 20° N
Horizon 10° N
Rigel
LEPUS
COLUMBA
Canopus
CARINA
M42
M35
ORION
Betelgeuse
MONOCEROS
Sirius
M41
CANIS MAJOR
Adhara
PUPPIS
GEMINI
CANIS MINOR
Procyon

NOVEMBER
Southern latitudes

Facing North

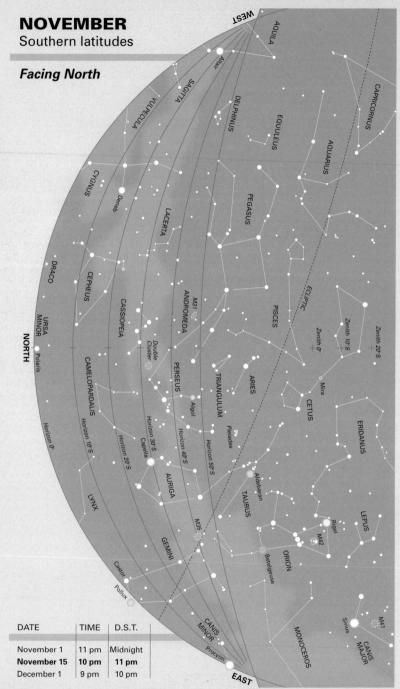

DATE	TIME	D.S.T.
November 1	11 pm	Midnight
November 15	**10 pm**	**11 pm**
December 1	9 pm	10 pm

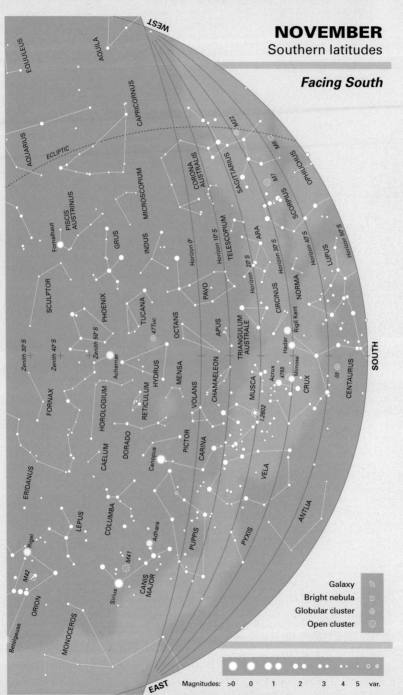

NOVEMBER
Southern latitudes

Facing South

WEST

EQUULEUS

AQUILA

CAPRICORNUS

AQUARIUS

ECLIPTIC

M22

M8

OPHIUCHUS

CORONA
AUSTRALIS

SAGITTARIUS

M7

SCORPIUS

MICROSCOPIUM

PISCIS
AUSTRINUS

Fomalhaut

GRUS

INDUS

TELESCOPIUM

ARA

Horizon 0°

Horizon 10° S

Horizon 20° S

Horizon 30° S

Horizon 40° S

Horizon 50° S

SCULPTOR

PHOENIX

TUCANA

47Tuc

PAVO

CIRCINUS

NORMA

LUPUS

OCTANS

APUS

Zenith 30° S

Zenith 40° S

Zenith 50° S

Achernar

HYDRUS

RETICULUM

MENSA

CHAMAELEON

TRIANGULUM
AUSTRALE

Rigil Kent

Hadar

SOUTH

FORNAX

HOROLOGIUM

VOLANS

Acrux

4755

Mimosa

ω

MUSCA

CENTAURUS

CAELUM

DORADO

PICTOR

CARINA

I.2602

CRUX

ERIDANUS

LEPUS

COLUMBA

Canopus

Adhara

VELA

PYXIS

ANTLIA

PUPPIS

M41

Rigel

M42

Sirius

CANIS
MAJOR

ORION

Betelgeuse

MONOCEROS

EAST

Galaxy
Bright nebula
Globular cluster
Open cluster

Magnitudes: >0   0   1   2   3   4   5   var.

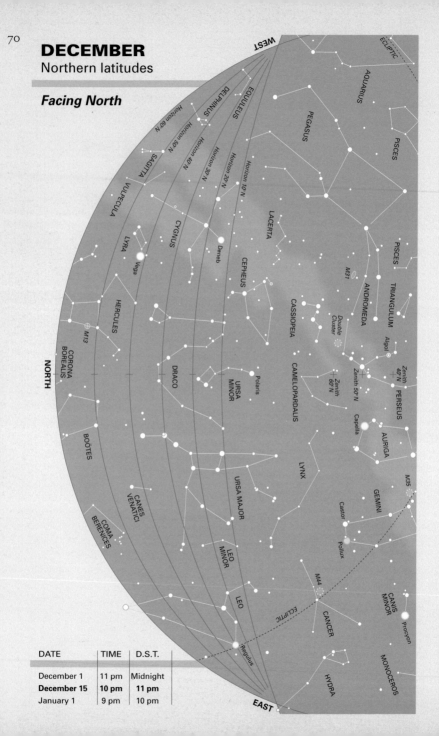

DECEMBER
Northern latitudes

Facing North

WEST

ECLIPTIC

AQUARIUS

PISCES

PEGASUS

Horizon 60 N
Horizon 50 N
Horizon 40 N
Horizon 30 N
Horizon 20 N
Horizon 10 N

DELPHINUS

EQUULEUS

SAGITTA

VULPECULA

LACERTA

PISCES

LYRA

Vega

CYGNUS

Deneb

CEPHEUS

CASSIOPEIA

M31

ANDROMEDA

TRIANGULUM

Double
Cluster

Algo

Zenith
40° N

PERSEUS

HERCULES

M13

CAMELOPARDALIS

Zenith 50° N

Zenith
60° N

Capella

AURIGA

CORONA
BOREALIS

DRACO

URSA
MINOR

Polaris

NORTH

LYNX

M35

GEMINI

BOÖTES

CANES
VENATICI

URSA MAJOR

Castor

Pollux

COMA
BERENICES

LEO
MINOR

M44

CANCER

CANIS
MINOR

Procyon

LEO

ECLIPTIC

Regulus

HYDRA

MONOCEROS

EAST

DATE	TIME	D.S.T.
December 1	11 pm	Midnight
December 15	**10 pm**	**11 pm**
January 1	9 pm	10 pm

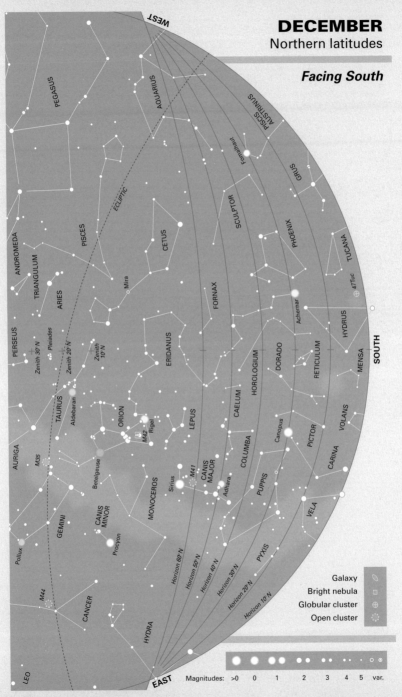

DECEMBER
Northern latitudes

Facing South

WEST

EAST

SOUTH

PEGASUS

AQUARIUS

PISCIS AUSTRINUS

Fomalhaut

GRUS

ANDROMEDA

TRIANGULUM

ARIES

PISCES

ECLIPTIC

CETUS

SCULPTOR

PHOENIX

TUCANA

47 Tuc

PERSEUS

Mira

FORNAX

ERIDANUS

Achernar

HYDRUS

Zenith 30° N

Pleiades

Zenith 20° N

Zenith 10° N

HOROLOGIUM

DORADO

RETICULUM

MENSA

TAURUS

Aldebaran

ORION

LEPUS

CAELEM

COLUMBA

Canopus

PICTOR

VOLANS

M42   Rigel

AURIGA

M35

Betelgeuse

Sirius

M41

CANIS MAJOR

Adhara

PUPPIS

CARINA

GEMINI

CANIS MINOR

MONOCEROS

VELA

Pollux

Procyon

PYXIS

M44

CANCER

Horizon 60° N

Horizon 50° N

Horizon 40° N

Horizon 30° N

Horizon 20° N

Horizon 10° N

LEO

HYDRA

Galaxy
Bright nebula
Globular cluster
Open cluster

Magnitudes: >0   0   1   2   3   4   5   var.

DECEMBER
Southern latitudes

Facing North

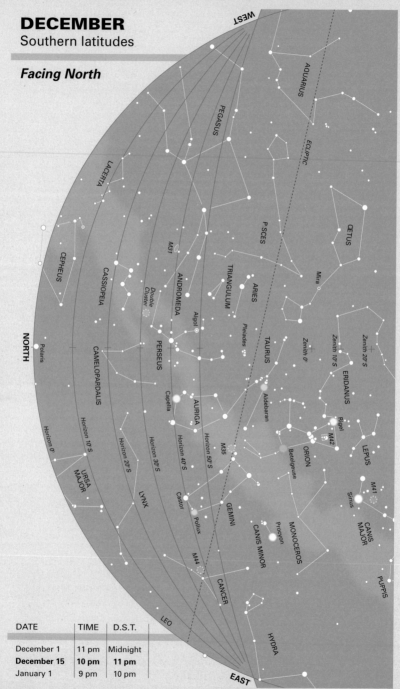

WEST

AQUARIUS

PEGASUS

ECLIPTIC

LACERTA

CETUS

P SCES

Mira

CEPHEUS

M31

CASSIOPEIA

Double Cluster

ANDROMEDA

TRIANGULUM

ARIES

Algol

Pleiades

Zenith 10° S

Zenith 20° S

ERIDANUS

NORTH

Polaris

CAMELOPARDALIS

PERSEUS

TAURUS

Zenith 0°

Rigel

Capella

Aldebaran

M42

LEPUS

AURIGA

Betelgeuse

ORION

Horizon 0°

Horizon 10° S

Horizon 20° S

Horizon 30° S

Horizon 40° S

Horizon 50° S

M35

URSA MAJOR

LYNX

Castor

GEMINI

M41

Sirius

CANIS MAJOR

Pollux

CANIS MINOR

MONOCEROS

Procyon

M44

CANCER

PUPPIS

LEO

HYDRA

EAST

DATE	TIME	D.S.T.
December 1	11 pm	Midnight
December 15	**10 pm**	**11 pm**
January 1	9 pm	10 pm

DECEMBER
Southern latitudes

Facing South

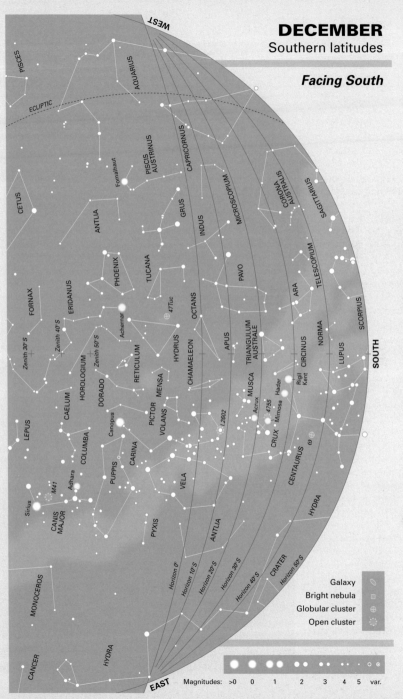

WEST

PISCES

AQUARIUS

ECLIPTIC

CETUS

Fomalhaut

PISCIS AUSTRINUS

CAPRICORNUS

ANTLIA

GRUS

MICROSCOPIUM

CORONA AUSTRALIS

SAGITTARIUS

FORNAX

PHOENIX

TUCANA

INDUS

PAVO

TELESCOPIUM

ERIDANUS

Achernar

47 Tuc

OCTANS

ARA

SCORPIUS

Zenith 30° S

Zenith 40° S

Zenith 50° S

RETICULUM

HYDRUS

CHAMAELEON

APUS

TRIANGULUM AUSTRALE

CIRCINUS

NORMA

SOUTH

CAELUM

HOROLOGIUM

DORADO

MENSA

MUSCA

Hadar

Rigil Kent

LUPUS

LEPUS

COLUMBA

Canopus

PICTOR

VOLANS

I.2602

Acrux

4755

Mimosa

CRUX

ω

CENTAURUS

PUPPIS

CARINA

VELA

Sirius

M41

Adhara

CANIS MAJOR

PYXIS

ANTLIA

HYDRA

MONOCEROS

CRATER

Horizon 0°

Horizon 10° S

Horizon 20° S

Horizon 30° S

Horizon 40° S

Horizon 50° S

CANCER

HYDRA

Galaxy

Bright nebula

Globular cluster

Open cluster

Magnitudes: >0 0 1 2 3 4 5 var.

EAST

Andromeda

Andromeda represents the daughter of Queen Cassiopeia who was chained to a rock as a sacrifice to the sea monster Cetus until saved by Perseus, whom she subsequently married. The constellation originated in ancient times. Despite its fame, Andromeda is not particularly striking: its brightest stars are only 2nd magnitude. Its most prominent feature is a crooked line of four stars extending from the Square of Pegasus; the first of these stars marks a corner of the Square, although it is actually part of Andromeda. This star, known both as Alpheratz and Sirrah, marks the head of the chained Andromeda; another star in the line, Mirach, represents her waist and a third, Almaak, is her chained foot. The most celebrated object in the constellation is the Andromeda Galaxy, M31, a spiral galaxy like our own Milky Way; it is the most distant object visible to the naked eye. Two stars leading from Mirach, β (beta) Andromedae, act as a guide to it.

α (alpha) Andromedae, 0h 08m +29°.1, (Alpheratz or Sirrah), mag. 2.1, is a blue-white star 97 l.y. away.

β (beta) And, 1h 10m +35°.6, (Mirach), mag. 2.1, is a red giant 199 l.y. away.

γ (gamma) And, 2h 04m +42°.3, (Almaak or Almach), 355 l.y. away, is an outstanding triple star. Its two brightest components, of mags. 2.3 and 4.8, form one of the finest pairs for small telescopes: their colours are orange and blue. The fainter, blue star also has a close 6th-mag. blue-white companion that orbits it every 64 years. This fainter pair is closest around the year 2015 and is unresolvable by amateur telescopes for several years either side. ▶

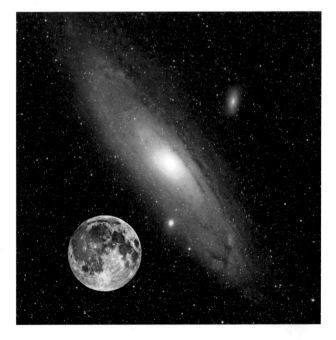

M31, the Andromeda Galaxy, is a magnificent spiral visible to the naked eye, with two smaller companion galaxies that can be seen in small telescopes: M32, superimposed on one of the spiral arms just below centre, and the larger but fainter M110 to the upper right. The full Moon has been added to the same scale to give an impression of the Andromeda spiral's full extent. (Galaxy image: Bill Schoening, Vanessa Harvey/REU program/ NOAO/AURA/NSF)

Andromeda · And · Andromedae

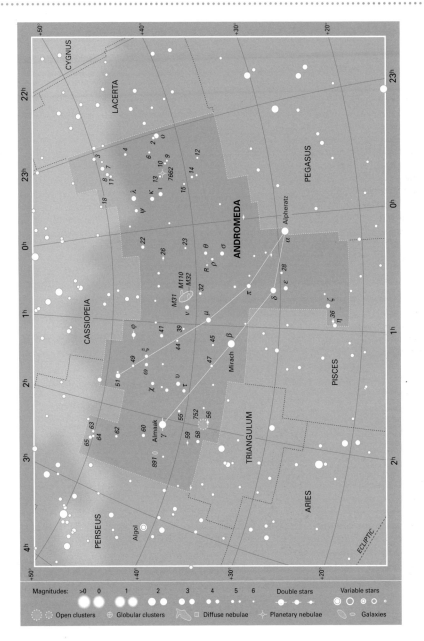

CYGNUS

LACERTA

PEGASUS

ANDROMEDA

Alpheratz

CASSIOPEIA

PISCES

PERSEUS

TRIANGULUM

ARIES

Mirach

Almaak

Algol

ECLIPTIC

Magnitudes: >0   0   1   2   3   4   5   6   Double stars   Variable stars

Open clusters   Globular clusters   Diffuse nebulae   Planetary nebulae   Galaxies

δ (delta) And, 0h 39m +30°.9, mag. 3.3, is an orange giant 101 l.y. away.

μ (mu) And, 0h 57m +38°.5, mag. 3.9, is a white star 136 l.y. away.

π (pi) And, 0h 37m +33°.7, 660 l.y. away, is a blue-white star of mag. 4.3 with a mag. 8.9 companion visible in small telescopes.

υ (upsilon) And, 1h 37m, +41°.4, mag. 4.1, is a yellow-white main-sequence star 44 l.y. away. Astronomers have found that it is accompanied by three planets, the first known multi-planet system around a star other than our own Sun.

56 And, 1h 56m +37°.3, is a yellow giant star of mag. 5.7, 320 l.y. away, with an orange giant companion, mag. 5.9 and 990 l.y. away, easily split in binoculars. The two stars are found near the star cluster NGC 752 but are actually closer to us.

M31 (NGC 224), 0h 43m +41°.3, the Andromeda Galaxy, is a spiral galaxy 2.5 million l.y. away. It is visible to the naked eye as an elliptical fuzzy patch, and becomes more prominent when seen through binoculars or a telescope with low magnification (too high a power reduces the contrast and renders the fainter parts of the galaxy less visible). Dark lanes can be seen in the spiral arms surrounding the nucleus. But the full extent of the galaxy becomes apparent only on long-exposure photographs (visual observers see just its brightest, central portion). If the entire Andromeda Galaxy were bright enough to be seen by the naked eye, it would appear five or six times the diameter of the full Moon, as demonstrated in the composite image on page 74. M31 is accompanied by two small satellite galaxies, the equivalents of our Magellanic Clouds but both elliptical rather than irregular in shape. The brighter of these, M32 (NGC 221), is visible in small telescopes as a fuzzy, 8th-mag. star-like glow ½° south of M31's core. The second companion, M110 (also known as NGC 205), is larger but visually more elusive, and over 1° northwest of M31.

NGC 752, 1h 58m +37°.7, is a widely spread cluster of about 60 stars of 9th mag. and fainter, 1200 l.y. away, visible in binoculars and easily resolved in telescopes.

NGC 7662, 23h 26m +42°.6, is one of the brightest and easiest planetary nebulae to see with a small telescope. At low powers it appears as a fuzzy, 9th-mag. blue-green star, but magnifications of ×100 or so reveal its slightly elliptical disk. Larger apertures show a central hole; the central star is a difficult object for amateur telescopes. NGC 7662 lies about 4000 l.y. away.

Antlia – The Air Pump

An obscure southern constellation introduced on a map published in 1756 by the French astronomer Nicolas Louis de Lacaille to commemorate the air pump invented by the French physicist Denis Papin. Lacaille, the first person to map the naked-eye stars in the southern skies completely (see page 216), introduced 14 new constellations to fill gaps between existing ones. Most of these new figures are unremarkable, as is Antlia.

α (alpha) Antliae, 10h 27m −31°.1, mag. 4.3, the constellation's brightest star, is an orange giant 366 l.y. distant.

δ (delta) Ant, 10h 30m −30°.6, 481 l.y. away, is a blue-white star of mag. 5.6 with a mag. 9.6 companion visible in small telescopes.

ζ¹ ζ² (zeta¹ zeta²) Ant, 9h 31m −31°.9, 373 l.y. away, is a wide pair of stars of mags. 5.8 and 5.9, visible in binoculars. Small telescopes show that ζ¹ Ant is itself double, with components of mags. 6.2 and 7.0.

▶

Antlia · Ant · Antliae

NGC 2997: Knots of young blue stars and pink gas as well as lanes of dark dust are visible in the curling arms of this handsome 9th-magnitude spiral galaxy in central Antlia, presented virtually face-on to us. Moderate-sized amateur telescopes are required to see it. (M. Bessell, R. Sutherland and M. Buxton, RSAA – Australian National University)

Apus – The Bird of Paradise

A faint constellation near the south celestial pole, one of the 12 new figures introduced in the 1590s by the Dutch navigators Pieter Dirkszoon Keyser and Frederick de Houtman during voyages to the southern hemisphere. This one represents the bird of paradise, native to New Guinea.

α (alpha) Apodis, 14h 48m −79°.0, mag. 3.8, is an orange giant star 411 l.y. away.

β (beta) Aps, 16h 43m −77°.5, mag. 4.2, is an orange giant 158 l.y. away.

γ (gamma) Aps, 16h 33m −78°.9, mag. 3.9, is an orange giant 160 l.y. away.

δ¹ δ² (delta¹ delta²) Aps, 16h 20m −78°.7, is a naked-eye or binocular pair of red and orange giant stars of mags. 4.7 and 5.3, 765 and 663 l.y. away.

θ (theta) Aps, 14h 05m −76°.8, 328 l.y. away, is a red giant that varies semi-regularly between 5th and 7th mags. every 4 months or so.

Apus · Aps · Apodis

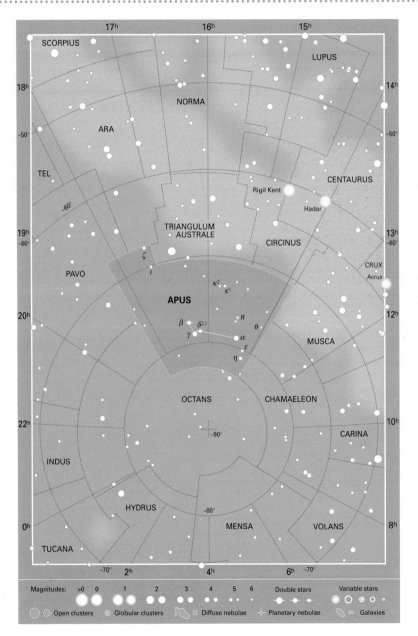

Aquarius – The Water Carrier

Aquarius is one of the most ancient constellations. The Babylonians saw in this area of sky the figure of a man pouring water from a jar. In Greek mythology the constellation represents Ganymede, a shepherd boy carried off by Zeus to Mount Olympus, where he became wine-waiter to the gods.

The most prominent part of Aquarius is the Y-shaped asterism of four stars representing the Water Jar itself, centred on the star ζ (zeta) Aquarii. Aquarius is in an area of 'watery' constellations that includes Pisces, Cetus and Capricornus. The Sun is within the boundaries of Aquarius from late February to early March. Aquarius will one day contain the vernal equinox, the point at which the Sun crosses into the northern celestial hemisphere each year. This astronomically important point, from which the coordinate of right ascension is measured, will move into Aquarius from neighbouring Pisces in AD 2597 because of the effect of precession. Hence the so-called Age of Aquarius is a long way off yet.

Three main meteor showers radiate from Aquarius each year. The first, the Eta Aquarids, is the richest, reaching a maximum of 35 meteors per hour around May 5–6. The Delta Aquarids have a double radiant: the southerly one produces about 20 meteors per hour around July 29, while the northerly radiant (which is actually near the Circlet of Pisces) peaks at about 10 an hour on August 6. A weaker stream, the Iota Aquarids, produces a maximum of 8 meteors per hour, also on August 6. Each shower is named after the bright star closest to its radiant.

α (alpha) Aquarii, 22h 06m −0°.3, (Sadalmelik, from the Arabic for 'the lucky stars of the king'), mag. 2.9, is a yellow supergiant 760 l.y. away.

β (beta) Aqr, 21h 32m −5°.6, (Sadalsuud, from the Arabic for 'luckiest of the lucky stars'), mag. 2.9, is a yellow supergiant 610 l.y. away.

γ (gamma) Aqr, 22h 22m −1°.4, (Sadachbia), mag. 3.9, is a blue-white star 158 l.y. away.

δ (delta) Aqr, 22h 55m −15°.8, (Skat), mag. 3.3, is a blue-white star 160 l.y. away.

ε (epsilon) Aqr, 20h 48m −9°.5, (Albali), mag. 3.8, is a blue-white star 102 l.y. away.

ζ (zeta) Aqr, 22h 29m −0°.0, 103 l.y. away, is a celebrated binary consisting of twin white stars of mags. 4.3 and 4.5 orbiting each other every 590 years. The two stars are gradually moving apart as seen from Earth and hence are becoming increasingly easy to divide in small telescopes. ▶

NGC 7009 is popularly known as the Saturn Nebula because of the faint 'handles' either side of it, like Saturn's rings, that can be seen through large telescopes and in photographs. The name Saturn Nebula was given to it in the 19th century by Lord Rosse. (SAAO)

Aquarius · Aqr · Aquarii

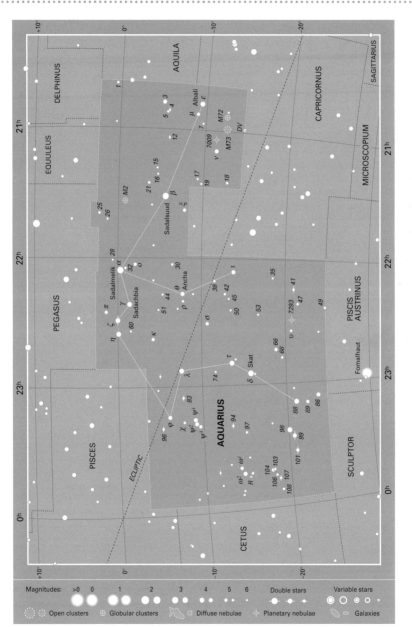

Magnitudes: >0 0 1 2 3 4 5 6   Double stars   Variable stars

Open clusters   Globular clusters   Diffuse nebulae   Planetary nebulae   Galaxies

M2 (NGC 7089), 21h 34m −0°.8, is a mag. 6.5 globular cluster easily visible in binoculars or small telescopes, but requiring 100 mm aperture to resolve the brightest individual stars. This rich and highly concentrated globular lies 37,000 l.y. away.

M72 (NGC 6981), 20h 54m −12°.5, is a 9th-mag. globular cluster 56,000 l.y. away, much smaller and less impressive than M2.

NGC 7009, 21h 04m −11°.4, is a famous planetary nebula, 3000 l.y. away, known as the Saturn Nebula because of its resemblance to that planet when seen in large telescopes (see the photograph on page 80). But in most amateur telescopes, of 75 mm aperture or more, it appears as merely an 8th-mag. blue-green ellipse of similar apparent size to the globe of Saturn. Its central star is of mag. 11.5.

NGC 7293, 22h 30m −20°.8, is the nearest planetary nebula to the Sun, about 700 l.y. away, and is commonly known as the Helix Nebula. It is the largest planetary nebula in apparent size, covering ¼° of sky, half the apparent size of the Moon. Its popular name arises because on early photographs it appeared to consist of two overlapping turns of a spiral, although that

NGC 7293, the Helix Nebula, appears as a colourful ring on photographs, but visually is only a pale grey disk. (Nik Szymanek)

structure is less apparent on modern colour images which instead emphasize its ring shape (see the photograph above). Despite its size the Helix Nebula is quite faint and is best found with binoculars or very low power on a telescope, when it appears as a circular misty patch, not as impressive as its large size would suggest.

Aquila – The Eagle

A constellation dating from ancient times, representing the bird that in Greek mythology carried the thunderbolts of Zeus. According to Greek myth, Zeus sent an eagle (or, in a variant of the tale, turned himself into an eagle) to abduct the shepherd boy Ganymede, represented in the sky by the neighbouring constellation of Aquarius.

Aquila's brightest star, Altair, forms one corner of a giant stellar triangle (known in the northern hemisphere as the Summer Triangle) that is completed by Deneb in Cygnus and Vega in Lyra. The name Altair comes from the Arabic *al-nasr al-tair*,

meaning 'the flying eagle'. Two fainter stars, β (beta) and γ (gamma) Aquilae, stand like sentinels either side of it; these are called Alshain and Tarazed, both from the Persian *shahin-i tarazu*, a translation of an Arabic name meaning 'the balance' which referred jointly to these two stars and Altair. Aquila lies in the Milky Way and contains rich starfields, particularly towards Scutum in the south. It is an abundant area for novae.

α (alpha) Aquilae, 19h 51m +8°.9, (Altair, 'the flying eagle'), mag. 0.76, is a white star 17 l.y. away, among the closest naked-eye stars to us. ▶

Aquila · Aql · Aquilae

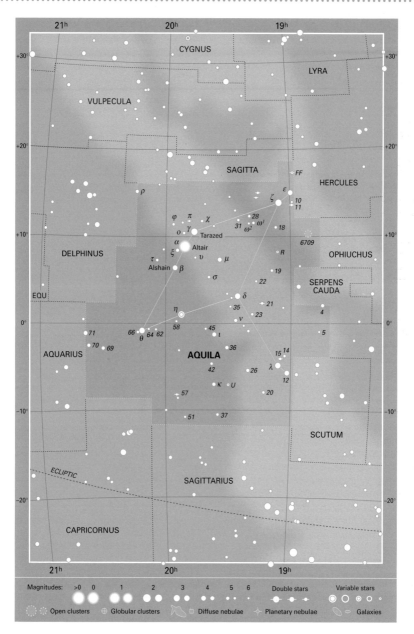

β (beta) Aql, 19h 55m +6°.4, (Alshain), mag. 3.7, is a yellow star 45 l.y. away.

γ (gamma) Aql, 19h 46m +10°.6, (Tarazed), mag. 2.7, is an orange giant star 460 l.y. away.

ζ (zeta) Aql, 19h 05m +13°.9, mag. 3.0, is a blue-white star 83 l.y. away.

η (eta) Aql, 19h 52m +1°.0, a yellow-white super-giant 1200 l.y. away, is one of the brightest Cepheid variable stars. Its brightness ranges from mag. 3.5 to 4.4 with a period of 7.2 days.

15 Aql, 19h 05m −4°.0, is an orange giant star of mag. 5.4, 325 l.y. away, with a purplish mag. 7.0 companion, 550 l.y. away, easily visible in small telescopes.

57 Aql, 19h 55m −8°.2, is an easy double for small telescopes, consisting of a bluish star of mag. 5.7 with a mag. 6.5 companion, both about 350 l.y. away.

R Aql, 19h 06m +8°.2, 690 l.y. away, is a red giant variable of Mira type, about 400 times the diameter of the Sun, ranging from 6th to 12th mag. every 9 months or so.

FF Aql, 18h 58m +17°.4, is a yellow-white super-giant Cepheid variable that ranges from mag. 5.2 to 5.7 every 4.5 days. It lies about 2500 l.y. away.

NGC 6709, 18h 52m +10°.3, is a loosely scattered cluster of some 40 stars of mags. 9 to 11, about 3000 l.y. away.

Ara – The Altar

This constellation, although faint and relatively little-known, originated with the Greeks, who visualized it as the altar on which the gods of Olympus swore an oath of allegiance before their war against the Titans, who at that time ruled the Universe. After their victory, the Olympian gods divided up the Universe between them. Zeus, the greatest of them all, was allotted the sky and placed the altar of the gods among the stars in lasting memory of their victory. Old star atlases depict Ara as the altar on which Centaurus, the centaur, is about to sacrifice Lupus, the wolf. Ara lies in a rich part of the Milky Way, south of Scorpius.

α (alpha) Arae, 17h 32m −49°.9, mag. 2.8, is a blue-white star 242 l.y. away.

β (beta) Ara, 17h 25m −55°.5, mag. 2.8, is an orange supergiant 600 l.y. away.

γ (gamma) Ara, 17h 25m −56°.4, mag. 3.3, is a blue supergiant 1140 l.y. away.

δ (delta) Ara, 17h 31m −60°.7, mag. 3.6, is a blue-white star 187 l.y. away.

ζ (zeta) Ara, 16h 59m −56°.0, mag. 3.1, is an orange giant star 574 l.y. away.

NGC 6193, 16h 41m −48°.8, is a 5th-mag. open cluster of about 30 stars 4200 l.y. away, about half the apparent size of the full Moon. The brightest member is a blue-white star of mag. 5.6 which small telescopes show has a companion of mag. 6.9. Around the cluster is an irregularly shaped patch of faint nebulosity, NGC 6188, which shows up well only on photographs.

NGC 6397, 17h 41m −53°.7, is a 6th-mag. globular cluster that appears as a fuzzy star through binoculars; under good conditions it is visible to the naked eye. Small telescopes resolve the brightest stars in its outer regions, which extend across at least half the apparent diameter of the full Moon. It is one of the closer globulars to us, 10,500 l.y. away.

Ara · Ara · Arae

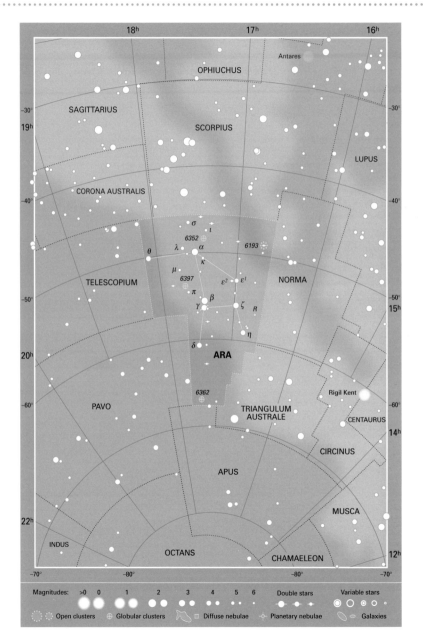

Magnitudes: >0 0 1 2 3 4 5 6   Double stars   Variable stars

Open clusters   Globular clusters   Diffuse nebulae   Planetary nebulae   Galaxies

Aries – The Ram

A constellation lying between Taurus and Andromeda, whose origin dates back to ancient times. Aries represents the ram of Greek legend whose golden fleece was sought by Jason and the Argonauts. Despite its faintness, Aries has assumed great importance in astronomy because, over 2000 years ago, it contained the point where the Sun passes from south to north across the celestial equator each year. This point, the vernal equinox, is still sometimes known as the First Point of Aries, although it no longer lies in Aries, having moved into neighbouring Pisces in 67 BC as a result of the slow wobble of the Earth in space known as precession. It is now headed for Aquarius (see diagram below). Currently, the Sun passes through Aries from late April to mid-May.

α (alpha) Arietis, 2h 07m +23°.5, (Hamal, from the Arabic for 'lamb'), mag. 2.0, is an orange giant star 66 l.y. away.

β (beta) Ari, 1h 55m +20°.8, (Sheratan, from the Arabic for 'two'), mag. 2.6, is a blue-white star 60 l.y. away.

γ (gamma) Ari, 1h 54m +19°.3, (Mesartim), 204 l.y. away, is a striking double consisting of twin white stars of mags. 4.7 and 4.6, clearly visible through small telescopes even under low magnification.

ε (epsilon) Ari, 2h 59m +21°.3, 290 l.y. away, is a challenging double star for apertures of 100 mm or more. High magnification reveals a tight pair of white stars of mags. 5.2 and 5.5.

λ (lambda) Ari, 1h 58m +23°.6, 133 l.y. away, is a white star of mag. 4.8 with a yellow mag. 7.3 companion visible in small telescopes or even good binoculars.

π (pi) Ari, 2h 49m +17°.5, 600 l.y. away, is a blue-white star of mag. 5.3, with a close mag. 8.5 companion, difficult to distinguish in the smallest telescopes.

Movement of the so-called First Point of Aries over 800 years. At present it lies in Pisces and is approaching Aquarius. (Wil Tirion)

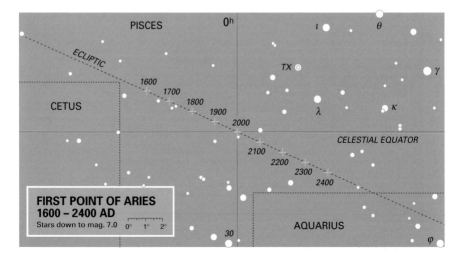

Aries · Ari · Arietis

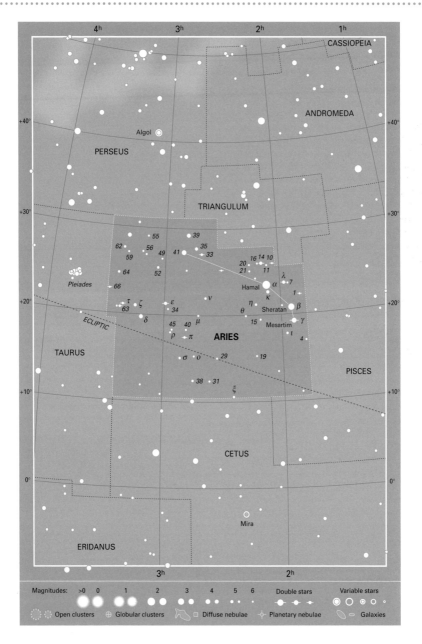

Auriga – The Charioteer

A large and prominent constellation, usually identified by the ancient Greeks as Erichthonius, a legendary king of Athens and a skilled charioteer. Auriga's leading star is Capella, sixth-brightest in the whole sky and the most northerly first-magnitude star. In legend this star represented the she-goat Amaltheia which suckled the infant Zeus; the stars ζ (zeta) and η (eta) Aurigae are supposedly her kids and were visualized as being carried in the charioteer's arms. The star marking the charioteer's foot was originally shared with Taurus the Bull, but it is now assigned exclusively to Taurus as β (beta) Tauri. However, it once bore the alternative designation of γ (gamma) Aurigae and is shown on the chart to complete the classical figure.

α (alpha) Aurigae, 5h 17m +46°.0, (Capella, 'she-goat'), mag. 0.08, lies 42 l.y. away. It is a spectroscopic binary, consisting of two yellow giant stars orbiting every 104 days, although they do not eclipse each other.

β (beta) Aur, 6h 00m +44°.9, (Menkalinan, 'shoulder of the charioteer'), 82 l.y. away, is an eclipsing variable of mag. 1.9 consisting of two blue-white stars that orbit every 3.96 days, causing two dips in brightness of 0.1 mag. on each orbit.

ε (epsilon) Aur, 5h 02m +43°.6, a white supergiant about 2000 l.y. away, is an eclipsing binary of exceptionally long period. Normally it appears at mag. 3.0, but every 27 years it sinks to mag. 3.8 as it is eclipsed by a dark companion, remaining at minimum for a year. One theory is that ε Aurigae's companion is a binary star enveloped in a disk of matter. Its next eclipse starts in late 2009.

ζ (zeta) Aur, 5h 02m +41°.1, 790 l.y. away, is a famous eclipsing binary of contrasting stars: an orange giant orbited every 972 days by a smaller blue companion. During eclipses, ζ Aurigae's brightness drops from mag. 3.7 to 4.0.

θ (theta) Aur, 6h 00m +37°.2, 173 l.y. away, is a blue-white star of mag. 2.6. It has a yellowish companion of mag. 7.1 which, because of its closeness and relative faintness, needs at least 100 mm aperture and high magnification to distinguish. This is a tough double for steady nights.

ψ¹ (psi¹) Aur, 6h 25m +49°.3, is a variable orange supergiant that ranges between mags. 4.8 and 5.7 with no set period. Its distance is uncertain.

4 Aur, 4h 59m +37°.9, 159 l.y. away, is a double star of mags. 5.0 and 8.1, divisible through small telescopes.

14 Aur, 5h 15m +32°.7, 270 l.y. away, is a white star of mag. 5.0 with a mag. 7.9 companion, 82 l.y. away, split in small telescopes.

RT Aurigae, 6h 29m +30°.5, a yellow-white supergiant, is a Cepheid variable ranging between mag. 5.0 and 5.8 every 3.7 days. It lies about 1600 l.y. away. ▶

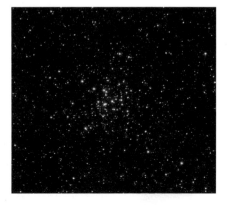

M36 is the most prominent of the three Messier clusters in Auriga when viewed through small instruments. It displays crooked chains of stars and a triangular centre. (Peter Wienerroither)

Auriga · Aur · Aurigae

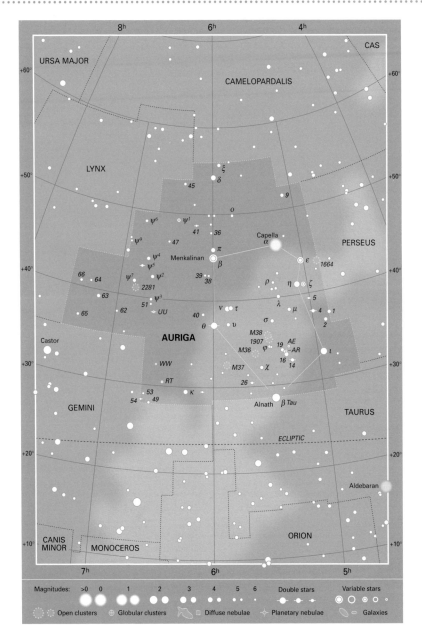

UU Aur, 6h 37m +38°.4, is a semi-regular variable, ranging between 5th and 7th mags. with a rough period of 234 days. It is a giant star appearing deep red in colour, around 2000 l.y. away.

M36 (NGC 1960), 5h 36m +34°.1, is a small, bright open cluster of about 60 stars, visible in binoculars and resolvable into stars in small tele-scopes (see the picture on the previous page). It appears about one-third the diameter of the full Moon and is the most prominent of Auriga's three Messier clusters. M36 lies 3900 l.y. away.

M37 (NGC 2099), 5h 52m +33°.5, is the largest and richest of the three Messier clusters in Auriga, containing about 150 stars. In binoculars the cluster appears as a hazy, unresolved patch about half the diameter of the full Moon, but a 100-mm telescope resolves it into a sparkling field of faint stardust, with a brighter orange star at the centre. It lies 4200 l.y. away.

M38 (NGC 1912), 5h 29m +35°.8, is a large, scattered cluster of about 100 faint stars, visible in binoculars, with a noticeable cross-shape when seen through a telescope. Its distance is 3900 l.y. Half a degree south of it lies the small fuzzy blob of NGC 1907, a much smaller and fainter cluster 4200 l.y. away.

NGC 2281, 6h 49m +41°.1, is a binocular cluster of about 30 stars, 1500 l.y. away. Through a tele-scope the stars appear to be arranged in a crescent, with four brighter stars forming a diamond shape.

Boötes – The Herdsman

An ancient constellation, representing a herdsman driving a bear (Ursa Major) around the sky; the herdsman is often depicted holding the leash of the hunting dogs, Canes Venatici. The name of the con-stellation's brightest star, Arcturus, actually means 'bear-keeper' in Greek. In Greek mythology Boötes represents Arcas, the son of Zeus and the nymph Callisto (neigh-bouring Ursa Major represents Callisto herself, who was turned into a bear by Zeus's jealous wife Hera).

Arcturus is the brightest star in the north-ern half of the sky, and is easily identified: the curving handle of the Big Dipper or Plough points to it. Arcturus forms the base of a large 'Y' shape completed by ε (epsilon) Boötis, γ (gamma) Boötis and α (alpha) Coronae Borealis. The other stars of the con-stellation are much fainter than Arcturus, but include many doubles of interest.

The year's most abundant meteor shower, the Quadrantids, radiates from the northern part of Boötes, an area of sky that was once occupied by the now-abandoned constel-lation of Quadrans Muralis, the Mural Quadrant (hence the shower's name). The Quadrantid meteors reach a peak of about 100 per hour on January 3–4 each year, although they are not as bright as other rich showers such as the Perseids and Geminids.

α (alpha) Boötis, 14h 16m +19°.2, (Arcturus), mag. −0.05, is the fourth-brightest star in the entire sky. It is an orange giant, about 27 times the diameter of the Sun, lying 37 l.y. away; its ruddy colour is noticeable to the naked eye, and is more striking with optical aid. Arcturus has a mass similar to our Sun's, and it is believed that the Sun will swell up to become a red giant like Arcturus 5000 million years from now.

β (beta) Boo, 15h 02m +40°.4, (Nekkar, corrupted from the Arabic for 'ox-driver', referring to the whole constellation), mag. 3.5, is a yellow giant 219 l.y. away.   ▶

Boötes · Boo · Boötis

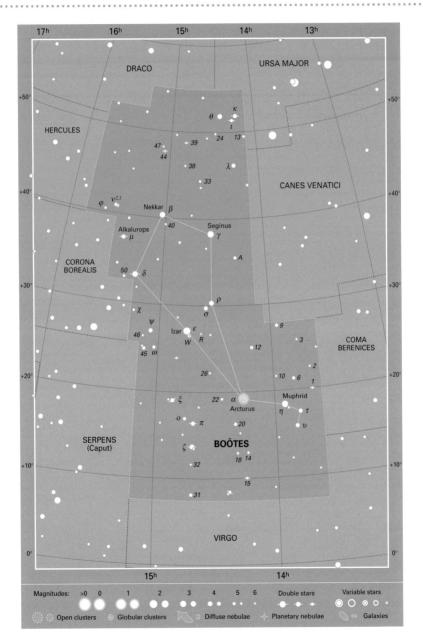

γ (gamma) Boo, 14h 32m +38°.3, (Seginus), mag. 3.0, is a white star 85 l.y. away.

δ (delta) Boo, 15h 16m +33°.3, mag. 3.5, is a yellow giant 117 l.y. away. It has a wide binocular companion of mag. 7.8.

ε (epsilon) Boo, 14h 45m +27°.1, (Izar, 'girdle' or 'loincloth'), 210 l.y. away, is a celebrated double star: an orange giant primary of mag. 2.5 with a blue main-sequence companion of mag. 4.6. This close double of contrasting colours requires a telescope of at least 75 mm at ×100 power or more, because the bright primary tends to overwhelm its fainter companion; but its appearance when split has led to the alternative name Pulcherrima, meaning 'most beautiful'.

ι (iota) Boo, 14h 16m +51°.4, 97 l.y. away, is a wide double star of mags. 4.8 and 7.5.

κ (kappa) Boo, 14h 13m +51°.8, is an easy double star for small telescopes, consisting of unrelated components of mags. 4.5 and 6.6, distances 155 and 196 l.y.

μ (mu) Boo, 15h 24m +37°.4, (Alkalurops, 'club' or 'staff'), 121 l.y. distant, is an attractive triple star. To the naked eye it appears as a blue-white star of mag. 4.3, but binoculars reveal a wide com-

panion of mag. 6.5. Telescopes of 75 mm aperture with high magnification show that this companion actually consists of two close stars of mags. 7.0 and 7.6; they orbit each other every 260 years.

ν¹ ν² (nu¹ nu²) Boo, 15h 31m +40°.8, is an unrelated binocular duo: ν¹ is an orange giant of mag. 5.0, 870 l.y. distant, while ν² is a white star also of mag. 5.0, 430 l.y. away.

π (pi) Boo, 14h 41m +16°.4, 317 l.y. away, is a double star with blue-white components of mags. 4.9 and 5.8, visible in small telescopes.

ξ (xi) Boo, 14h 51m +19°.1, 22 l.y. away, is a showpiece double for small telescopes, consisting of yellow and orange stars of mags. 4.7 and 7.0, orbiting each other every 150 years.

44 Boo, 15h 04m +47°.7, 42 l.y. away, is a complex double–variable star. To the naked eye it appears as a yellow star of mag. 4.8. In fact, it is a binary of mags. 5.3 and 6.1 orbiting every 206 years. Until the year 2011 the two can be split in apertures of 100 mm, but thereafter they will become more difficult as they move together until, around their closest in 2019–20, they will be indivisible in amateur telescopes. The fainter star is itself an eclipsing binary with a period of 6.4 hours and a range of 0.6 mag.

Caelum – The Chisel

An obscure, almost irrelevant constellation at the foot of Eridanus, representing a pair of engraving tools, or burins. It was one of the figures representing instruments of the arts and sciences introduced in the 1750s by the French astronomer Nicolas Louis de Lacaille during his mapping of the southern sky.

α (alpha) Caeli, 4h 41m −41°.9, mag. 4.4, is a white main-sequence star 66 l.y. away.

β (beta) Cae, 4h 42m −37°.1, mag. 5.0, is a white star 90 l.y. away.

γ (gamma) Cae, 5h 04m −35°.5, mag. 4.6, is an orange giant 185 l.y. away. It has a close mag. 8.1 companion, difficult to see in the smallest apertures because of the brightness contrast.

δ (delta) Cae, 4h 31m −45°.0, mag. 5.1, is a blue-white star 710 l.y. away.

Caelum · Cae · Caeli

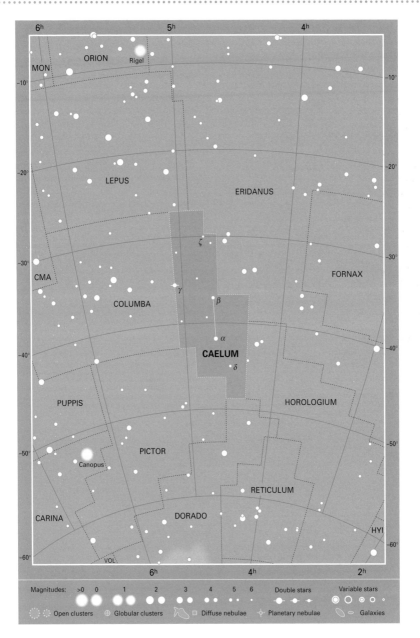

Camelopardalis – The Giraffe

A faint and obscure constellation of the north polar region of the sky, sometimes written as Camelopardus, an obsolete variant of the name. It was invented in 1613 by the Dutch theologian and astronomer Petrus Plancius. The German astronomer Jacob Bartsch wrote that it represented the camel on which Rebecca rode into Canaan for her marriage to Isaac. But Camelopardalis is a giraffe not a camel, so its true meaning remains unclear.

α (alpha) Camelopardalis, 4h 54m +66°.3, mag. 4.3, is a highly luminous blue supergiant star approximately 5000 l.y. away, exceptionally distant for a naked-eye star.

β (beta) Cam, 5h 03m +60°.4, at mag. 4.0 the brightest star in the constellation, is a yellow supergiant 1000 l.y. distant. It has a wide mag. 8.6 companion star visible in small telescopes or even good binoculars.

11 Cam, 5h 06m +59°.0, mag. 5.2, forms an easy binocular pairing with 12 Cam, mag. 6.1. Both lie at a similar distance, about 650 l.y., but are too widely separated to be a genuine double.

Σ 1694 (Struve 1694), 12h 49m +83°.4, 300 l.y. away, is a pair of blue-white stars, mags. 5.4 and 5.9, easily split in small telescopes. In some old catalogues the star was listed as 32 Cam.

NGC 1502, 4h 08m +62°.3, is a small 6th-mag. open star cluster with about 45 members visible in binoculars or a small telescope, somewhat triangular in shape and with an easy 7th-mag. double star at its centre. It lies 3000 l.y. away. Note a chain of 15 or more stars called Kemble's Cascade which runs for 2½° from NGC 1502 towards Cassiopeia, parallel to the Milky Way (see illustration below). The brightest star in the chain is of 5th magnitude, about halfway along.

NGC 2403, 7h 37m +65°.6, is an 8th-mag. spiral galaxy ¼° long, visible as an elliptical glow in a 100-mm telescope. It lies about 12 million l.y. away.

Kemble's Cascade is a chain of faint stars visible in binoculars and small telescopes near the open cluster NGC 1502. (Wil Tirion)

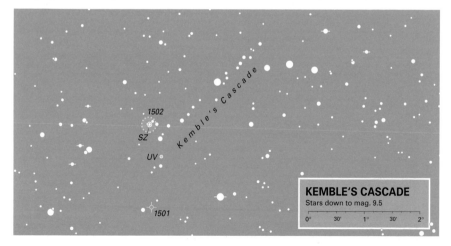

KEMBLE'S CASCADE
Stars down to mag. 9.5

| 0° | 30' | 1° | 30' | 2° |

Camelopardalis · Cam · Camelopardalis

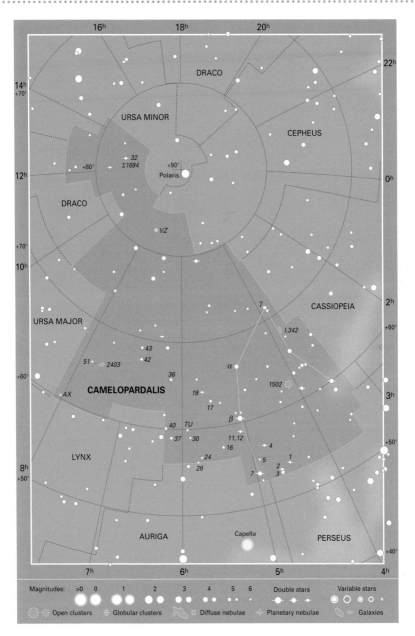

Cancer – The Crab

Cancer represents the crab that attacked Hercules when he was fighting the multi-headed Hydra; the luckless crab was crushed underfoot by mighty Hercules but was subsequently elevated to the heavens.

In ancient times, the Sun reached its most northerly point in the sky each year while it was in Cancer. The date on which the Sun is farthest north of the Earth's equator, on or around June 21, is the northern summer solstice; on this day the Sun appears overhead at noon at latitude 23½° north on Earth. This latitude came to be known as the Tropic of Cancer, a name it retains today even though, because of the effect of precession, the Sun now lies in the constellation of Taurus on that date.

With only two stars brighter than mag. 4.0, Cancer is the faintest of the 12 constellations of the zodiac, but it nevertheless contains much of interest, notably the star cluster Praesepe, the Manger. Praesepe is flanked by a pair of 4th- and 5th-magnitude stars, Asellus Borealis and Asellus Australis, the northern and southern donkeys, visualized as feeding at the stellar manger.

α (alpha) Cancri, 8h 58m +11°.9, (Acubens, 'the claw'), mag. 4.3, is a white star 174 l.y. away. It has a 12th-mag. companion visible with telescopes of 75 mm aperture and over.

β (beta) Cnc, 8h 17m +9°.2, mag. 3.5, an orange giant 290 l.y. away, is the brightest star in the constellation.

γ (gamma) Cnc, 8h 43m +21°.5, (Asellus Borealis, 'northern donkey'), mag. 4.7, is a white star 158 l.y. away.

δ (delta) Cnc, 8h 45m +18°.2, (Asellus Australis, 'southern donkey'), mag. 3.9, is an orange giant star 136 l.y. away.

ζ (zeta) Cnc, 8h 12m +17°.6, 83 l.y. away, is an interesting multiple star. A small telescope reveals two yellow stars of mags. 5.0 and 6.2; they form a genuine binary with an estimated orbital period of 1100 years. Larger telescopes split the brighter component into a tight binary of mags. 5.6 and 6.0, orbital period 59.5 years. These two stars are currently moving apart; at their widest, around the year 2018, 100 mm should be sufficient to split them, but 150 mm will be needed before about 2010.

ι (iota) Cnc, 8h 47m +28°.8, 298 l.y. away, is a yellow giant of mag. 4.0 with a blue-white mag. 6.6 companion just visible in binoculars, or easily seen through a small telescope.

M44 (NGC 2632), 8h 40m +20°.0, Praesepe ('manger'), commonly called the Beehive Cluster, is a swarm of about 50 stars of 6th mag. and fainter, visible as a misty patch to the naked eye and best seen through binoculars. The brightest member of this open cluster, ε (epsilon) Cancri, is mag. 6.3. Praesepe sprawls over 1½° of sky, three times the apparent diameter of the Moon. The distance to the cluster's centre is 577 l.y. according to the Hipparcos measurements.

M67 (NGC 2682), 8h 50m +11°.8, is a smaller and denser open cluster than M44, visible as a Moon-sized misty ellipse in binoculars or small telescopes. Apertures of at least 75 mm are needed to resolve the brightest of its 200 or so individual stars, which are of 10th mag. and fainter. It lies 2500 l.y. away.

Cancer · Cnc · Cancri

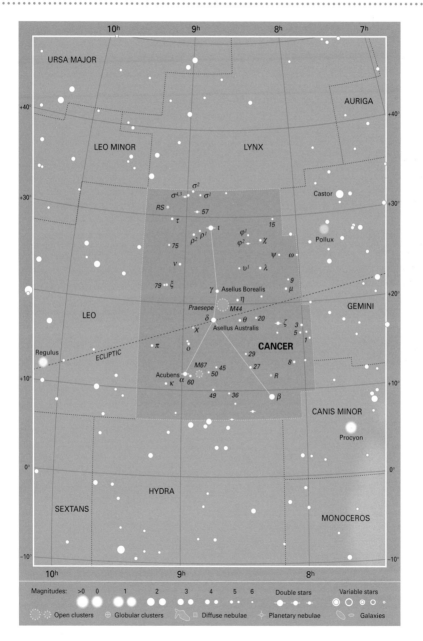

Canes Venatici – The Hunting Dogs

A constellation introduced in 1687 by the Polish astronomer Johannes Hevelius, consisting of a sprinkling of faint stars below Ursa Major. It represents two dogs, Asterion ('little star') and Chara ('joy'), held on a leash by neighbouring Boötes as they pursue the Great Bear around the pole. Canes Venatici contains numerous galaxies, the most famous being M51, the Whirlpool, a beautiful face-on spiral (pictured on page 290). It was the first galaxy in which spiral form was detected, by Lord Rosse in 1845 with his 72-inch (1.8-m) reflector at Birr Castle, Ireland.

α (alpha) Canum Venaticorum, 12h 56m +38°.3, is popularly called Cor Caroli, meaning 'Charles's heart', a reference to the executed King Charles I of England; it is reputed, doubtless apocryphally, to have shone particularly brightly in 1660 on the arrival of Charles II in England at the Restoration of the monarchy. It is a double star of mags. 2.9 and 5.6, easily split in small telescopes. Both stars are white, but various observers have reported subtle shades of colour in them when viewed through a telescope. The brighter star is the standard example of a rare class of stars with strong and variable magnetic fields; its brightness fluctuates slightly but not enough to be noticeable to the eye. Cor Caroli is 110 l.y. away.

β (beta) CVn, 12h 34m +41°.4, (Chara, 'joy'), mag. 4.2, is the only other star of any prominence in the constellation. It is a yellow main-sequence star similar to the Sun, 27 l.y. away.

Y CVn, 12h 45m +45°.4, 710 l.y. away, is a semi-regular variable supergiant of deep red colour sometimes known as La Superba. Its range is about mag. 5.0 to 6.5 and its period is approximately 160 days.

M3 (NGC 5272), 13h 42m +28°.4, is a rich globular cluster located midway between Cor Caroli and Arcturus, regarded as one of the finest globulars in the northern sky. At 6th mag. it is on the naked-eye limit, but can be picked up easily as a hazy star in binoculars or a small telescope; a 6th-mag. star nearby acts as a guide. In small telescopes the cluster appears as a condensed ball of light with a faint outer halo. Apertures of 100 mm or more are needed to resolve individual stars in its outer regions. M3 is 32,000 l.y. away.

M51 (NGC 5194), 13h 30m +47°.2, the Whirlpool Galaxy, is an 8th-mag. spiral galaxy about 26 million l.y. away with a smaller satellite galaxy, NGC 5195, apparently lying at the end of one of its arms; in reality, this companion lies slightly behind M51, having brushed past it some time in the past 100 million years or so. The Whirlpool can be seen in binoculars, appearing elongated. It is disappointing in small telescopes, which show a faint milky radiance around the starlike nuclei of the galaxy and its satellite; apertures of at least 250 mm are needed to see the spiral arms well. Nevertheless, M51 is well worth hunting for on clear, dark nights. (For a photograph of M51 and NGC 5195 see page 290.)

M63 (NGC 5055), 13h 16m +42°.0, is a 9th-mag. spiral galaxy visible in small telescopes as an elliptical haze with a somewhat mottled texture. It is popularly known as the Sunflower Galaxy because of its appearance in large instruments.

M94 (NGC 4736), 12h 51m +41°.1, is a compact spiral galaxy presented nearly face-on. In amateur telescopes it looks like an 8th-mag. comet, with a fuzzy star-like nucleus surrounded by an elliptical halo. M94 is about 15 million l.y. away.

Canes Venatici · CVn · Canum Venaticorum

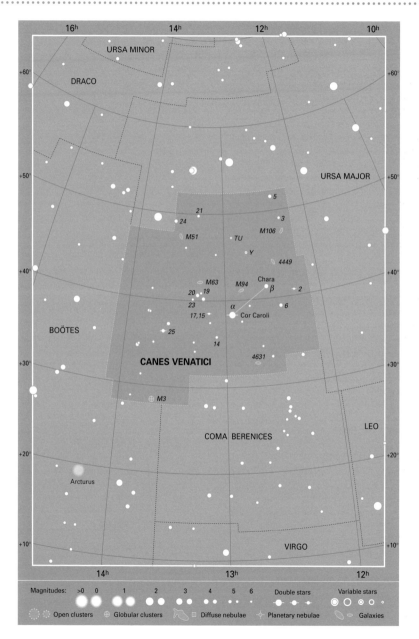

Canis Major – The Greater Dog

An ancient constellation, representing one of the two dogs (the other being Canis Minor) following at the heels of Orion. In other Greek legends it was identified as Laelaps, a dog so swift that no prey could outrun it. Laelaps was sent in pursuit of a fox that could never be caught. The chase would have continued for eternity had not Zeus plucked up the dog and placed it in the sky as Canis Major, although without the fox.

Canis Major contains many brilliant stars, making it one of the most prominent constellations; its leading star, Sirius, is the brightest star in the entire sky as seen from Earth. Sirius features in many legends, and the ancient Egyptians based their calendar on its yearly motion around the sky.

α (alpha) Canis Majoris, 6h 45m −16°.7, (Sirius, from the Greek meaning 'searing' or 'scorching'), mag. −1.44, is a brilliant white star 8.6 l.y. away, one of the Sun's closest neighbours; the Hipparcos satellite detected variability of 0.1 mag. or so in its brightness. It has a white dwarf companion of mag. 8.4 that orbits it every 50 years. The brilliance of Sirius overpowers this white dwarf so that even when the two stars are at their greatest separation, as between the years 2020 and 2025, telescopes of 200 mm aperture or more and steady atmospheric conditions are required to see it. The two were last at their closest in 1993.

β (beta) CMa, 6h 23m −18°.0, (Mirzam, 'the announcer', i.e. of Sirius), mag. 2.0, is a blue giant 500 l.y. away. It is a pulsating star whose variations, of a few hundredths of a magnitude every 6 hours, are undetectable to the naked eye.

δ (delta) CMa, 7h 08m −26°.4, (Wezen, 'the weight'), mag. 1.8, is a white supergiant star 1800 l.y. away.

ε (epsilon) CMa, 6h 59m −29°.0, (Adhara, 'the virgins'), mag. 1.5, is a blue giant 430 l.y. away. It has a mag. 7.4 companion which is difficult to see in small telescopes because of the glare from the primary.

η (eta) CMa, 7h 24m −29°.3, (Aludra), mag. 2.4, is a blue supergiant about 3200 l.y. away.

μ (mu) CMa, 6h 56m −14°.0, mag. 5.0, is a yellow giant about 900 l.y. away with a close 7th-mag. blue-white companion difficult to pick up in the smallest apertures because of the magnitude contrast.

ν¹ (nu¹) CMa, 6h 36m −18°.7, mag. 5.7, is a yellow giant 278 l.y. away with a mag. 8.1 companion, visible in small telescopes.

UW CMa, 7h 19m −24°.6, is an eclipsing binary of β (beta) Lyrae type that varies between mags. 4.8 and 5.3 in 4.4 days. It lies about 3000 l.y. away.

M41 (NGC 2287), 6h 47m −20°.7, is a large and bright open cluster of about 80 stars, easily visible through binoculars or a small telescope and, with a total magnitude of 4.5, detectable by the naked eye under good conditions – it was known to the ancient Greeks. A low-power view in a small telescope shows the individual stars grouped in bunches and curves, covering an area of sky equivalent to the apparent diameter of the Moon. The brightest stars in the cluster are 7th-mag. orange giants. M41 is 2100 l.y. away.

NGC 2362, 7h 19m −25°.0, is a compact cluster surrounding the mag. 4.4 blue supergiant star τ (tau) Canis Majoris, which is a genuine member. Small telescopes show about 60 stars in the cluster, which lies some 5200 l.y. away.

Canis Major · CMa · Canis Majoris

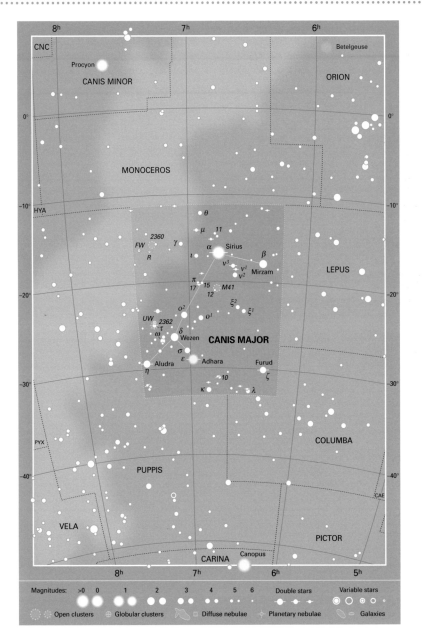

Canis Minor – The Lesser Dog

The second of the two dogs of Orion, the other being Canis Major. In one Greek legend, the constellation represents Maera, dog of Icarius, the man who was first taught to make wine by the god Dionysus. When some shepherds drank the wine they became intoxicated. Suspecting that Icarius had poisoned them they killed him. Maera the dog ran howling to Icarius's daughter Erigone and led her to her father's body. Both Erigone and the dog took their own lives where Icarius lay. Zeus placed their images among the stars as a reminder of the tragic affair. In this story, Icarius is identified with the constellation Boötes, Erigone is Virgo and Maera is Canis Minor.

Apart from its leading star Procyon, which is the eighth-brightest star in the sky, there are few objects of importance in Canis Minor. Procyon forms a prominent equilateral triangle with the bright stars Sirius (in Canis Major) and Betelgeuse (in Orion).

α (alpha) Canis Minoris, 7h 39m +5°.2, (Procyon, from the Greek meaning 'preceding the dog', referring to its rising before Sirius), mag. 0.40, is a yellow-white star 11.4 l.y. away, and therefore among the nearest stars to the Sun. Like Sirius, Procyon has a white dwarf companion, but this star, of mag. 10.7, is even more difficult to see than the companion of Sirius, requiring the use of large professional telescopes. Procyon's companion orbits it every 41 years.

β (beta) CMi, 7h 27m +8°.3, (Gomeisa), mag. 2.9, is a blue-white main-sequence star 170 l.y. away.

White dwarfs

By a remarkable coincidence both Sirius and Procyon, the brightest stars in Canis Major and Canis Minor respectively, are accompanied by tiny, faint stars known as white dwarfs. The existence of these companion stars was predicted in 1844 by the German astronomer Friedrich Wilhelm Bessel (1784–1846), who detected a slight wobble in the proper motions (*page 16*) of Sirius and Procyon. Bessel realized that this wobble was most probably caused by the presence of unseen companions orbiting around the visible stars.

The companion of Sirius, called Sirius B, was first seen in 1862 by the American astronomer Alvan G. Clark using a 47-cm (18½-inch) refractor, while the companion to Procyon (Procyon B) was first seen in 1896 by John M. Schaeberle with the 91-cm (36-inch) refractor at Lick Observatory.

But not until 1915 did astronomers realize the truly extraordinary nature of these stars. Observations showed that Sirius B was very hot, very small and very dense. In fact Sirius B has the mass of the Sun packed into a sphere less than 1 per cent of the Sun's diameter (that is, their diameter is about the same as that of the Earth). The resulting density of Sirius B is over 100,000 times that of water.

We now know that a white dwarf is a star at the end of its life; it is the shrunken remnant of a once-proud star like the Sun whose central nuclear fires have burnt out. The cause of the immense densities of white dwarfs is the inexorable pull of gravity, which squeezes the electrons of the dying star as closely together as is physically possible.

Canis Minor · CMi · Canis Minoris

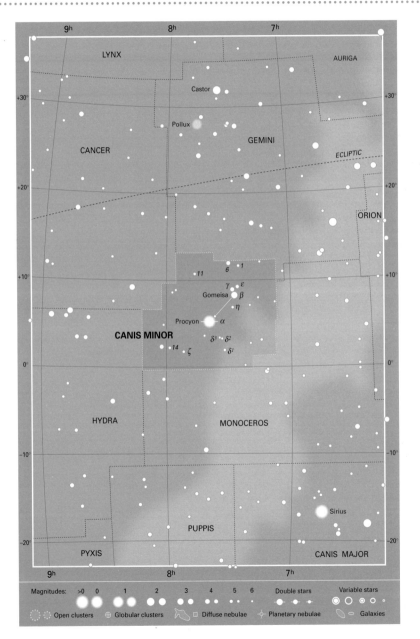

Capricornus – The Sea Goat

Capricornus is depicted as a goat with a fish's tail. Amphibious creatures feature prominently in ancient legends, and the origin of Capricornus certainly dates back to ancient times. In Greek legend, the constellation represented the goat-headed god Pan who jumped into a river to escape the approach of the monster Typhon, turning his lower half into a fish.

Before about 130 BC the Sun lay in Capricornus when it reached its farthest point south of the equator each year; this point, known as the northern winter solstice, currently occurs on December 21 or 22 (the date can vary from year to year). The latitude on Earth at which the Sun appears overhead at noon on that date, 23½° south, became known as the Tropic of Capricorn. Because of precession, the northern winter solstice has since moved from Capricornus into the neighbouring constellation Sagittarius (and will reach Ophiuchus in the year 2269), but the Tropic of Capricorn retains its name.

Capricornus is the smallest constellation of the zodiac. The Sun is within its boundaries from late January to mid-February.

α^1 α^2 (alpha^1 alpha^2) Capricorni, 20h 18m −12°.5, (Algedi or Giedi, both from the Arabic meaning 'the kid' in reference to the constellation as a whole) is a multiple star, consisting of an unrelated yellow supergiant of mag. 4.3 and a yellow giant of mag. 3.6, 690 and 109 l.y. away respectively, visible separately with the naked eye or binoculars. Telescopes reveal that each star is itself double. The fainter of the pair, α^1, has a wide mag. 9.2 companion visible in small telescopes; α^2 has its own companion of mag. 11. Telescopes of at least 100 mm aperture show that this faint companion is itself composed of two 11th-mag. stars. α Capricorni is therefore a fascinating hybrid system.

β (beta) Cap, 20h 21m −14°.8, (Dabih, from the Arabic meaning 'the lucky stars of the slaughterer'), 340 l.y. away, is a golden-yellow giant star of mag. 3.1 with a wide, blue-white companion of mag. 6.1, visible through binoculars or small telescopes.

γ (gamma) Cap, 21h 40m −16°.7, (Nashira), mag. 3.7, is a white giant star 139 l.y. away.

δ (delta) Cap, 21h 47m −16°.1, (Deneb Algedi, 'the kid's tail'), mag. 2.9, is the brightest star in the constellation. It is an eclipsing binary of β (beta) Lyrae type, varying by a barely perceptible 0.2 mag. over 24.5 hours. It lies 39 l.y. away.

π (pi) Cap, 20h 27m −18°.2, 670 l.y. away, is a blue-white star of mag. 5.1 with a close mag. 8.3 companion visible with a small telescope.

M30 (NGC 7099), 21h 40m −23°.2, is a mag. 7.5 globular cluster 30,000 l.y. away, visible in small telescopes and resolvable in 100 mm aperture, notably centrally condensed with finger-like chains of stars extending northwards.

Capricornus · Cap · Capricorni

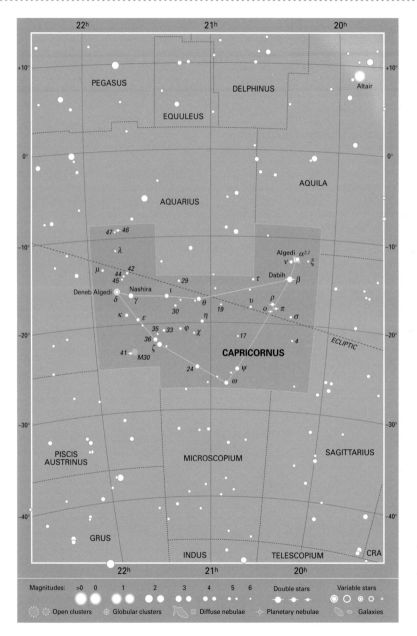

Carina – The Keel

This constellation was originally part of the extensive Argo Navis, the ship of the Argonauts, until that figure was subdivided into three parts in 1763 by the French celestial cartographer Nicolas Louis de Lacaille. The other two parts are Puppis and Vela; Carina is the smallest of the three. As a part of Argo Navis, Carina originated in ancient Greek times and is associated with the legend of Jason and the Argonauts and their quest for the Golden Fleece. Carina lies in the Milky Way, providing rich starfields and clusters for binoculars. The stars ι (iota) and ε (epsilon) Carinae, together with κ (kappa) and δ (delta) Velorum, form a shape known as the False Cross, sometimes confused with the real Southern Cross.

α (alpha) Carinae, 6h 24m −52°.7, (Canopus), mag. −0.62, the second-brightest star in the sky, is a white giant 313 l.y. away; the Hipparcos satellite found it to be variable by 0.1 mag. or so. It is named after the helmsman of the Greek King Menelaus, and appropriately enough is now used by spacecraft as a guide for navigation.

β (beta) Car, 9h 13m −69°.7, (Miaplacidus), mag. 1.7, is a blue-white star 111 l.y. away.

ε (epsilon) Car, 8h 23m −59°.5, mag. 1.9, is an orange giant star 630 l.y. away.

η (eta) Car, 10h 45m −59°.7, 7500 l.y. away, is a peculiar nova-like variable star embedded in the nebula NGC 3372 (see separate entry). In the past, η Carinae has fluctuated erratically in brightness, reaching a maximum of mag. −1 in 1843 when it was temporarily the second-brightest star in the sky; it subsequently settled at around 6th mag., but brightened to 5th mag. in 1998. The star is estimated to be over 100 times more massive and 4 million times brighter than the Sun, with an unseen companion orbiting it every 5½ years. The binary pair is surrounded by a shell of dust and gas thrown off in the 1843 outburst.

θ (theta) Car, 10h 43m −64°.4, mag. 2.7, is a blue-white star 440 l.y. away, a member of the sparkling cluster IC 2602 (see page 108).

ι (iota) Car, 9h 17m −59°.3, mag. 2.2, is a white supergiant 690 l.y. away.

υ (upsilon) Car, 9h 47m −65°.1, 1600 l.y. away, is a double star consisting of two blue-white giants of mags. 3.0 and 6.0, divisible in small telescopes.

l Car, 9h 45m −62°.5, a yellow supergiant 1500 l.y. away, is the brightest Cepheid whose variations are large enough to be obvious to the naked eye. It rises and falls between mags. 3.3 and 4.2 every 35.5 days.

R Car, 9h 32m −62°.8, 416 l.y. away, is a red giant variable star of Mira type that ranges between 4th and 10th mags. with a period of 309 days.

S Car, 10h 09m −61°.5, about 1300 l.y. away, is a red giant variable similar to nearby R Car, varying from 5th to 10th mag. over 150 days.

NGC 2516, 7h 58m −60°.9, is a naked-eye open cluster of some 80 stars, 1100 l.y. away, as large as the full Moon. It is a sparkling sight in binoculars, which show it to be cross-shaped. Its brightest member is a mag. 5.2 red giant.

NGC 3114, 10h 03m −60°.1, is a widely scattered open cluster the same apparent size as the full Moon, containing stars of 6th mag. and fainter, about 3000 l.y. away. It is best seen in binoculars and small telescopes with low power.

NGC 3372, 10h 44m −59°.9, is a celebrated diffuse nebula easily visible to the naked eye as a brilliant patch of the Milky Way four Moon diameters wide, surrounding the erratic variable star η (eta) Carinae. The nebula shines from the light of brilliant young stars born within it. Binoculars and small telescopes show jewelled star clusters and swirls of glowing gas alternating with ▶

Carina · Car · Carinae

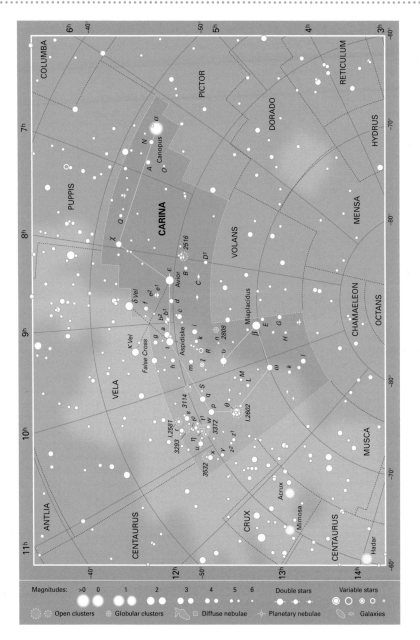

dark lanes. Most famous is a dark notch, called the Keyhole because of its distinctive shape, silhouetted against the nebula's brightest central portion near η Carinae itself. The whole NGC 3372 nebula lies about 7500 l.y. from us, the same distance as η Carinae. (For a photograph of NGC 3372, Eta Carinae and the Keyhole see page 279.)

NGC 3532, 11h 06m −58°.7, 1350 l.y. away, is an outstanding open cluster, visible to the naked eye as a brighter patch nearly 1° wide among rich Milky Way starfields and glorious in binoculars. It contains 150 or so stars of 7th mag. and fainter, including several orange giants, arranged in an elliptical shape with a star-free lane across its centre. The mag. 3.9 yellow-white supergiant x Carinae at the cluster's edge is not a member but a background object five times farther off.

IC 2602, 10h 43m −64°.4, is a large and brilliant open cluster containing a few dozen stars, popularly known as the Southern Pleiades, centred about 480 l.y. away. Its brightest members are visible to the naked eye, notably the blue-white star θ Carinae, mag. 2.7. The whole cluster appears twice as wide as the full Moon.

Cassiopeia

In Greek legend, Cassiopeia was the beautiful but boastful Queen of Ethiopia, wife of King Cepheus and mother of Andromeda. In the sky she is depicted sitting in a chair. The constellation is easily identifiable by the distinctive W-shape of its five brightest stars. Cassiopeia lies on the opposite side of the Pole Star from Ursa Major, in a rich part of the Milky Way.

Near the star κ (kappa) Cassiopeiae, at 0h 25.3m, +64° 09′, occurred the brilliant supernova outburst of 1572, which reached mag. −4. It is now known as Tycho's Star because it was observed by the great Danish astronomer Tycho Brahe. The remains of another supernova, which erupted around 1660 but which went unseen at the time, form the strongest radio source in the sky, Cassiopeia A; it lies at 23h 23.4m, +58° 50′, and is some 10,000 l.y. away.

α (alpha) Cassiopeiae, 0h 41m +56°.5, (Shedir, 'the breast'), mag. 2.2, is an orange giant star 229 l.y. away. It has a wide mag. 8.9 companion, which is unrelated.

β (beta) Cas, 0h 09m +59°.1, (Caph), mag. 2.3, is a white star 54 l.y. away.

γ (gamma) Cas, 0h 57m +60°.7, 613 l.y. away, is a remarkable blue-white variable star that rotates so rapidly that it is unstable, throwing off rings of gas from its equatorial region at irregular intervals. The ejections of gas cause it to vary unpredictably between mags. 3.0 and 1.6. Currently it hovers around mag. 2.2.

δ (delta) Cas, 1h 26m +60°.2, (Ruchbah, 'the knee'), mag. 2.7, is a blue-white star 99 l.y. away. It is an eclipsing binary of the Algol type that varies by about 0.1 mag. with the relatively long period of 2 years and 1 month.

ε (epsilon) Cas, 1h 54m +63°.7, mag. 3.3, is a blue-white star 442 l.y. away.

η (eta) Cas, 0h 49m +57°.8, 19 l.y. away, is a beautiful double star with yellow and red components of mags. 3.5 and 7.5, visible in small telescopes. They form a true binary with a period of 480 years.   ▶

Cassiopeia · Cas · Cassiopeiae

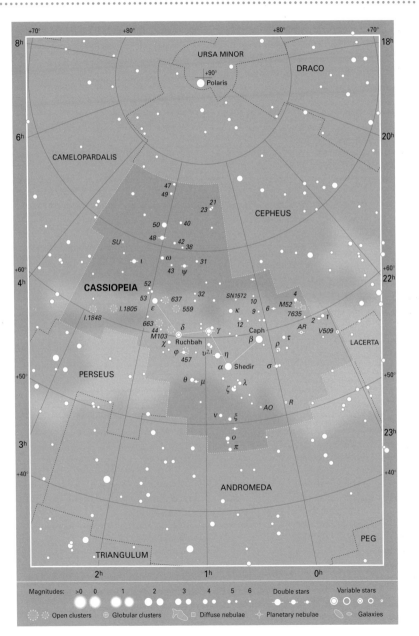

ι (iota) Cas, 2h 29m +67°.4, 142 l.y. away, is a mag. 4.5 white star with a wide mag. 8.4 companion visible through a 60-mm telescope. With an aperture of 100 mm and high magnification, the brighter star is seen to have a closer mag. 6.9 yellow companion of its own, making this an impressive triple.

ρ (rho) Cas, 23h 54m +57°.5, is a yellow-white supergiant, one of the most luminous stars known, giving out as much light as half a million Suns. It is a semi-regular pulsating variable, ranging between mag. 4.1 and 6.2 every 320 days or so. Its distance is not known accurately, but is probably over 10,000 l.y.

σ (sigma) Cas, 23h 59m +55°.8, 1500 l.y. away, is a close pair of mags. 5.0 and 7.3 appearing green and blue, in striking contrast to the warmer hues of η Cas. An aperture of 75 mm and high power will split the pair.

ψ (psi) Cas, 1h 26m +68°.1, 193 l.y. away, is a mag. 4.7 orange giant with a wide 9th-mag. companion visible in a small telescope. High powers reveal that this companion is itself a close binary.

M52 (NGC 7654), 23h 24m +61°.6, is an open cluster of about 100 stars, 5200 l.y. away, visible as a misty patch in binoculars. It is somewhat kidney-shaped, with an 8th-mag. orange star embedded at one edge, like a poorer version of the Wild Duck Cluster (M11 in Scutum). M52 can be resolved into stars with 75 mm aperture.

M103 (NGC 581), 1h 33m +60°.7, is a small, elongated group of about 25 faint stars, 8200 l.y. away. What appears to be the brightest member, a double of 7th and 10th mags. near the cluster's northern tip, is in fact a foreground object.

NGC 457, 1h 19m +58°.3, is a loose open cluster of about 80 stars, 10,000 l.y. away, seemingly arranged in chains. It has been given various imaginative names inspired by its shape, such as the Dragonfly Cluster, the ET Cluster and the Owl Cluster. The mag. 5.0 white supergiant φ (phi) Cas on its southern outskirts is probably a true member.

NGC 663, 1h 46m +61°.2, is a prominent binocular cluster of about 80 stars appearing half the size of the full Moon, 8200 l.y. away.

Centaurus – The Centaur

A large and rich constellation representing a centaur, the mythical beast that was half man, half horse. Reputedly Centaurus depicts the scholarly centaur Chiron, the tutor of many Greek gods and heroes, who was raised to the sky after being accidentally struck by a poisoned arrow from Hercules.

A line from α (alpha) through β (beta) Centauri points to Crux, the Southern Cross (see the photograph on page 120). Currently α and β Centauri are about 4½° apart, but the rapid proper motion of α Cen is taking it towards β Cen, and about 4000 years from now the two will be little more than ½° (one Moon diameter) apart in our skies, making a stunning naked-eye pairing. Of particular

note in Centaurus is the closest star to the Sun, Proxima Centauri, an 11th-mag. red dwarf which is an outlying companion of α Centauri.

One of the strongest radio sources in the sky, Centaurus A, is associated with the unusual galaxy NGC 5128. Centaurus lies in a prominent part of the Milky Way and contains more naked-eye stars than any other constellation: 281 brighter than magnitude 6.5, according to the Hipparcos catalogue.

α (alpha) Centauri, 14h 40m −60°.8, (Rigil Kentaurus, abbreviated as Rigil Kent, 'foot of the centaur', or Toliman) lies 4.4 l.y. away. To the naked eye it shines at mag. −0.28, the third-

▶

Centaurus · Cen · Centauri

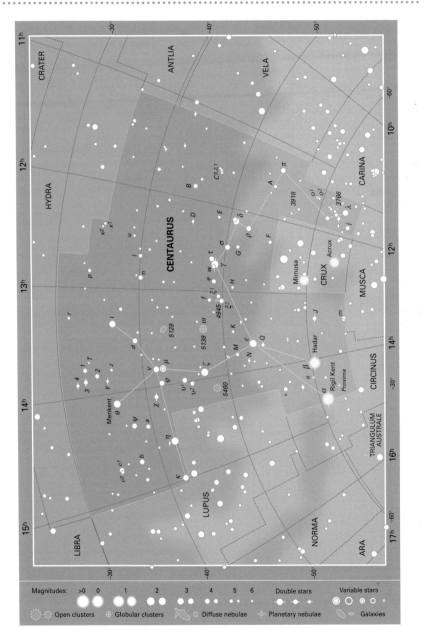

brightest star in the sky. But the smallest of telescopes reveals that it consists of twin yellow stars of mags. −0.01 and 1.35; the brighter of these is very similar to the Sun. They orbit each other every 80 years and are always divisible in amateur telescopes, although at their closest, around the years 2037–38, they will need 75 mm aperture. Also associated with α Centauri is an 11th-mag. red dwarf called Proxima Centauri, lying 2° away and therefore not even in the same telescopic field of view (see the finder chart below.) This star is estimated to take as long as a million years to orbit its two brilliant companions. At present, Proxima Centauri is about 0.2 l.y. closer to us than the two other members of α Centauri. Proxima Centauri is a flare star, suddenly brightening by as much as one magnitude for several minutes at a time.

β (beta) Cen, 14h 04m −60°.4, (Hadar or Agena), mag. 0.6, is a blue giant 525 l.y. away. It is in fact a very close double with a 4th-mag. companion, divisible only in larger apertures.

Finder chart for 11th-mag. Proxima Centauri, showing its proper motion over 200 years. Proxima lies 2° from Alpha Centauri , the bright star at upper left of the diagram. (Wil Tirion)

γ (gamma) Cen, 12h 42m −49°.0, 130 l.y. away, is a close double with blue-white components each of mag. 2.9, orbiting each other every 85 years. Together they shine as a star of mag. 2.2. As seen from Earth they are currently moving together, requiring 220 mm aperture to separate in 2005. At their closest, between 2010 and 2020, amateur telescopes will be unable to split them, but they will become divisible in 220 mm again by 2025, and in 150 mm from 2030.

3 Cen, 13h 52m −33°.0, 298 l.y. away, is a blue-white star of mag. 4.6 with a companion of mag. 6.1, forming a striking pair in small telescopes.

R Cen, 14h 17m −59°.9, is a red giant variable of Mira type, which varies between mags. 5.3 and 11.8 in about 18 months. It lies about 2100 l.y. away.

ω (omega) Cen (NGC 5139), 13h 27m −47°.5, is the largest and brightest globular cluster in the sky – so prominent that it was labelled as a star on early charts. It appears as a hazy star of mag. 3.7 to the naked eye, noticeably elliptical in shape and as large as the full Moon. It is inherently the most luminous of all globulars, with the light output of a million Suns. Small telescopes or even binoculars

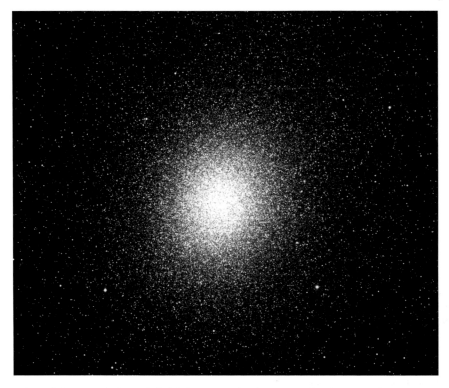

ω (omega) Centauri, the finest of all globular star clusters, appears noticeably elliptical in shape. It has a true diameter of about 150 light years. (NOAO/AURA/NSF)

begin to resolve its outer regions into stars, and it is a showpiece for all apertures. Its brilliance and large apparent size are due in part to its relative closeness, 17,000 l.y., which places it among the nearest globular clusters to us.

NGC 3766, 11h 36m −61°.6, is a naked-eye open cluster of about 100 stars of mag. 7 and fainter, resolvable in binoculars, 6300 l.y. away.

NGC 3918, 11h 50m −57°.2, is an 8th-mag. planetary nebula 2600 l.y. away, discovered in the 19th century by John Herschel and called by him the Blue Planetary. It is similar in appearance to the planet Uranus, but three times the apparent diameter. Its central star, of 11th mag., should be detectable in modest amateur instruments.

NGC 5128, 13h 25m −43°.0, is a peculiar 7th-mag. galaxy, visually the brightest galaxy outside the Local Group, known to radio astronomers as Centaurus A. In long-exposure photographs such as the one on page 295 it appears as a giant elliptical galaxy with an encircling band of dust. It is thought to result from a merger between an elliptical and a spiral galaxy. Under good skies it is visible in binoculars as a rounded glow, but at least 100 mm aperture is necessary to trace its outline and the dark bisecting lane of dust. NGC 5128 is 12 million l.y. away.

NGC 5460, 14h 08m −48°.3, is a large, 6th-mag. open cluster of about 40 stars visible in binoculars or small telescopes. It lies 2500 l.y. away.

Cepheus

An ancient constellation representing the mythological King Cepheus of Ethiopia, the husband of Cassiopeia and father of Andromeda, who are themselves depicted by nearby constellations. Cepheus is replete with double and variable stars, including the celebrated δ (delta) Cephei, prototype of the Cepheid variables which are used as 'standard candles' for distance-finding in space. This star's fluctuations in light output were discovered in 1784 by the English amateur astronomer John Goodricke.

α (alpha) Cephei, 21h 19m +62°.6, (Alderamin), mag. 2.5, is a white star 49 l.y. away.

β (beta) Cep, 21h 29m +70°.6, (Alfirk, meaning 'flock', i.e. of sheep), 595 l.y. away, is both a double and a variable star. A small telescope shows that this blue giant, of mag. 3.2, has a mag. 7.9 companion. β Cephei is the prototype of a class of pulsating variable stars (also known as β Canis Majoris stars) with periods of a few hours and tiny brightness fluctuations. Over a period of 4.6 hours or so β Cephei varies by 0.1 mag., an amount indistinguishable to the naked eye but detectable by sensitive instruments.

γ (gamma) Cep, 23h 39m +77°.6, (Errai, 'the shepherd'), mag. 3.2, is an orange star 45 l.y. away.

δ (delta) Cep, 22h 29m +58°.4, 980 l.y. away, is a famous pulsating variable star, the prototype of the classic Cepheid variables. This yellow supergiant varies between mags. 3.5 and 4.4 in 5 days 9 hours, changing in size between about 40 and 46 times the Sun's diameter as it does so – see page 287 for a comparison chart. Less well known is that δ Cephei is also an attractive double star for binoculars or the smallest telescopes, with a wide, bluish mag. 6.3 companion.

μ (mu) Cep, 21h 44m +58°.8, 2800 l.y. away, is a famous red star, called the Garnet Star by William Herschel because of its striking tint, which is notable in binoculars. μ Cephei is a red super-giant, a prominent example of a class of variable stars known as semi-regular variables. It varies between mags. 3.4 and 5.1 with a period of around 2 years.

ξ (xi) Cep, 22h 04m +64°.6, 102 l.y. away, is a double star of mags. 4.4 and 6.5 visible in small telescopes. The components are blue-white and yellow and form a true binary with an estimated orbital period of nearly 4000 years.

o (omicron) Cep, 23h 19m +68°.1, 211 l.y. away, is an orange giant of mag. 4.9 with a close mag. 7.1 companion for telescopes of 60 mm aperture and above. They orbit each other about every 1500 years.

T Cep, 21h 10m +68°.5, 685 l.y. away, is a red giant variable of the Mira type, about 500 times the diameter of the Sun, ranging between mags. 5.2 and 11.3 in around 13 months.

VV Cep, 21h 57m +63°.6, is an enormous red supergiant that varies semi-regularly between mags. 4.8 and 5.4. It is also an eclipsing binary with the unusually long period of 20.3 years, but the light dip during eclipse is too slight to be noticeable to the naked eye. Its estimated distance is more than 2000 l.y. and its diameter is probably over 1000 Suns, one of the largest stars known.

Cepheus · Cep · Cephei

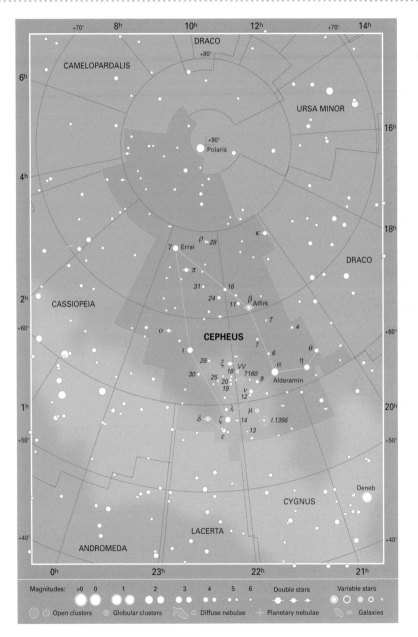

Cetus – The Whale

An ancient constellation depicting the sea monster that was about to devour Andromeda before she was rescued by Perseus. In the sky, Cetus is found basking on the banks of Eridanus, the River. The constellation is large but not prominent; nevertheless it contains several stars of particular interest, notably o (omicron) Ceti and τ (tau) Ceti.

One faint but famous star is UV Ceti, position 1h 38.8m, −17° 57′, which consists of a pair of 13th-mag. red dwarfs 8.7 l.y. away. One of these is the prototype of a class of erratic variables known as flare stars; these are red dwarfs that undergo sudden increases in light output lasting only a few minutes. The outbursts of the flare star component of UV Ceti can take it from its normal level of 13th mag. to as bright as 7th mag. ▶

Finder chart for the variable star Mira, also known as o (omicron) Ceti. The numbers against the surrounding stars are their magnitudes with the decimal points omitted, to prevent confusion with faint stars. These stars can be used to estimate the magnitude of Mira. (Wil Tirion)

41
δ Cet
53
CELESTIAL EQUATOR
53
54
59
56
71
95
63 90 93
72 Mira 88
84 73
81
80 69
65
77
82

MIRA – O (omicron) Ceti
Stars down to mag. 9.5
0°   1°   2°

70
55

Cetus · Cet · Ceti

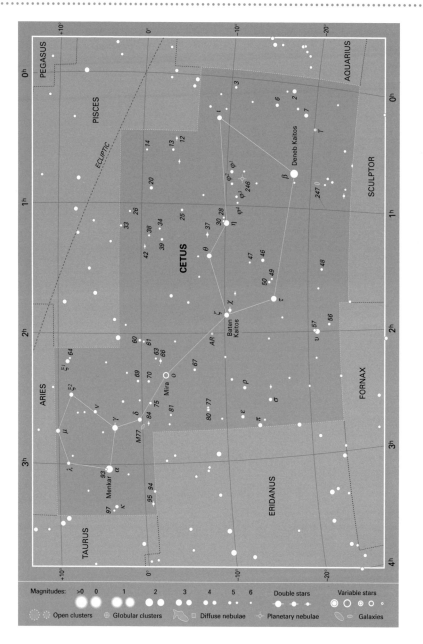

α (alpha) Ceti, 3h 02m +4°.1, (Menkar, 'nose'), mag. 2.5, is a red giant star 220 l.y. away. Binoculars show a wide mag. 5.6 blue-white apparent companion, 93 Ceti, which is actually unrelated to it, lying twice as far away.

β (beta) Cet, 0h 44m −18°.0, (Deneb Kaitos, 'tail of the whale', or Diphda), mag. 2.0, the brightest star of the constellation, is an orange giant 96 l.y. away.

γ (gamma) Cet, 2h 43m +3°.2, 82 l.y. away, is a close double star needing telescopes of at least 60 mm aperture and high power to split. The stars are of mags. 3.5 and 6.6, the colours yellow and bluish.

o (omicron) Cet, 2h 19m −3°.0, (Mira, 'the amazing one'), 420 l.y. away, is the prototype of a famous class of red giant long-period variable stars. Mira itself varies between about 3rd and 9th mags. (although it can become as bright as 2nd mag.) in an average of 332 days, changing in diameter from

about 400 to 500 times the size of the Sun as it does so. Mira's light variations were first noted in 1596 by the Dutch astronomer David Fabricius, making it the first variable star (other than novae) to be discovered. The finder chart on page 116 shows various comparison stars down to mag 9.5 from which Mira's brightness can be estimated.

τ (tau) Cet, 1h 44m −15°.9, mag. 3.5, is a yellow main-sequence star, is one of the nearest stars to us, lying 11.9 l.y. away. Its main claim to fame is that, of all the nearby single stars, it is the one most like the Sun, although we do not yet know whether it also has planets.

M77 (NGC 1068), 2h 43m −0°.0, is a small, softly glowing 9th-mag. face-on spiral galaxy with a 10th-mag. starlike nucleus. Apertures of 100 mm show the brighter patches in the spiral arms. M77 is the brightest example of a Seyfert galaxy, a close relative of the quasars, and it is a radio source. It lies roughly 50 million l.y. away. (See photograph on page 294.)

Chamaeleon – The Chameleon

A faint and unremarkable constellation between Carina and the south celestial pole, representing the colour-changing lizard. It was one of the constellations representing exotic animals introduced by the Dutch navigators Pieter Dirkszoon Keyser and Frederick de Houtman at the end of the 16th century.

α (alpha) Chamaeleontis, 8h 19m −76°.9, mag. 4.1, is a white star 63 l.y. away.

β (beta) Cha, 12h 18m −79°.3, mag. 4.2, is a blue-white star 271 l.y. away.

γ (gamma) Cha, 10h 35m −78°.6, mag. 4.1, is a red giant 413 l.y. away.

δ¹ δ² (delta¹ delta²) Cha, 10h 45m −80°.5, consists of a wide pair of unrelated stars, both clearly seen in binoculars: δ¹ Cha, mag. 5.5, is an orange giant 354 l.y. away and δ² Cha, mag. 4.4, is a blue star 364 l.y. away.

NGC 3195, 10h 09m −80°.9, is a faint planetary nebula of similar apparent size to the planet Jupiter, needing at least 100 mm aperture to be seen well.

Chamaeleon · Cha · Chamaeleontis

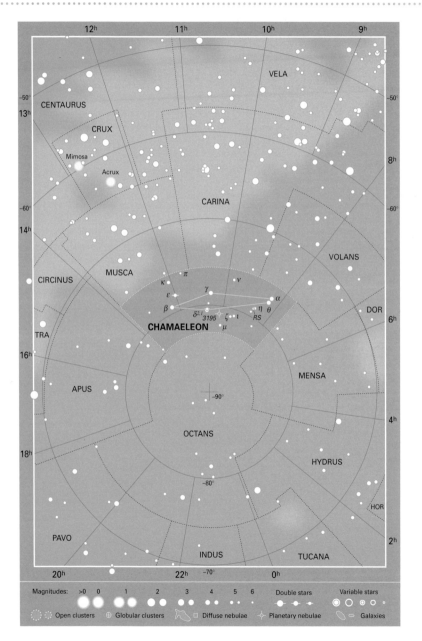

Circinus – The Compasses

Another of the small and obscure southern constellations introduced in 1756 by the French astronomer Nicolas Louis de Lacaille. It represents a pair of dividing compasses as used by draughtsmen and surveyors, and is appropriately placed in the sky next to Norma, the Set Square. It is overshadowed by the brilliance of neighbouring Centaurus.

α (alpha) Circini, 14h 43m −65°.0, 53 l.y. away, is a white main-sequence star of mag. 3.2 with a companion of mag. 8.5 easily visible in small telescopes.

γ (gamma) Cir, 15h 23m −59°.3, 500 l.y. away, consists of a very close pair of blue and yellow stars of mags. 5.1 and 5.5, requiring an aperture of at least 150 mm and high magnification to see separately. Their orbital period is calculated to be 270 years.

Circinus is most easily found by reference to the prominent stars Alpha and Beta Centauri (seen left of centre in the photograph below), which point towards the familiar figure of Crux, the Southern Cross, at upper right. Alpha Circini is the star below and to the right of Alpha Centauri. Long-exposure photographs such as this one emphasize the darkness of the Coalsack Nebula in Crux against the star background of the Milky Way. (M. Bessell, R. Sutherland and M. Buxton, RSAA – Australian National University)

Circinus · Cir · Circini

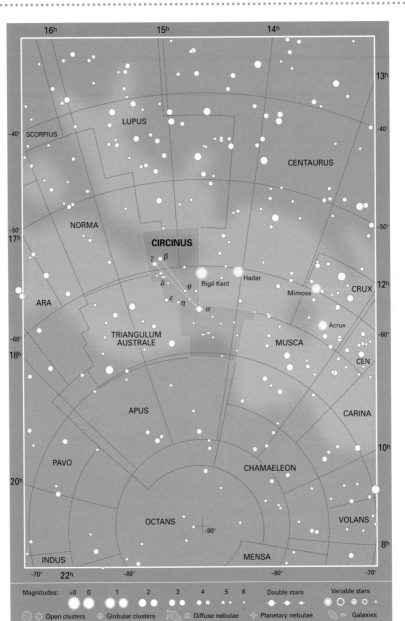

Columba – The Dove

A constellation representing the dove sent out from Noah's Ark to find dry land after the Biblical Flood. The Dutchman Petrus Plancius formed it in 1592 from some stars adjacent to Canis Major that had not previously been part of any constellation. Columba is placed next to Puppis, the stern of the ship Argo, so it might alternatively be imagined as the dove that the Argonauts sent ahead to help them pass safely between the Clashing Rocks (the Symplegades) at the mouth of the Black Sea. Columba contains little of interest for amateur telescopes.

α (alpha) Columbae, 5h 40m −34°.1, (Phact, 'ring dove'), mag. 2.7, is a blue-white star 268 l.y. away.

β (beta) Col, 5h 51m −35°.8, (Wazn), mag. 3.1, is an orange giant 86 l.y. away.

NGC 1851, 5h 14m −40°.1, is a 7th-mag. globular cluster, visible in small telescopes but requiring moderate apertures to resolve its brightest stars. It lies 35,000 l.y. away.

Vanished constellations

Antinous, Felis, Globus Aerostaticus, Harpa Georgii, Machina Electrica, Officina Typographica, Quadrans Muralis, Robur Carolinum, Taurus Poniatovii, Telescopium Herschelii and Turdus Solitarius – these are all constellations which once graced the sky but which you will not find on any chart today. Where did they go?

During the golden age of celestial mapping, in the 17th and 18th and centuries, various astronomers tried to leave their mark on the sky by adding new constellations to those of the ancient Greeks. Constellations depicted on an atlas by one astronomer could be completely ignored by everyone else.

Some constellations were blatant attempts to flatter their monarchs or wealthy patrons. A 17th-century German astronomer, Julius Schiller, produced an atlas in which all the constellations were based on characters from the Bible – for example, the familiar constellations of the zodiac were changed to represent the 12 apostles. Such attempts to insert politics or religion into the skies did not succeed.

Over 100 different constellations were included in the immense *Uranographia* star atlas published by the German astronomer Johann Elert Bode in 1801, but during the following century common usage gave rise to the selection of 88 constellations that were officially adopted by the International Astronomical Union in 1922. Unlike the map of the Earth, which is continually affected by political changes, the map of the sky is now fixed.

Noctua, the night owl, perches on the tail of Hydra the water snake, on a colourful set of 19th-century constellation cards called Urania's Mirror, but has now flown from modern star charts. (Ian Ridpath)

Columba · Col · Columbae

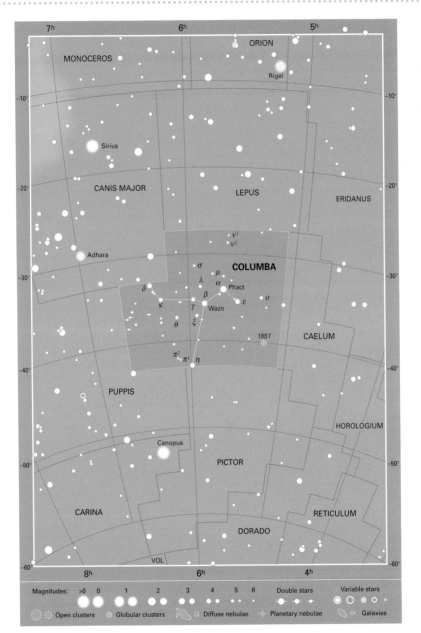

Coma Berenices – Berenice's Hair

This faint constellation represents the flowing locks of Queen Berenice of Egypt, who cut off her hair in gratitude to the gods for the safe return of her husband Ptolemy III Euergetes from battle. Although the legend dates from Greek times, this group of stars was regarded as part of Leo until 1551 when the Dutch cartographer Gerardus Mercator made them into a separate constellation. The main part of the queen's severed tresses is represented by the extensive Coma Star Cluster (Melotte 111). Coma Berenices also contains another type of cluster – a cluster of galaxies. The Coma Cluster of galaxies lies about 280 million l.y. away, so its members are too faint for all but the largest amateur telescopes. But the constellation also contains some brighter galaxies, members of the nearer Virgo Cluster, the brightest of which are visible in amateur telescopes. The north pole of our Galaxy lies in Coma Berenices.

α (alpha) Comae Berenices, 13h 10m +17°.5, (Diadem), mag. 4.3, is a tight binary 47 l.y. away, consisting of twin yellow-white stars of mag. 5.1 that orbit each other every 26 years. Even at their widest, around the year 2010, they are at the limit of resolution of a 220-mm telescope.

β (beta) Com, 13h 12m +27°.9, mag. 4.2, the brightest star in the constellation, is a yellow main-sequence star 30 l.y. away.

γ (gamma) Com, 12h 27m +28°.3, mag. 4.4, is an orange giant 170 l.y. away, which appears to be a member of the Coma Star Cluster but is actually a foreground star.

24 Com, 12h 35m +18°.4, is a beautiful coloured double star for small telescopes, consisting of an orange giant of mag. 5.0, 610 l.y. away, and an unrelated blue-white companion of mag 6.6.

35 Com, 12h 53m +21°.2, 324 l.y. away, is a tight binary star with yellow and white components of mags. 5.1 and 7.2 orbiting every 360 years and divisible in 150-mm apertures. Small telescopes show a wider 9th-mag. companion.

FS Com, 13h 06m +22°.6, 572 l.y. away, is a red giant that varies semi-regularly between mags. 5.3 and 6.1 every two months or so.

Coma Star Cluster (Melotte 111), 12h 25m +26°, is a scattered group of about 50 stars best seen in binoculars. The cluster's brightest members, of 5th mag., form a noticeable V-shape extending for several degrees south of γ (gamma) Com, which is not actually a member but a foreground star. The brightest true member seems to be 12 Com, mag. 4.8. The distance to the cluster's centre is 288 l.y.

M53 (NGC 5024), 13h 13m +18°.2, is an 8th-mag. globular cluster 56,000 l.y. away, visible in small telescopes as a rounded, hazy patch.

M64 (NGC 4826), 12h 57m +21°.7, is a famous spiral galaxy, called the Black Eye Galaxy because of a dark cloud of dust silhouetted against its nucleus. This dark dust lane shows up well in apertures above 150 mm; observers with smaller instruments must content themselves simply with locating this 9th-mag. galaxy some 20 million l.y. away, closer than the Virgo Cluster and not a member of it.

M85 (NGC 4382), 12h 25m +18°.2, is a 9th-mag. elliptical galaxy in the Virgo Cluster, 55 million l.y. away. Small telescopes show a brighter, starlike centre.

M88 (NGC 4501), 12h 32m +14°.4, is a 10th-mag. spiral galaxy in the Virgo Cluster, 55 million l.y. away. It is presented at an angle to us, so that it appears elliptical.

M99 (NGC 4254), 12h 19m +14°.6, is a 10th-mag. spiral galaxy 55 million l.y. away in the Virgo Cluster, presented face-on so that it appears almost circular.

▶

Coma Berenices · Com · Comae Berenices

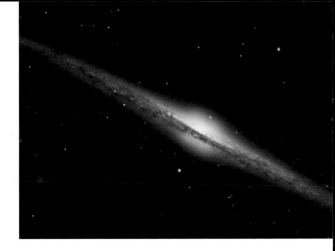

NGC 4565, a classically elegant spiral galaxy in Coma Berenices seen almost exactly edge-on. A dark lane of dust crosses its centre. NGC 4565 is estimated to be over 100,000 light-years in diameter, similar to our own Milky Way. (Bruce Hugo and Leslie Gaul/Adam Block/ NOAO/AURA/NSF)

M100 (NGC 4321), 12h 23m +15°.8, is a 9th-mag. Virgo Cluster spiral seen face-on, similar to M99 but larger. The Hubble Space Telescope has measured its distance accurately at 52.5 million l.y.

NGC 4565, 12h 36m +26°.0, is a 10th-mag. spiral galaxy seen edge-on. It is the most famous of the edge-on spirals, and is pictured above. Apertures of 100 mm show its cigar-shaped body with a central bulge and starlike core, but larger instruments are needed to trace the dark band (actually a dust lane) that splits it lengthwise. NGC 4565 is not a member of the Virgo Cluster but only about half as far, around 30 million l.y. away.

Corona Australis – The Southern Crown

The southern counterpart of the Northern Crown (Corona Borealis). Corona Australis has been known since the time of the Greek astronomer Ptolemy in the 2nd century AD who visualized it not as a crown but as a wreath. In one legend it represents the crown placed in the sky by Bacchus when he res-cued his dead mother from the Underworld; alternatively, it could simply have slipped from the head of the centaur Sagittarius, at whose feet it lies. Although faint, it is a dis-tinctive figure, and is situated on the edge of the Milky Way.

α (alpha) Coronae Australis, 19h 09m −37°.9, mag. 4.1, is a blue-white main-sequence star 130 l.y. away.

β (beta) CrA, 19h 10m −39°.3, mag. 4.1, is an orange giant 510 l.y. away.

γ (gamma) CrA, 19h 06m −37°.1, 58 l.y. away, consists of a pair of near-identical yellow-white stars, mags. 4.9 and 5.0, orbiting every 122 years, forming a tight double for small telescopes. The stars were closest together during the 1990s, but since 2000 have become progressively easier to split with 100 mm aperture.

κ (kappa) CrA, 18h 33m −38°.6, is a pair of unrelated blue-white stars of mags. 5.7 and 6.3, distances about 1700 and 490 l.y., easily divisible in small telescopes.

λ (lambda) CrA, 18h 44m −38°.3, 202 l.y. away, is a mag. 5.1 blue-white star with a wide mag. 9.7 companion, visible in small telescopes.

NGC 6541, 18h 08m −43°.7, is a 7th-mag. globular cluster 22,000 l.y. away, visible in binoculars or small telescopes.

Corona Australis · CrA · Coronae Australis

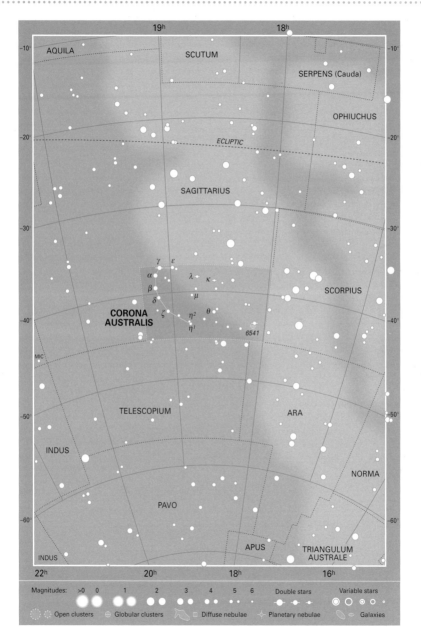

Corona Borealis – The Northern Crown

An ancient constellation, representing the jewelled crown worn by Ariadne when she married Bacchus and cast by him into the sky to mark the happy event. It consists of an arc of seven stars, all but one of 4th magnitude, the exception being Alphekka (a word derived from the Arabic name for the constellation) which is of 2nd magnitude. Alphekka is set in the crown like a central gemstone, as its alternative name Gemma implies. Corona Borealis contains a famous cluster of about 400 galaxies more than 1000 million light years away. Being so very distant, the galaxies are no brighter than 16th magnitude and are thus far beyond the reach of amateur telescopes.

α (alpha) Coronae Borealis, 15h 35m +26°.7, (Alphekka or Gemma), mag. 2.2, is a blue-white main-sequence star 75 l.y. distant. It is an eclipsing binary of the Algol type, but its variation every 17.4 days is only 0.1 mag., too slight to be noticeable to the naked eye.

ζ (zeta) CrB, 15h 39m +36°.6, 470 l.y. away, is a pair of blue-white stars, mags. 5.0 and 6.0, visible in small telescopes.

ν^1 ν^2 (nu^1 nu^2) CrB, 16h 22m +33°.8, is a wide binocular pair of red and orange giants of mags. 5.2 and 5.4, both about 550 l.y. from us but moving in different directions, so they probably do not constitute a true binary.

σ (sigma) CrB, 16h 15m +33°.9, 71 l.y. away, is a pair of yellow stars for small telescopes, of mags. 5.6 and 6.6. They form a genuine binary with an estimated orbital period of about 900 years.

R CrB, 15h 49m +28°.2, is a remarkable yellow supergiant star lying within the arc of the Crown, halfway between the stars α (alpha) and ι (iota). It usually appears about 6th mag., but occasionally and unpredictably drops in a matter of weeks to as faint as mag. 15, from where it may take many months to regain its former brightness. Recent catastrophic declines in the star's brilliance were seen in 1983, 1995 and 1999, although less extreme fades are more frequent and can occur at any time. These sudden dips in the light of R Coronae Borealis are believed to be due to the accumulation of carbon particles (i.e. soot) in its atmosphere. R Coronae Borealis is estimated to lie over 7000 l.y. away.

S CrB, 15h 21m +31°.4, is a variable of Mira type, ranging from 6th to 14th mag. in just under a year.

T CrB, 16h 00m +25°.9, is another spectacular variable star, known as the Blaze Star, which performs in almost the opposite way to R Coronae Borealis. It is a recurrent nova, usually slumbering at around mag. 11 but which can suddenly and unpredictably brighten to mag. 2. Its last recorded outburst was in 1946, and the previous one was 80 years before that. It is not known when it may erupt again.

Corona Borealis · CrB · Coronae Borealis

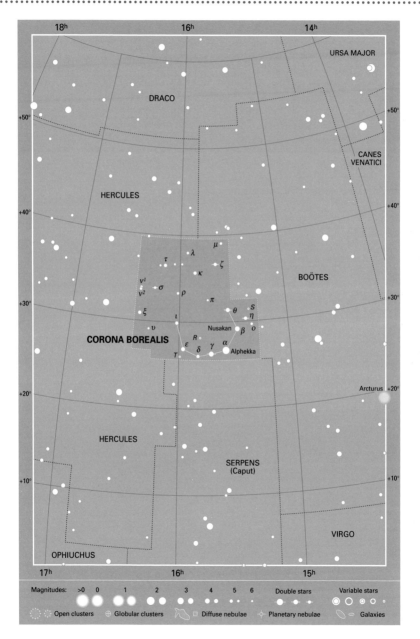

Corvus – The Crow

In Greek legend, Corvus is associated with the neighbouring constellations Crater, the Cup, and Hydra, the Water Snake. The crow is said to have been sent by Apollo to fetch water in a cup, but along the way it dallied to eat figs. When the crow returned to Apollo it carried the water snake in its claws, claiming that this creature had been blocking the spring and was the cause of its delay. Apollo, seeing through the lie, banished the the trio to the sky, where the Crow and the Cup lie on the back of Hydra. For its misdeed the crow was condemned to suffer from eternal thirst, which is why crows croak so harshly; in the sky, the cup is just out of the thirsty crow's reach. In another legend, a snow-white crow brought Apollo the bad news that his lover Coronis had been unfaithful to him. In his anger, Apollo turned the crow black. Apollo and crows are closely linked in legend, for during the war waged by the giants on the gods, Apollo turned himself into a crow. The four brightest stars in Corvus, γ (gamma), β (beta), δ (delta) and ε (epsilon), form a quadrilateral shape somewhat reminiscent of the Keystone of Hercules.

α (alpha) Corvi, 12h 08m −24°.7, (Alchiba), mag. 4.0, is a white star 48 l.y. away.

β (beta) Crv, 12h 34m −23°.4, mag. 2.7, is a yellow giant star 140 l.y. away.

γ (gamma) Crv, 12h 16m −17°.5, (Gienah, 'wing'), mag. 2.6, the brightest star in the constellation, is a blue-white giant 165 l.y. away.

δ (delta) Crv, 12h 30m −16°.5, (Algorab, 'the raven'), is a wide double star for small telescopes. The brighter component, visible to the naked eye, is a mag. 2.9 blue-white star 88 l.y. away, which is accompanied by a 9th-mag. star often described as purplish in colour.   ▶

Just over the northern border of Corvus into Virgo lies the Sombrero Galaxy, M104, a spiral galaxy with a bulging core and a dark central lane of dust. See page 260. (ESO)

Corvus · Crv · Corvi

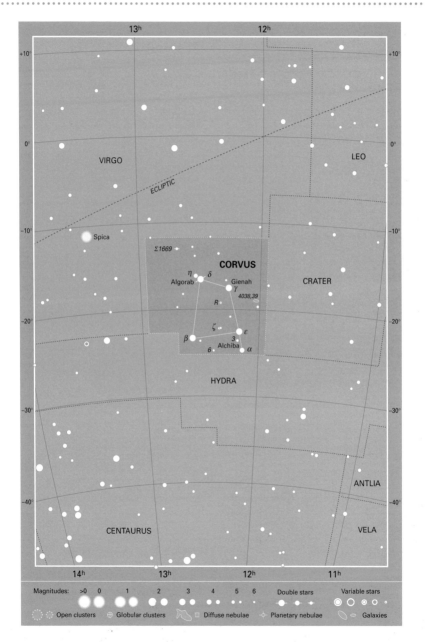

ε (epsilon) Crv, 12h 10m −22°.6, mag. 3.0, is an orange giant star 303 l.y. away.

Σ 1669 (Struve 1669), 12h 41m −13°.0, is a neat pair of white stars 280 l.y. away, appearing to the naked eye as a single star of mag. 5.2 but divisible by small telescopes into near-identical components of mags. 5.9 and 6.0.

NGC 4038–9, 12h 02m −18°.9, the Antennae, are a pair of spiral galaxies of 10th mag. that collided a few hundred million years ago. Streams of gas and stars thrown off in the collision, visible on photographs such as the one below, stretch for 250,000 l.y. or more and form the antennae that give the galaxies their popular name. The galaxies lie 63 million l.y. away.

NGC 4038–9, the peculiar pair of interacting galaxies known as the Antennae, lie in Corvus near the border with Crater. (M. Bessell, R. Sutherland and M. Buxton, RSAA – Australian National University)

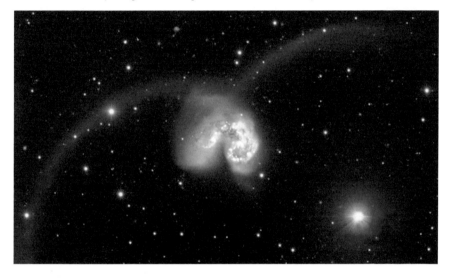

Crater – The Cup

An ancient constellation representing the chalice of Apollo, associated in Greek legend with neighbouring Corvus and Hydra (see page 130). The stars of Crater are faint and the constellation contains no objects of particular interest.

α (alpha) Crateris, 11h 00m −18°.3, (Alkes, 'the cup'), mag. 4.1, is an orange giant star 174 l.y. away.

β (beta) Crt, 11h 12m −22°.8, mag. 4.5, is a blue-white star 266 l.y. away.

γ (gamma) Crt, 11h 25m −17°.7, mag. 4.1, is a white star 84 l.y. away. It has a mag. 9.6 companion visible in small telescopes.

δ (delta) Crt, 11h 19m −14°.8, mag. 3.6, the constellation's brightest star, is an orange giant lying 195 l.y. away.

Crater · Crt · Crateris

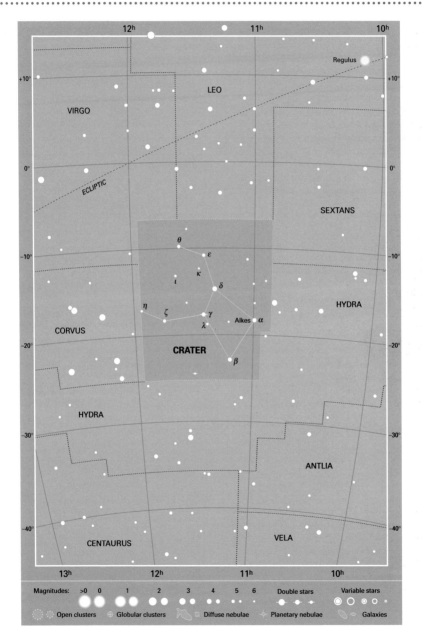

Crux – The Southern Cross

The smallest constellation in the sky, but one of the most celebrated and distinctive. Crux was visible from the Mediterranean area in ancient times, so its stars were known to Greek astronomers; the effects of precession have since carried it below the horizon from such northerly latitudes. The Greeks regarded it as part of Centaurus, but it was made a separate constellation by various seamen and astronomers in the 16th century. It can be useful for direction-finding at night, since its long axis, from γ (gamma) via α (alpha) Crucis, points towards the south celestial pole.

Crux lies in a dense and brilliant part of the Milky Way, and the famous dark nebula known as the Coalsack is particularly striking in silhouette against the starry background. α (alpha) Crucis is the most southerly first-magnitude star. α (alpha) and β (beta) Crucis are, by a small margin, the bluest of all first-magnitude stars.

α (alpha) Crucis, 12h 27m −63°.1, (Acrux), 321 l.y. away, appears to the naked eye as a bluish star of mag. 0.8, but a small telescope divides it into a sparkling double with components of mags. 1.3 and 1.8. There is also a wider 5th-mag. companion that can be seen with binoculars.

β (beta) Cru, 12h 48m −59°.7, (Mimosa), mag. 1.3, is a blue giant 353 l.y. away. It is a variable star of the β Cephei type, fluctuating by less than 0.1 mag. every 5 hours, too small to be noticeable to the eye.

γ (gamma) Cru, 12h 31m −57°.1, (Gacrux), mag. 1.6, is a red giant star 88 l.y. away. There is a very wide mag. 6.5 unrelated companion, three times as distant, visible in binoculars.

δ (delta) Cru, 12h 15m −58°.7, mag. 2.8, the faintest of the four main stars in the Cross, is a blue-white star 364 l.y. away.

ε (epsilon) Cru, 12h 21m −60°.4, mag. 3.6, is an orange giant 228 l.y. away.

ι (iota) Cru, 12h 46m −61°.0, 125 l.y. away, is an orange giant of mag. 4.7 with a mag. 9.5 companion visible in small telescopes.

μ (mu) Cru, 12h 55m −57°.2, is a wide pair of blue-white stars of mags. 4.0 and 5.1, divisible by the smallest telescopes or even good binoculars. They both lie about 370 l.y away.

NGC 4755, 12h 54m −60°.3, the Jewel Box or κ (kappa) Crucis Cluster, is one of the finest open clusters in the sky, visible to the naked eye as a hazy 4th-mag. star. Binoculars resolve the brightest individual members, which are blue supergiants of 6th and 7th mag.; three of these form a chain across the cluster like a mini Orion's belt, with an 8th-mag. red supergiant next to the middle one. The star κ Crucis itself, of mag. 5.9, is the most southerly member of this chain. A small telescope brings at least 50 stars into view. John Herschel gave this cluster its popular name when he likened it to a piece of multicoloured jewellery. The distance of the Jewel Box is just under 5000 l.y.

The Coalsack Nebula is a dark, pear-shaped dust cloud silhouetted against the Milky Way, covering nearly 7° by 5° of sky and spilling over into neighbouring Centaurus and Musca. It lies an estimated 600 l.y. away. The Coalsack has no NGC or other identification number. For a photograph of the Coalsack, see page 120.

Crux · Cru · Crucis

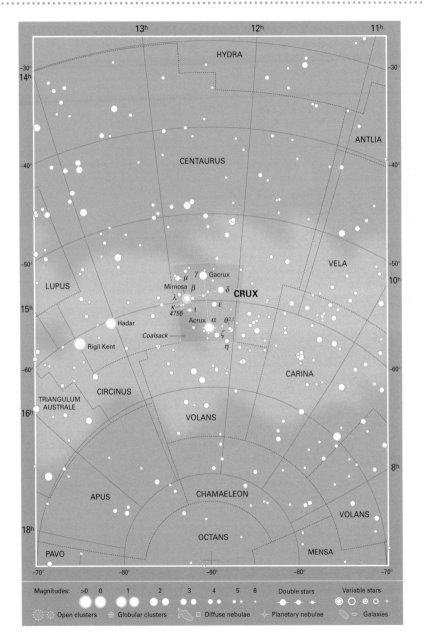

Cygnus – The Swan

Cygnus represents a swan flying down the Milky Way. In Greek mythology, the swan was the guise in which the god Zeus visited Leda, wife of King Tyndareus of Sparta; the result of their union was Pollux, one of the heavenly twins. The flying swan's tail is marked by the star Deneb (α Cygni), its beak by Albireo (β Cygni), and its wings by δ (delta) and ε (epsilon) Cygni. These stars form a distinctive cross-shape, so the constellation is also referred to as the Northern Cross; it happens to be far larger than the more famous Southern Cross. Cygnus lies in a rich part of the Milky Way, which is split here by a dark lane of dust known as the Cygnus Rift or the Northern Coalsack. Deneb, the constellation's brightest star, forms one corner of the so-called Summer Triangle, completed by Altair and Vega.

Among the fascinating objects in Cygnus is an X-ray source called Cygnus X-1, thought to be a black hole orbiting a 9th-mag. blue supergiant star; it lies at 19h 58.4m, +35° 12′, near η (eta) Cygni. Near γ (gamma) Cygni, at 19h 59.5m, +40° 44′, is Cygnus A, a powerful radio source caused by two distant galaxies in collision. Long-exposure photographs of the region between ε (epsilon) Cygni and the border with Vulpecula reveal beautiful swirls of gas known as the Veil Nebula, the brightest part of which, NGC 6992, is detectable in amateur instruments (see page 138).

α (alpha) Cygni, 20h 41m +45°.3, (Deneb, 'tail'), mag. 1.2, is a blue-white supergiant star about 3200 l.y. away.

β (beta) Cyg, 19h 31m +28°.0, (Albireo), 386 l.y. away, is one of the sky's showpiece doubles. It consists of gloriously contrasting amber and blue-green stars, like a celestial traffic light. The brighter star, of mag. 3.1, is an orange giant, and its blue-green companion is mag. 5.1. They can be separated through good binoculars and are a beautiful sight in any amateur telescope.

γ (gamma) Cyg, 20h 22m +40°.3, (Sadr, 'breast'), mag. 2.2, is a yellow-white supergiant about 1500 l.y. away.

δ (delta) Cyg, 19h 45m +45°.1, 171 l.y. away, is a blue-white giant of mag. 2.9 with a close mag. 6.6 companion, visible in a telescope of 100 mm aperture or above at high magnification. The stars have an orbital period of nearly 800 years.

ε (epsilon) Cyg, 20h 46m +34°.0, (Gienah, 'wing'), mag. 2.5, is an orange giant 72 l.y. away.

μ (mu) Cyg, 21h 44m +28°.7, 73 l.y. away, is a pair of white stars of mags. 4.8 and 6.2 orbiting each other about every 790 years. They are currently closing slowly and should remain divisible in 100 mm apertures until about 2020, although 150 mm will probably be necessary when they are closest, between 2043 and 2050. A wide mag. 6.9 binocular companion is an unrelated background star.

o[1] (omicron[1]) Cyg, 20h 14m +46°.7, also known as 31 Cygni, forms perhaps the most beautiful binocular double in the heavens with 30 Cygni. The stars are orange and turquoise, mags. 3.8 and 4.8, 1400 l.y. and 720 l.y. away, like a wider version of Albireo. A small telescope, or binoculars held steadily, shows a closer blue companion of mag. 7.0 to the brighter (orange giant) star.

χ (chi) Cyg, 19h 51m +32°.9, 350 l.y. away, is a red giant long-period variable of Mira type that varies every 400 days or so. At best it reaches mag. 3.3, brighter than any other star of its type except Mira itself, although R Hydrae can rival it. At its faintest χ Cyg drops to 14th mag. Its diameter is about 300 Suns.

▶

Cygnus · Cyg · Cygni

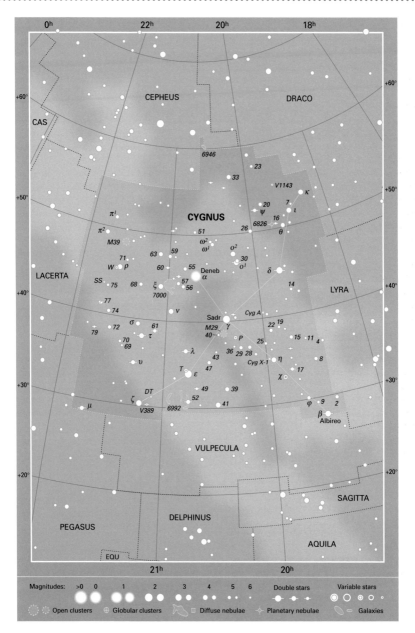

ψ (psi) Cyg, 19h 56m +52°.4, 289 l.y. away, is a pair of white stars of mags. 5.0 and 7.5 divisible in small to moderate apertures.

61 Cyg, 21h 07m +38°.7, 11.4 l.y. away, is a showpiece pair of orange dwarf stars of mags. 5.2 and 6.1, orbiting each other in about 680 years, divisible in a small telescope or even binoculars. In addition to being among the closest stars to Earth, 61 Cygni was the first star to have its parallax measured, by the German astronomer Friedrich Wilhelm Bessel in 1838.

P Cyg, 20h 18m +38°.0, is an erratically variable blue supergiant that normally resides around 5th mag., although in the year 1600 it reached a peak brightness of 3rd mag. Evidently the star is so large and luminous that it is close to instability. Its average brightness has increased gradually since the

18th century as it evolves into a red supergiant. Its distance is uncertain, but is probably several thousand light years.

W Cyg, 21h 36m +45°.4, 618 l.y. away, is a red giant that varies semi-regularly between 5th and 8th mags. with a rough period of 130 days.

M39 (NGC 7092), 21h 32m +48°.4, is a large, loose cluster of about 30 stars of 7th mag. and fainter, arranged in a triangle. It is visible in binoculars, and is even detectable with the naked eye under ideal conditions. It lies 950 l.y. away.

NGC 6826, 19h 45m +50°.5, is an 8th-mag. planetary nebula about 3200 l.y. away, known as the Blinking Planetary because it appears to blink on and off. It is visible as a pale blue disk in 75-mm telescopes, but 150 mm is needed to show it

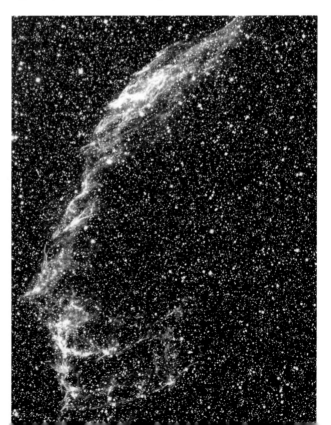

The Veil Nebula:
Drifting apart over thousands of years, the delicate traceries of the Veil Nebula are the tenuous remains of a star that shattered itself in a supernova explosion some 5000 years ago. Its brightest part, shown here, is NGC 6992, which can be glimpsed in binoculars under clear skies.
The arc of NGC 6992 extends across more than two Moon diameters of sky, equivalent to a true length of about 25 light years.
(Nigel Sharp, REU program/ AURA/NOAO/NSF)

The North America Nebula, NGC 7000, is a glowing cloud of gas shaped like the continent of North America, with a particularly distinctive Gulf of Mexico formed by an obscuring cloud of dark dust. It lies near the star ξ (xi) Cygni, an orange supergiant of mag. 3.7 seen at left. To the right is a smaller and fainter patch of nebulosity, IC 5070, known as the Pelican from its supposed resemblance to that bird. (Philip Perkins)

to advantage. At its centre is a 10th-mag. star. Looking alternately at this star and away again produces the 'blinking' effect. NGC 6826 lies less than 1° from the wide double star 16 Cygni, which consists of two 6th-mag. creamy-white stars 70 l.y. away.

NGC 6992, 20h 56m +31°.7, is the brightest part of the Veil Nebula, the remnant of a supernova explosion about 5000 years ago. Under ideal con-

ditions NGC 6992 can be seen in binoculars as a faint arc. Another part of the nebula, NGC 6960, can be located in wide-angle telescopes with low power near the 4th-mag. star 52 Cyg, but the complete Veil Nebula, also known as the Cygnus Loop, is apparent only on long-exposure photographs, spanning nearly 3° of sky. Its distance is 1400 l.y.

NGC 7000, 20h 59m +44°.3, the North America Nebula, can be seen as a hook-shaped brightening in the Milky Way with the naked eye or binoculars in good skies. Despite its large size, 2° at its widest, the nebula can be difficult to detect because of its low surface brightness. Long-exposure photographs such as the one above clearly show its shape, like the continent of North America. The nebula is about 1500 l.y. away, half the distance to the star Deneb. The nebula is believed to be lit up by an extremely hot 6th-mag. star that lies within it.

Delphinus – The Dolphin

A constellation which originated in Greek times, celebrating the long-standing relationship between humans and these intelligent aquatic mammals. In legend, dolphins were the messengers of the sea-god Poseidon, who sent a dolphin to bring the sea nymph Amphitrite to his underwater palace where he made her his wife. In another story, dolphins were credited with saving the life of Arion, a musician and poet, who was attacked on a ship. According to this tale, Arion jumped overboard to escape his attackers and was carried on the back of a dolphin to Greece, where he later identified the robbers who were sentenced to death. In this legend, the nearby constellation Lyra represents Arion's lyre.

Delphinus has a distinctive shape, its four main stars forming a diamond known as Job's Coffin. Its two brightest stars are called Sualocin and Rotanev, which backwards read Nicolaus Venator, the Latinized form of the name Niccolò Cacciatore (1780–1841), who was assistant and successor to the Italian astronomer Giuseppe Piazzi at Palermo Observatory. These names first appeared in the Palermo Catalogue of 1814, and it is usually supposed that Cacciatore himself was responsible, although it is equally possible that Piazzi may have applied the names to honour his appointed successor.

Like the small neighbouring constellations Vulpecula and Sagitta, Delphinus lies in a rich area of the Milky Way and has become a favourite hunting-ground for novae.

α (alpha) Delphini, 20h 40m +15°.9, (Sualocin), mag. 3.8, is a blue-white main-sequence star 241 l.y. away.

β (beta) Del, 20h 38m +14°.6, (Rotanev), mag. 3.6, the constellation's brightest member, is a white star 97 l.y. away. It is a binary with an orbital period of 27 years, but the components are too close to be divided in all but the largest amateur telescopes.

γ (gamma) Del, 20h 47m +16°.1, 102 l.y. away, is a showpiece double star consisting of golden and yellow-white stars of mags. 4.3 and 5.1, neatly separated in a small telescope. In the same telescopic field of view, but about 25 l.y. more distant, lies a faint binary, Σ 2725 (Struve 2725), consisting of stars of mags. 7.5 and 8.3.

The naming of variable stars

In addition to the normal system of naming stars in a constellation (page 7), stars that vary in brightness have their own system of nomenclature. Stars that were already named when their variability was discovered, such as δ (delta) Cephei, β (beta) Persei or ο (omicron) Ceti, retain their existing designation. Other variable stars are denoted by a system of one or two capital letters or, where that is insufficient, by the letter V and a number. The sequence starts with the letters R to Z, which covers the first nine variables in a constellation. Then double letters are applied, from RR to RZ. Next the lettering runs from SS to SZ, and so on until ZZ is reached. Then the sequence goes from AA to AZ, BB to BZ, ending with QZ, at which point 334 variable stars have been named (the letter J is omitted). Further variables are designated V335, V336, and so on. Novae, too, are allocated variable-star designations, such as the nova that erupted in Delphinus in 1967, which became known as HR Delphini.

Delphinus · Del · Delphini

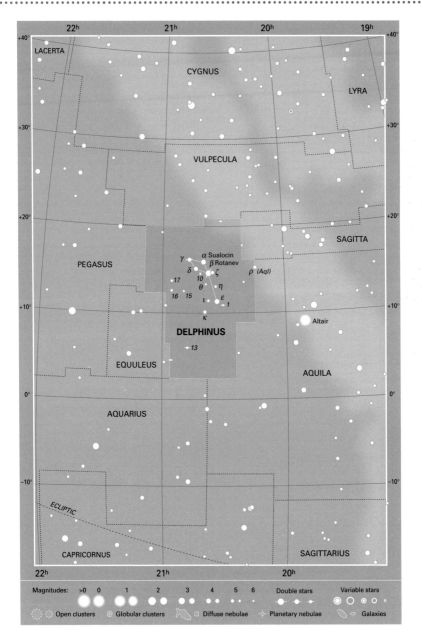

Dorado – The Goldfish

This constellation, also known as the Swordfish, was introduced at the end of the 16th century by the Dutch navigators Pieter Dirkszoon Keyser and Frederick de Houtman. Its most notable feature is the Large Magellanic Cloud (LMC), the larger of the two satellite galaxies that accompany our Milky Way. In 1987 the first supernova visible to the naked eye since 1604, Supernova 1987A, erupted in the LMC, at 5h 35m, −69°.3.

α (alpha) Doradus, 4h 34m −55°.0, mag. 3.3, is a blue-white star 176 l.y. away.

β (beta) Dor, 5h 34m −62°.5, a yellow-white supergiant 1040 l.y. away, is one of the brightest Cepheid variables, ranging from mag. 3.5 to 4.1 every 9 days 20 hours.

R Dor, 4h 37m −62°.1, 204 l.y. away, is a red giant that varies semi-regularly between mags. 4.8 and 6.6 with a period of around 11 months.

Large Magellanic Cloud (LMC), 5h 24m −69°, is a mini-galaxy 170,000 l.y. away, a satellite of the Milky Way, containing perhaps 10,000 million stars. To the naked eye it appears as a fuzzy elongated patch 6° in diameter, 12 times the apparent size of the Moon; its actual diameter is about 25,000 l.y. Binoculars and telescopes show individual stars, nebulae (notably the Tarantula Nebula, described below) and clusters.

NGC 2070, 5h 39m −69°.1, the Tarantula Nebula, is a looping cloud of hydrogen gas about 1000 l.y. in diameter in the Large Magellanic Cloud. To the naked eye the nebula appears as a fuzzy star, also known as 30 Doradus. The nebula's popular name, the Tarantula, comes from its spider-like shape. At the centre of the nebula is a cluster of supergiant stars, called R136, the light from which makes the nebula glow. The Tarantula is larger and brighter than any nebula in the Milky Way. If it were as close to us as the Orion Nebula, the Tarantula would fill the whole constellation of Orion, and would cast shadows on Earth.

The Large Magellanic Cloud: Loops of glowing gas can be seen among the star swarms. The pink knot of gas at left of centre is the Tarantula Nebula, NGC 2070. (Eckhard Slawik/ESA)

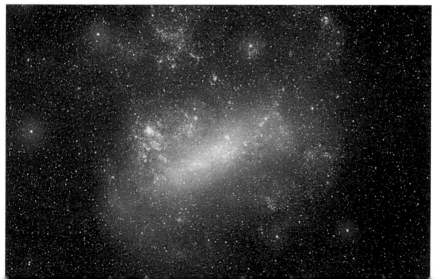

Dorado · Dor · Doradus

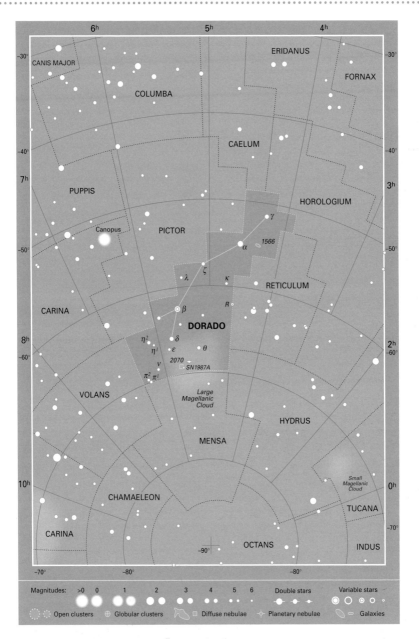

Draco – The Dragon

Dragons feature in many ancient legends, so it is not surprising to find such a monster in the sky. This one is said to be Ladon, the dragon slain by Hercules as a prelude to stealing the golden apples from the garden of the Hesperides. In the sky, one foot of Hercules rests upon the dragon's head, while the creature's body lies coiled around the north celestial pole. Although Draco is one of the largest and most ancient constellations it is indistinct, with no stars brighter than 2nd magnitude. Draco contains the north pole of the ecliptic, i.e. one of the two points 90° from the plane of the Earth's orbit, at 18h 00m, +66½°.

α (alpha) Draconis, 14h 04m +64°.4, (Thuban, 'serpent's head'), mag. 3.7, is a blue-white giant star 309 l.y. away. It was the pole star in about 2800 BC, but has lost that place to Polaris because of the effect of precession (see page 16).

β (beta) Dra, 17h 30m +52°.3, (Rastaban, 'serpent's head'), mag. 2.8, is a yellow giant 362 l.y. distant.

γ (gamma) Dra, 17h 57m +51°.5, (Etamin or Eltanin, 'the serpent'), mag. 2.2, is an orange giant 148 l.y. away, and the brightest star in the constellation. From observations of this star the English astronomer James Bradley discovered the effect known as the aberration of starlight in 1728.

μ (mu) Dra, 17h 05m +54°.5, (Alrakis), 88 l.y. away, is a close double star with matching cream-coloured components of mag. 5.6 and 5.7 which orbit every 670 years. The two stars are currently moving apart and becoming progressively easier to split in small apertures, although high magnification will be needed.

ν (nu) Dra, 17h 32m +55°.2, is a pair of identical white stars of mag. 4.9, easily visible in the smallest of telescopes and regarded as one of the finest binocular pairs. They lie 100 l.y. away.

o (omicron) Dra, 18h 51m +59°.4, 322 l.y. away, is a mag. 4.6 orange giant star with a mag. 7.8 companion visible in small telescopes.

ψ (psi) Dra, 17h 42m +72°.1, 72 l.y. away, is a mag. 4.6 yellow-white star with a yellow mag. 5.8 companion visible in small telescopes or even binoculars.

16–17 Dra, 16h 36m +52°.9, is a wide pair of blue-white stars of mags. 5.1 and 5.5, 400 l.y. away, easily found in binoculars. Telescopes of 60 mm aperture with high magnification split the brighter star into a binary of mags. 5.4 and 6.5, making this a striking triple system.

39 Dra, 18h 24m +58°.6, 188 l.y away, is an impressive triple system. The two brightest members, at mags. 5.0 and 7.4, appear in binoculars as a wide blue-and-yellow pair. A telescope of 60 mm aperture with high magnification reveals that the brighter star has a closer companion of mag. 8.0.

40–41 Dra, 18h 00m +80°.0, is an easy pair of orange dwarf stars for small telescopes, mags. 5.7 and 6.1. They lie about 170 l.y. away.

NGC 6543, 17h 59m +66°.6, is a 9th-mag. planetary nebula 3500 l.y. away, one of the brightest such nebulae, showing in amateur telescopes as an irregular blue-green disk like an out-of-focus star. It is now known as the Cat's Eye Nebula from its appearance in Hubble Space Telescope photographs (see page 277).

Draco · Dra · Draconis

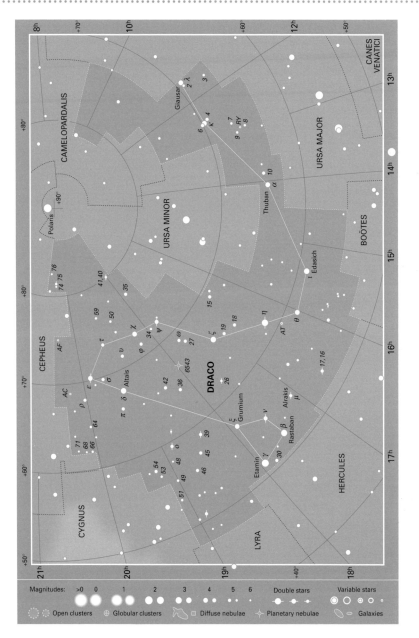

Equuleus – The Little Horse

The second-smallest constellation in the sky, Equuleus seems to have originated with the Greek astronomer Ptolemy in the second century AD, as it appears for the first time among the 48 constellations listed in his book the *Almagest*. Only the head of the horse is shown, next to the much larger horse Pegasus, and there are no legends that identify it.

α (alpha) Equulei, 21h 16m +5°.2, (Kitalpha, 'the section of the horse'), mag. 3.9, is a yellow giant 186 l.y. away.

γ (gamma) Equ, 21h 10m +10°.1, is a mag. 4.7 white star 115 l.y. away. Through binoculars an apparent mag. 6.1 companion, 6 Equ, can be seen, but this is an unrelated background star.

1 Equ, 20h 59m +4°.3, also known as ε (epsilon) Equ, is a triple star 197 l.y. away. In small telescopes it appears as a white-and-yellow pair of mags. 5.4 and 7.4. The brighter component is also a close binary, of mags. 6.0 and 6.3; these stars orbit each other every 101 years. Currently they are closing together, and by 2015 will be too close to separate in amateur telescopes.

NGC 1300 in Eridanus (see page 148) is a classic example of a barred spiral galaxy with tightly wound arms, requiring quite large amateur telescopes to be seen. (José Alfonso López Aguerri, M. Prieto, C. Muñoz-Tuñón and A. M. Varela/Instituto de Astrofísica de Canarias)

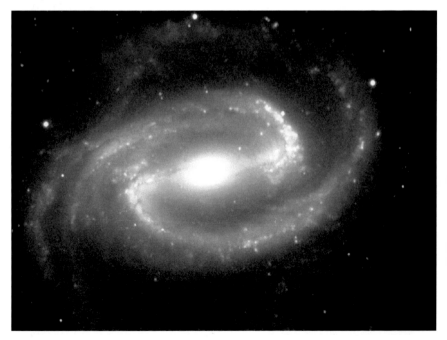

Equuleus · Equ · Equulei

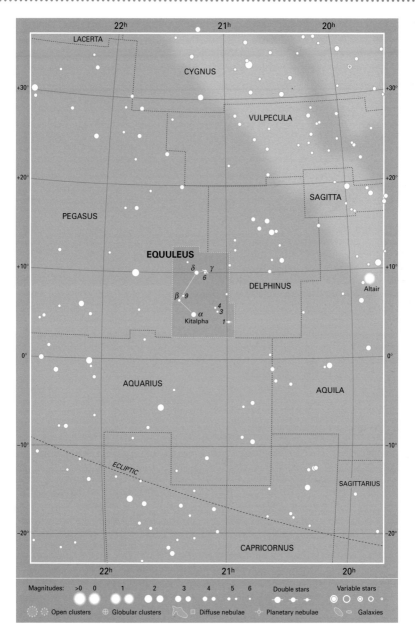

Eridanus – The River

An extensive constellation, the sixth-largest in the sky, but often overlooked because of its lack of any distinctive shape. It meanders from Taurus in the north to Hydrus in the south. In mythology, Eridanus was the river into which Phaethon fell after trying to drive the chariot of his father, Helios the Sun god. But it also supposedly represents a real river. Early mythologists identified it with the Nile, but later Greek writers said it was the Po in Italy.

Originally Eridanus included the stars of what is now Fornax and it stretched only as far as θ (theta) Eridani, which was given the name Achernar, from the Arabic meaning 'the river's end'. In more recent times Eridanus has been extended to nearly 60° south (below the horizon from Greece), and the title Achernar has been transferred to another star farther south. The star originally called Achernar is now known as Acamar, a name that comes from the same Arabic original as Achernar

The constellation contains several interesting galaxies, all too distant and too faint to be picked up easily in amateur telescopes. One celebrated example is NGC 1300, located at 3h 19.7m, −19° 25′; photographs such as the one on page 146 show that it is a beautiful 10th-mag. barred spiral. It lies 69 million l.y. away.

α (alpha) Eridani, 1h 38m −57°.2, (Achernar, 'the river's end'), mag. 0.5, is a blue-white main-sequence star 144 l.y. away.

β (beta) Eri, 5h 08m −5°.1, (Cursa, 'the footstool', referring to its position under the foot of Orion), mag. 2.8, is a blue-white star 89 l.y. away.

ε (epsilon) Eri, 3h 33m −9°.5, mag. 3.7, an orange main-sequence star 10.5 l.y. away, is among the most Sun-like of the nearby stars. It is orbited every 7 years by a planet of similar mass to Jupiter.

θ (theta) Eri, 2h 58m −40°.3, (Acamar), 161 l.y. away, is a striking pair of blue-white stars of mags. 3.2 and 4.3, divisible in small telescopes.

o² (omicron²) Eri, 4h 15m −7°.7, 16 l.y. away, also known as 40 Eridani, is a remarkable triple star. A small telescope shows that the mag. 4.4 orange primary, a star similar to the Sun, has a wide mag. 9.5 white dwarf companion, the most easily seen white dwarf in the sky. Small telescopes reveal that the white dwarf has an 11th-mag. companion which is a red dwarf, thereby completing a most interesting trio. The white dwarf and red dwarf orbit each other every 250 years, and will remain easy to split until the latter half of the 21st century.

32 Eri, 3h 54m −3°.0, 290 l.y. away, is a beautiful double star for small telescopes, consisting of a yellow giant of mag. 4.8 and a blue-green mag. 6.1 companion.

39 Eri, 4h 14m −10°.3, 206 l.y. away, is an orange giant of mag. 4.9 with an 8th-mag. companion divisible in small telescopes.

p Eri, 1h 40m −56°.2, 27 l.y. away, is a beautiful wide duo of orange stars, mags. 5.8 and 5.9, with an orbital period of about 500 years.

NGC 1535, 4h 14m −12°.7, is a small 9th-mag. planetary nebula about 2000 l.y. away. Small telescopes show it, but apertures of 150 mm are needed to appreciate its blue-grey disk.

Eridanus · Eri · Eridani

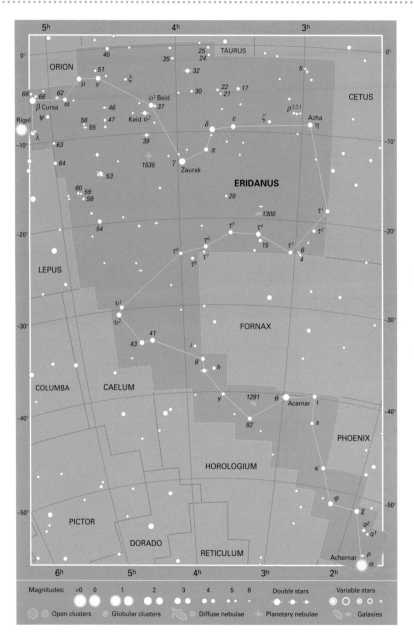

Fornax – The Furnace

A barren constellation introduced in the 1750s by Nicolas Louis de Lacaille, originally under the name of Fornax Chemica, the Chemical Furnace. It contains a dwarf member of our Local Group of galaxies, called the Fornax Dwarf, some 450,000 l.y. from the Milky Way but too faint to see in amateur telescopes. Fornax also contains a compact cluster of galaxies about 60 million l.y. away, the brightest member of which is the 9th-mag. peculiar galaxy NGC 1316, also known as the radio source Fornax A, located at 3h 22.7m, −37° 12′. Another well-known member of the Fornax cluster is NGC 1365, pictured below.

α (alpha) Fornacis, 3h 12m −29°.0, 46 l.y. away, is a binary consisting of a yellow-white main-sequence star of mag. 3.9 with a deeper yellow mag. 6.5 companion of suspected variability. The pair orbit each other every 270 years or so and will remain divisible in small telescopes throughout the 21st century.

β (beta) For, 2h 49m −32°.4, mag. 4.5, is a yellow giant star 169 l.y. away.

NGC 1097, 2h 46m −30°.6, is a 9th-mag. barred spiral galaxy with a bright nucleus, visible in medium-sized telescopes. It lies about 50 million l.y. away.

NGC 1365, on the border of Fornax and Eridanus at 3h 33.6m, −36° 08′, is a striking barred spiral galaxy of 10th mag. with widely flung arms. It is a member of the Fornax cluster of galaxies, 60 million l.y. away. (Patrick Woudt, Univ. of Cape Town/South African Astronomical Observatory)

Fornax · For · Fornacis

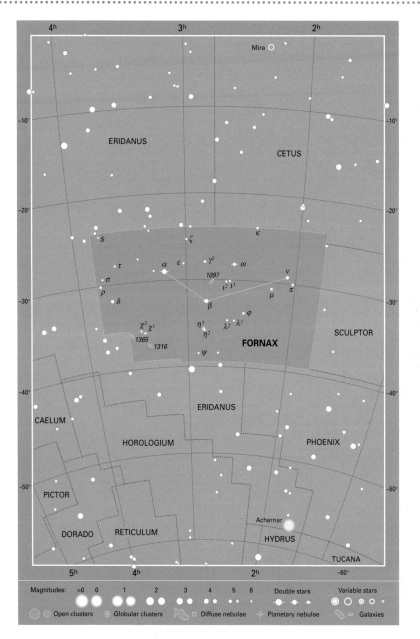

Gemini – The Twins

A constellation that dates from ancient times, representing a pair of twins. We know them as Castor and Pollux, members of the Argonauts' crew and of mixed parentage: both were sons of Queen Leda of Sparta, but by different fathers. Castor's father was her husband King Tyndareus, while the father of Pollux was the god Zeus. The twins were the protectors of mariners, appearing in ships' rigging as the electrical phenomenon now known as St Elmo's fire. In the sky, the stars Castor and Pollux provide a useful yardstick for measuring angular distances – they are exactly 4°.5 apart.

Gemini is a member of the zodiac, and the Sun passes through the constellation from late June to late July. Each year the Geminid meteors, one of the year's richest and brightest showers, radiate from a point near Castor, reaching a maximum around December 13–14, when up to 100 meteors per hour may be seen.

α (alpha) Geminorum, 7h 35m +31°.9, (Castor), 52 l.y. away, is an astounding multiple star, consisting of six separate components. To the naked eye Castor appears as a blue-white star of mag. 1.6, somewhat fainter than Pollux. A 60-mm telescope with high magnification splits Castor into two components of mags. 1.9 and 3.0, which orbit each other every 450 years. Separation of these stars is increasing, and they will become progressively easier to split throughout the 21st century, after which they will start to close up again. Both stars are spectroscopic binaries. Small telescopes also show a wide red dwarf companion to Castor; this is itself an eclipsing binary of Algol type, varying between mags. 9.3 and 9.8 in 19.5 hours, completing the six-star system.

β (beta) Gem, 7h 45m +28°.0, (Pollux), mag. 1.2, is the brightest star in the constellation. Some astronomers have speculated that Pollux, being labelled β Geminorum, was once fainter than Castor and has since brightened, or Castor has

faded. But the truth is that Johann Bayer, who allotted the Greek letters in 1603, did not distinguish carefully which star was the brighter, and so caused unnecessary confusion. Pollux is an orange giant 34 l.y. away.

γ (gamma) Gem, 6h 38m +16°.4, (Alhena), mag. 1.9, is a blue-white star 105 l.y. away.

δ (delta) Gem, 7h 20m +22°.0, (Wasat), 59 l.y. away, is a creamy-white star of mag. 3.5 with an orange dwarf companion of mag. 8.2. The brightness contrast makes the pair difficult in telescopes below about 75 mm aperture. Their estimated orbital period is over 1000 years.

ε (epsilon) Gem, 6h 44m +25°.1, mag. 3.1, is a yellow supergiant about 900 l.y. away. Binoculars, or a small telescope, reveal a wide companion of mag. 9.2.

ζ (zeta) Gem, 7h 04m +20°.6, 1200 l.y. away, is both a variable star and a binocular double. A yellow supergiant, it is a Cepheid variable, ▶

NGC 2392, popularly known as the Eskimo or Clown Face Nebula, is a planetary nebula visible as a blue disk in amateur telescopes. See page 154. (Tom Tekach/Flynn Haase/NOAO/AURA/NSF)

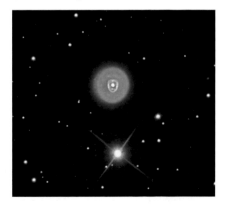

Gemini · Gem · Geminorum

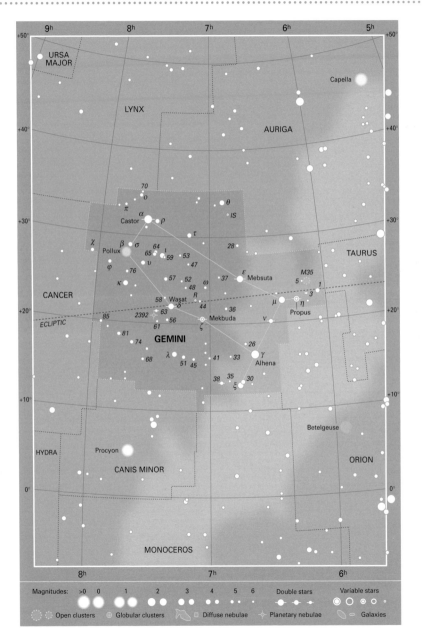

fluctuating between mags. 3.6 and 4.2 every 10.2 days. Binoculars or small telescopes reveal a wide mag. 7.6 companion, which is unrelated.

η (eta) Gem, 6h 15m +22°.5, 350 l.y. away, is another double–variable. A red giant star, it fluctuates in semi-regular manner between mags. 3.1 and 3.9 in about 233 days. It has a close 6th-mag. companion that orbits it every 500 years or so and requires a large telescope to distinguish it from the primary's glare.

κ (kappa) Gem, 7h 44m +24°.4, 143 l.y. away, is a mag. 3.6 yellow giant, with an 8th-mag. companion made difficult in small telescopes because of the extreme brightness difference.

ν (nu) Gem, 6h 29m +20°.2, mag. 4.1, is a blue giant 500 l.y. away with a wide 8th-mag. companion visible in binoculars or small telescopes.

38 Gem, 6h 55m +13°.2, 91 l.y. away, is a double star for small telescopes, with white and yellow components of mags. 4.8 and 7.8.

M35 (NGC 2168), 6h 09m +24°.3, is a large, 5th-mag. open cluster visible to the naked eye or through binoculars, noticeably elongated in shape. It consists of about 200 stars covering a similar area of sky as the Moon, and is 2800 l.y. away. Even a small telescope at low magnification will show the members of this outstanding cluster to be arranged in curving chains. Nearby in the sky, but actually about 10,000 l.y. farther off, is NGC 2158, a very rich open cluster that appears as a small, faint patch of light requiring at least 100 mm aperture to distinguish.

NGC 2392, 7h 29m +20°.9, is an 8th-mag. planetary nebula known as the Eskimo or Clown Face Nebula because it looks somewhat like a face surrounded by a fringe when seen through a large telescope. A small telescope shows it as a blue-green ellipse about the same apparent size as the disk of Saturn, with a central star of 10th mag. (see page 152 for a photograph). NGC 2392 lies 3000 l.y. away.

Grus – The Crane

One of the 12 constellations introduced at the end of the 16th century by the Dutch navigators Pieter Dirkszoon Keyser and Frederick de Houtman. It represents a water bird, the long-necked Crane, although it has also been depicted as a flamingo. The stars δ (delta) and μ (mu) Gruis are striking naked-eye doubles.

α (alpha) Gruis, 22h 08m −47°.0, (Alnair, 'the bright one'), mag. 1.7, is a blue-white star 101 l.y. away.

β (beta) Gru, 22h 43m −46°.9, is a red giant 170 l.y. away that varies between about mags. 2.0 and 2.3.

γ (gamma) Gru, 21h 54m −37°.4, mag. 3.0, is a blue giant 203 l.y. away.

δ^1 δ^2 (delta^1 delta^2) Gru, 22h 29m −43°.5, is a naked-eye pairing of two unrelated stars: δ^1, mag. 4.0, is a yellow giant 296 l.y. away; δ^2 is a mag. 4.1 red giant 325 l.y. away.

μ^1 μ^2 (mu^1 mu^2) Gru, 22h 16m −41°.3, is another naked-eye double of unrelated yellow giants that appear in the same line of sight by chance: μ^1 is of mag. 4.8, 262 l.y. away; μ^2 is of mag. 5.1, 240 l.y. away.

π^1 π^2 (pi^1 pi^2) Gru, 22h 23m −45°.9, is a binocular duo of unrelated stars. π^1 is a deep-red semi-regular variable that ranges between mags. 5.4 and 6.7 every 150 days or so; it lies 500 l.y. away. π^2 is a white giant of mag. 5.6, 132 l.y. away.

Grus · Gru · Gruis

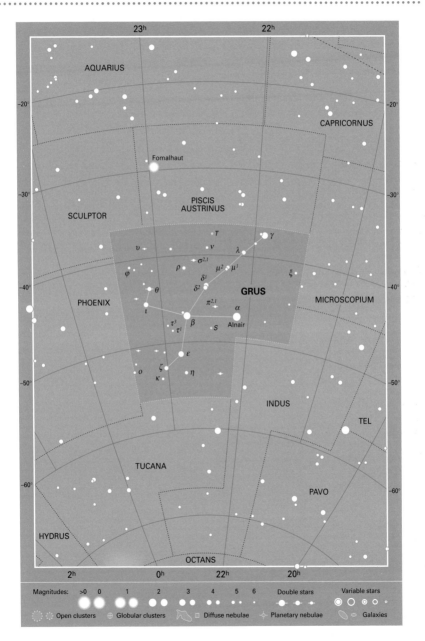

Hercules

Hercules represents the Greek mythological hero famous for his twelve labours. Originally, though, the constellation was visualized as an anonymous kneeling man, with one foot on the head of the celestial dragon, Draco, which adjoins to the north. Some legends identify the constellation with the ancient Sumerian superman, Gilgamesh. Despite being the fifth-largest constellation, Hercules is not prominent. But it is stocked with an abundance of double stars for users of small telescopes, plus one of the brightest and richest globular clusters, M13, which is easily found on one side of the so-called Keystone, a quadrilateral consisting of ε, ζ, η and π Herculis that marks the pelvis of Hercules.

α (alpha) Herculis, 17h 15m +14°.4, (Rasalgethi, 'the kneeler's head'), about 400 l.y. distant, is a red giant star some 400 times the Sun's diameter, making it one of the largest stars known. Like most red giants it is erratically variable, in this case fluctuat-ing from about mag. 3 to mag. 4. It is actually a double star, with a mag. 5.4 blue-green companion visible in small telescopes. The estimated orbital period of the pair is 3600 years.

β (beta) Her, 16h 30m +21°.5, (Kornephoros, 'club-bearer'), mag. 2.8, the brightest star in the constellation, is a yellow giant 148 l.y. away.

γ (gamma) Her, 16h 22m +19°.2, mag. 3.8, is a white giant star 195 l.y. away, with a wide, unrelated 10th-mag. companion visible in small telescopes.

δ (delta) Her, 17h 15m +24°.8, mag. 3.1, is a blue-white star 78 l.y. away. Small telescopes show a nearby mag. 8.2 star which is physically unrelated.

ζ (zeta) Her, 16h 41m +31°.6, 35 l.y. away, is a mag. 2.9 yellow-white star with a close mag. 5.5 orange companion that orbits it in 34.5 years. The two stars were closest in 2001 but have since ▶

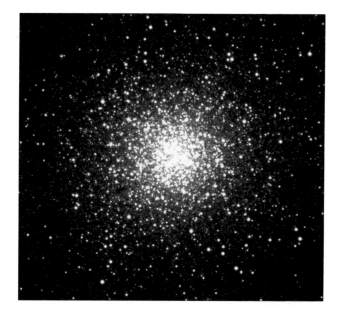

M13, the great globular cluster in Hercules, is a swarm of 300,000 stars appearing as a fuzzy ball in binoculars but which can be resolved by modest-sized amateur telescopes.
(Simon Tulloch and Daniel Folha, Isaac Newton Group of Telescopes, La Palma)

Hercules · Her · Herculis

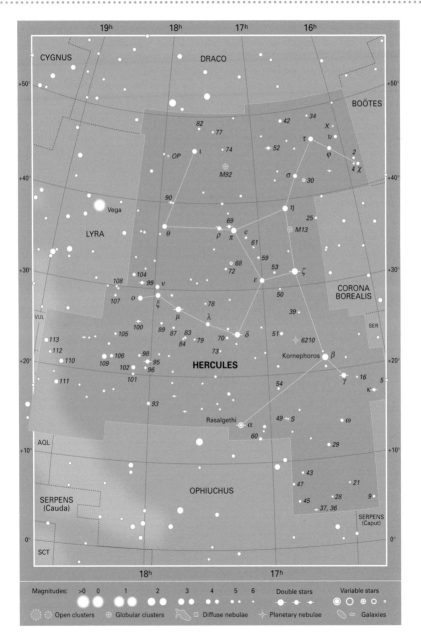

opened out, coming into the range of 220 mm apertures after 2010 and 150 mm apertures when widest around 2025. See page 282 for an orbit diagram.

κ (kappa) Her, 16h 08m +17°.0, is a mag. 5.0 yellow giant 388 l.y. away with an unrelated mag. 6.3 orange giant companion, 470 l.y. away, easily seen in small telescopes.

ρ (rho) Her, 17h 24m +37°.1, 402 l.y. away, is a pair of blue-white giants of mags. 4.5 and 5.5, visible in small telescopes.

30 Her, 16h 29m +41°.9, also known as g Her, is a red giant that varies semi-regularly between mags. 4.3 and 6.3 every 3 months or so. It lies 361 l.y. away.

68 Her, 17h 17m +33°.1, also known as u Her, 865 l.y. away, is an eclipsing binary of Beta Lyrae type varying between mag. 4.7 and 5.4 in just over 2 days.

95 Her, 18h 02m +21°.6, 470 l.y. away, is a double star for small telescopes, consisting of two giant stars of mags. 4.9 and 5.2, appearing silver and gold.

100 Her, 18h 08m +26°.1, is an easy duo for small telescopes consisting of identical blue-white stars of mag. 5.8, distances 165 and 230 l.y., like a pair of celestial cat's eyes.

M13 (NGC 6205), 16h 42m +36°.5, is a 6th-mag. globular cluster of 300,000 stars, the brightest of its kind in northern skies. It can be seen with the naked eye and is unmistakable in binoculars, spanning half the apparent width of the full Moon. The cluster lies 25,200 l.y. away and has a true diameter of over 100 l.y. Telescopes of 100 mm aperture resolve individual stars throughout the cluster, giving a mottled, sparkling effect.

M92 (NGC 6341), 17h 17m +43°.1, is a globular cluster only slightly inferior to its more famous neighbour, M13, which overshadows it. M92 is easily seen in binoculars as a fuzzy star. It is smaller and more condensed than M13, and needs a larger telescope to resolve its stars. It lies 29,000 l.y. away and has an estimated age of around 14 billion years, making it the oldest globular known.

NGC 6210, 16h 45m +23°.8, is a 9th-mag. planetary nebula which a telescope of 75 mm or larger shows as a blue-green ellipse. It is about 4000 l.y. away.

Horologium – The Pendulum Clock

One of the constellations representing mechanical instruments, in this case a pendulum clock used for timing observations, introduced in the 1750s by Nicolas Louis de Lacaille. As with so many of Lacaille's constellations, Horologium is faint and obscure.

α (alpha) Horologii, 4h 14m −42°.3, mag. 3.9, is an orange giant star 117 l.y. away.

β (beta) Hor, 2h 59m −64°.1, mag. 5.0, is a giant white star 314 l.y. away.

R Hor, 2h 54m −49°.9, is a red giant variable of the Mira type, ranging between extremes of mag. 4.7 and 14.3 in 13 months or so. It lies about 1000 l.y. away.

TW Hor, 3h 13m −57°.3, is a deep-red pulsating variable of semi-regular type, ranging between mag. 5.5 and 6.0 with an approximate period of 5 months. It lies about 1300 l.y. away.

NGC 1261, 3h 12m −55°.2, is an 8th-mag. globular cluster 44,000 l.y. away.

Horologium · Hor · Horologii

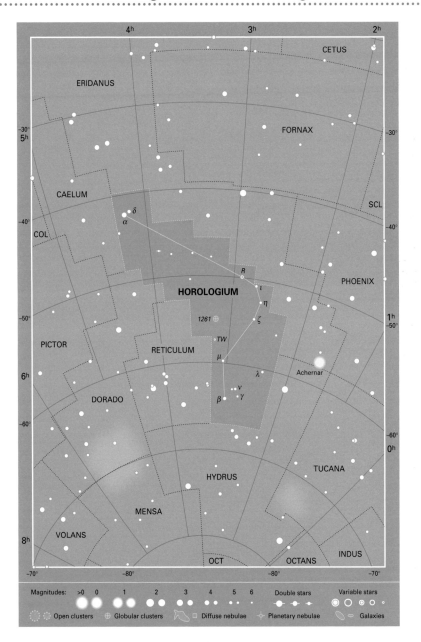

Hydra – The Water Snake

The largest constellation in the sky, but by no means easy to identify because of its faintness. Apart from its brightest star, Alphard, which marks the heart of the Water Snake, Hydra's only readily recognizable feature is its head, an attractive group of six stars. Hydra winds its way from the head in the northern celestial hemisphere, on the borders of Cancer, to the tip of its tail, south of the celestial equator adjacent to Libra and Centaurus, a total length of over 100°.

In mythology, Hydra is usually identified with the multi-headed monster slain by Hercules. Another legend links it with the adjoining constellations of Corvus the Crow and Crater the Cup, which are found on its back. In this story the bird returned to the god Apollo with Hydra in its claws as an excuse for its aborted mission to fetch water in the cup.

α (alpha) Hydrae, 9h 28m −8°.7, (Alphard, 'the solitary one'), mag. 2.0, is an orange giant 177 l.y. away.

β (beta) Hya, 11h 53m −33°.9, mag. 4.3, is a blue-white star 365 l.y. away.

γ (gamma) Hya, 13h 19m −23°.2, mag. 3.0, is a yellow giant 132 l.y. away.

δ (delta) Hya, 8h 38m +5°.7, mag. 4.1, is a blue-white star 179 l.y. away.

ε (epsilon) Hya, 8h 47m +6°.4, 135 l.y. away, is a beautiful but difficult double star of contrasting colours, needing high power on a telescope of at least 75 mm aperture. The stars are yellow and blue, of mags. 3.4 and 6.7, and form a genuine binary with an estimated orbital period of nearly 1000 years. ▶

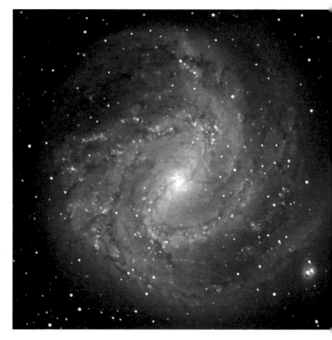

M83, a face-on barred spiral galaxy in Hydra, is a swirl of stars and gas sometimes known as the Southern Pinwheel. Our own Galaxy might look like this if we could see it from outside. See also page 162.
(Adam Block/NOAO/ AURA/NSF)

Hydra · Hya · Hydrae

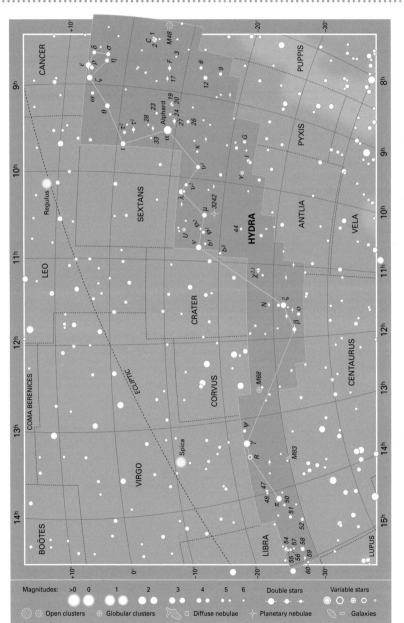

27 Hya, 9h 20m, −9°.6, mag. 4.8, is a white star 244 l.y. away with a wide mag. 7.0 companion visible in binoculars. The companion is 202 l.y. away and evidently unrelated. Small telescopes split this companion into components of mags. 7 and 11.

54 Hya, 14h 46m −25°.4, 99 l.y. away, is an easy double for small telescopes, consisting of yellow and purple stars of mags. 5.3 and 7.4.

R Hya, 13h 30m −23°.3, is a red giant variable star similar to Mira in Cetus that fluctuates between 4th and 10th mags. every 390 days. At its best it can reach mag. 3.5, making it one of the brightest Mira stars. Its distance is about 2000 l.y.

U Hya, 10h 38m −13°.4, 528 l.y. away, is a deep-red variable star that fluctuates semi-regularly between mags. 4.2 and 6.6 every 115 days.

M48 (NGC 2548), 8h 14m −5°.8, is a large open cluster of about 80 stars, 2000 l.y. away, just visible to the naked eye under clear skies and a fine sight in binoculars. It is somewhat triangular in shape, wider than the apparent size of the full Moon, and is well shown by small telescopes under low power.

M68 (NGC 4590), 12h 39m −26°.7, is an 8th-mag. globular cluster visible as a fuzzy star in binoculars and just resolved with apertures of 100 mm. It lies 31,000 l.y. away.

M83 (NGC 5236), 13h 37m −29°.9, is a large, face-on spiral galaxy of 8th mag., visible in a small telescope. It has a small, bright nucleus and signs of a central bar, similar to our own Galaxy. Its spiral arms can be traced with an aperture of 150 mm. (See photograph on page 160.) M83 has been the site of more known supernovae than any other Messier object, six in all. It is one of the brightest and closest galaxies outside the Local Group, about 15 million l.y. away. It is sometimes popularly termed the Southern Pinwheel.

NGC 3242, 10h 25m −18°.6, is a 9th-mag. planetary nebula of similar apparent size to the disk of Jupiter; it is popularly termed the Ghost of Jupiter. This often-overlooked object 2600 l.y. away is prominent enough to be picked up in small telescopes at low magnification as a blue-green disk, while larger instruments show a bright inner disk surrounded by a fainter halo.

Hydrus – The Lesser Water Snake

The Dutch navigators Pieter Dirkszoon Keyser and Frederick de Houtman introduced this constellation at the end of the 16th century as a smaller southern counterpart to the great Water Snake, Hydra. Lying between the two Magellanic Clouds, Hydrus almost bridges the gap between Eridanus and the south celestial pole. There is little to interest the casual observer.

α (alpha) Hydri, 1h 59m −61°.6, mag. 2.9, is a white main-sequence star 71 l.y. away.

β (beta) Hyi, 0h 26m −77°.3, mag. 2.8, is the constellation's brightest member, is a yellow star 24 l.y. away.

γ (gamma) Hyi, 3h 47m −74°.2, mag. 3.2, is a red giant star 214 l.y. away.

π^1 π^2 (pi^1 pi^2) Hyi, 2h 14m −67°.8, is a binocular pair of red and orange giants, unrelated: π^1 is mag. 5.6 and lies 740 l.y. away; π^2, mag. 5.7, is 468 l.y. distant.

Hydrus · Hyi · Hydri

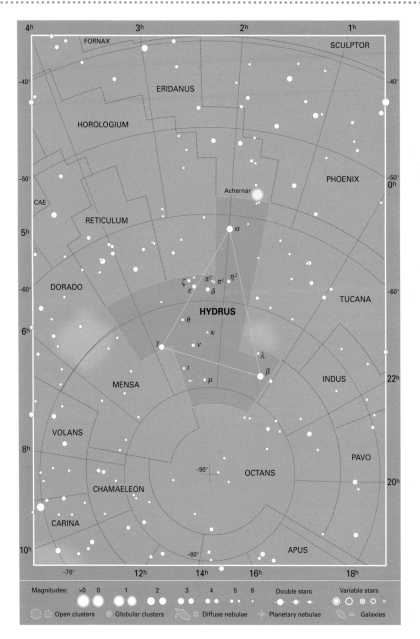

Indus – The Indian

A constellation representing a native hunter, introduced at the end of the 16th century by the Dutch navigators Pieter Dirkszoon Keyser and Frederick de Houtman. On old star charts he was depicted holding aloft a spear as though in pursuit of prey, but it is not known whether he is supposed to be a native of the East Indies or the Americas. None of its stars is brighter than 3rd magnitude and none of them have names.

α (alpha) Indi, 20h 38m −47°.3, mag. 3.1, is an orange giant star 101 l.y. away.

β (beta) Ind, 20h 55m −58°.5, mag. 3.7, is an orange giant 600 l.y. away.

δ (delta) Ind, 21h 58m −55°.0, mag. 4.4, is a white star 185 l.y. away.

ε (epsilon) Ind, 22h 03m −56°.8, mag. 4.7, is an orange dwarf somewhat smaller and cooler than the Sun. At a distance of 11.8 l.y., it is one of the Sun's closest neighbours.

θ (theta) Ind, 21h 20m −53°.4, 97 l.y. away, is a pair of white stars of mags. 4.5 and 7.0, divisible in a small telescope.

T Ind, 21h 20m −45°.0, is a deep-red variable star ranging semi-regularly between 5th and 7th magnitudes in 11 months or so. It lies about 1900 l.y. away.

Between Indus and Hydrus lies the Small Magellanic Cloud, an irregularly shaped splash of stars which is actually a satellite of our own Galaxy. See page 248. (NOAO/AURA/NSF)

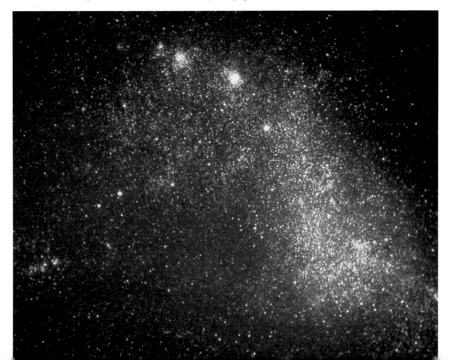

Indus · Ind · Indi

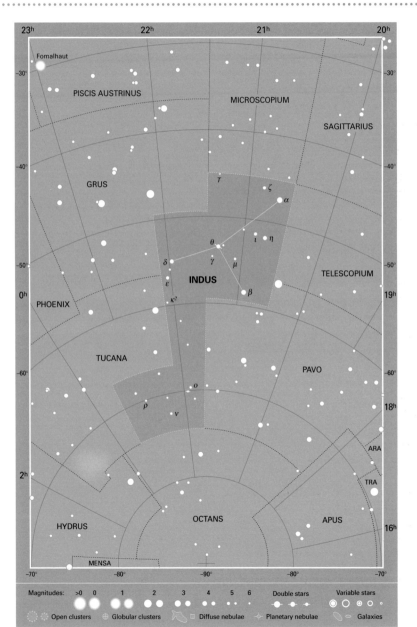

Lacerta – The Lizard

An inconspicuous constellation sandwiched between Cygnus and Andromeda, introduced in 1687 by the Polish astronomer Johannes Hevelius (see box below). An alternative figure that previously occupied this area was Sceptrum, the Sceptre and Hand of Justice, created in 1679 by the Frenchman Augustin Royer to commemorate King Louis XIV. In 1787 the German Johann Elert Bode called this region Frederick's Glory in honour of King Frederick the Great of Prussia. Both these alternatives have been discarded.

The constellation's most celebrated object is BL Lacertae, location 22h 02.7m, +42° 17′, originally thought to be a peculiar 14th-magnitude variable star. It is the prototype of a group of objects, the BL Lac objects or Lacertids, believed to be giant elliptical galaxies with variable centres, lying far off in the Universe and related to quasars and other galaxies with active nuclei. Three bright novae appeared within the boundaries of Lacerta during the 20th century.

α (alpha) Lacertae, 22h 31m +50°.3, mag. 3.8, is a blue-white main-sequence star 102 l.y. away.

β (beta) Lac, 22h 24m +52°.2, mag. 4.4, is a yellow giant 170 l.y. away.

NGC 7243, 22h 15m +49°.9, is a scattered open cluster of similar apparent size to the full Moon suitable for small telescopes, consisting of a few dozen stars of 8th mag. and fainter, 2500 l.y. away.

Johannes Hevelius (1611–1687)

Johannes Hevelius of Danzig (now Gdansk, Poland) was one of the finest observers of his day. His masterwork was a catalogue of 1564 stars, called *Catalogus Stellarum Fixarum*, completed just before his death in 1687 and published posthumously in 1690. The catalogue was accompanied by a set of sky charts, *Firmamentum Sobiescianum*, on which Hevelius introduced seven constellations still in use today: Canes Venatici, Lacerta, Leo Minor, Lynx, Scutum, Sextans and Vulpecula. Another important cartographic product of Hevelius was his map of the Moon, published in 1647 in his book *Selenographia*. It was the first major Moon map, and introduced the first system of lunar nomenclature. He named lunar formations after features on Earth: for instance, the large, bright crater we know as Copernicus he called Etna, while another prominent crater, Tycho, was Mount Sinai and Mare Imbrium was the Mediterranean. Only a few of the names given by Hevelius remain, such as the lunar Alps and Apennines; his names have mostly been discarded in favour of the system introduced by the Italian astronomer Giovanni Battista Riccioli (1598–1671), who named the craters after famous philosophers and astronomers. Fittingly enough, the craters commemorating Hevelius and Riccioli are found close together on the Moon.

Lacerta · Lac · Lacertae

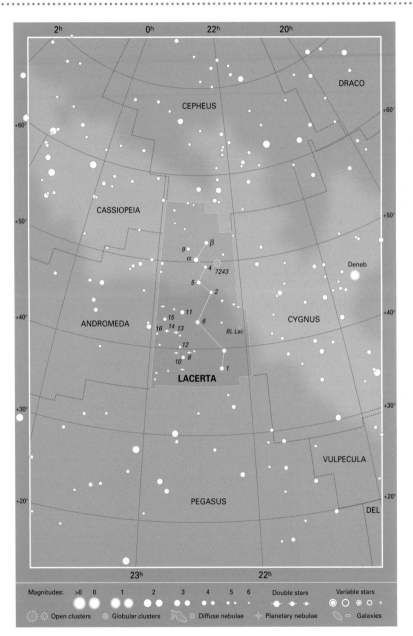

Leo – The Lion

One of the few constellations that looks like the figure it is supposed to represent – in this case, a crouching lion. The Lion's head is outlined by the so-called Sickle consisting of six stars from ε (epsilon) to α (alpha) Leonis; the Lion's body stretches out behind, its tail marked by β (beta) Leonis. Mythologically, this is the Lion reputedly slain by the hero Hercules as the first of his 12 labours. The Sun passes through the constellation from mid-August to mid-September.

Leo contains the third-nearest star to the Sun, CN Leonis (also called Wolf 359), a red dwarf 7.8 l.y. away. It is of magnitude 13.5, although it is a flare star and can occasionally brighten by up to a magnitude; it is located at 10h 56.5m, +7° 01′, near the border with Sextans. Leo contains numerous distant galaxies, the brightest of which are mentioned on page 170, plus three faint dwarf members of our Local Group, beyond the reach of amateur telescopes.

Every November, the Leonid meteors radiate from a point near γ (gamma) Leonis.

Usually the numbers are low, peaking at about 10 per hour on November 17–18, but the shower's activity increases dramatically at 33-year intervals when the parent comet, Tempel–Tuttle, returns to perihelion. High Leonid rates were last seen in 1999–2002.

α (alpha) Leonis, 10h 08m +12°.0, (Regulus, 'the little king'), mag. 1.4, is a blue-white main-sequence star 77 l.y. away. Binoculars or small telescopes will show a wide companion of mag. 7.7.

β (beta) Leo, 11h 49m +14°.6, (Denebola, 'the lion's tail'), mag. 2.1, is a blue-white star 36 l.y. away.

γ (gamma) Leo, 10h 20m +20°.0, (Algieba, 'the forehead'), 126 l.y. away, consists of a pair of golden-yellow giant stars of mags. 2.3 and 3.6, orbiting in about 500 years. They form an exceptionally handsome double in small telescopes, one of the finest in the sky. In binoculars, an unrelated mag. 4.8 yellowish foreground star, 40 Leonis, is visible nearby.   ▶

Trails of Leonid meteors sweep through the Big Dipper in this 43-second exposure, made from Kitt Peak, Arizona, during the great Leonid storm of 1966. On that occasion, thousands of meteors a minute were seen. (Dave McLean, NOAO/AURA/NSF)

Leo · Leo · Leonis

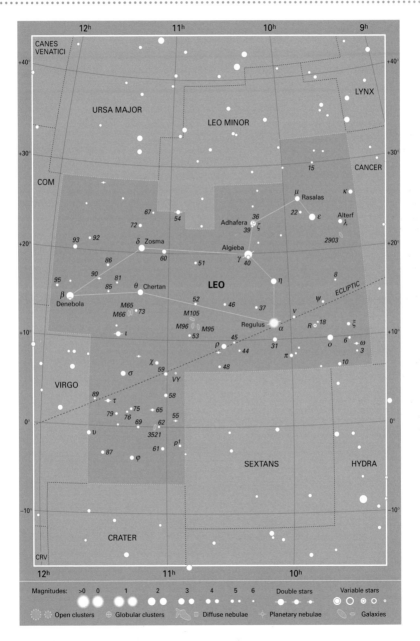

δ (delta) Leo, 11h 14m +20°.5, (Zosma), mag. 2.6, is a blue-white star 58 l.y. away.

ε (epsilon) Leo, 9h 46m +23°.8, mag. 3.0, is a yellow giant 251 l.y. away.

ζ (zeta) Leo, 10h 17m +23°.4, (Adhafera), mag. 3.4, is a giant white star 260 l.y. away. To the north of it, binoculars show 35 Leonis, an unrelated foreground star of mag. 6.0. To the south a wider third star, 39 Leonis, mag. 5.8 and also in the foreground, can be seen in binoculars, making this an optical triple.

ι (iota) Leo, 11h 24m +10°.5, 79 l.y. away, is a close and difficult double star. It appears to the naked eye as a yellow-white star of mag. 4.0, but actually consists of components of mags. 4.1 and 6.7 orbiting every 186 years. Currently the stars are moving apart; apertures of 100 mm should be sufficient to separate them until 2010 after which they become progressively easier, coming within range of all but the very smallest apertures when at their widest from 2053 to 2067.

τ (tau) Leo, 11h 28m +2°.9, 621 l.y. away, is a yellow giant of mag. 5.0 with an 8th-mag. companion visible in binoculars and small telescopes.

54 Leo, 10h 56m +24°.7, 289 l.y. away, is a double for small telescopes, consisting of blue-white components of mags. 4.5 and 6.3.

R Leo, 9h 48m +11°.4, is a red giant variable of Mira type, lying about 330 l.y. away. It appears strongly red when at maximum and is roughly 450 times larger than our Sun. R Leonis normally varies between 6th and 10th mags. with an average period of 310 days, but on occasion can become as bright as mag. 4.4.

M65, M66 (NGC 3623, NGC 3627), 11h 19m +13°.0, are a pair of spiral galaxies about 30 million l.y. away. At 9th mag. they can be detected in large binoculars under clear conditions, but at least 100 mm aperture at low power is required for their elongated shape and condensed centres to be clearly seen.

M95, M96 (NGC 3351, NGC 3368), 10h 44m +11°.7, 10h 47m +11°.8, are a pair of spiral galaxies of 10th and 9th mag. respectively, about 35 million l.y. away, visible as circular nebulosities in small telescopes. Larger apertures show that M95 has a central bar. About 1° away lies the smaller M105 (NGC 3379), 10h 48m +12°.6, a 9th-mag. elliptical galaxy at a similar distance.

Leo Minor – The Lesser Lion

Leo Minor lies between the larger and brighter constellations of Leo to the south and Ursa Major to the north. Johannes Hevelius, the Polish astronomer, introduced Leo Minor in 1687. There is little of interest in it. Surprisingly there is no star labelled α (alpha). This is the result of an error by the English astronomer Francis Baily, who assigned Greek letters to the stars of Leo

Minor in 1845 but failed to letter the brightest star through an oversight.

β (beta) Leonis Minoris, 10h 28m +36°.7, mag. 4.2, is a yellow giant star 146 l.y. away.

46 LMi, 10h 53m +34°.2, mag. 3.8, the brightest star in the constellation, is an orange giant 98 l.y. away.

Leo Minor · LMi · Leonis Minoris

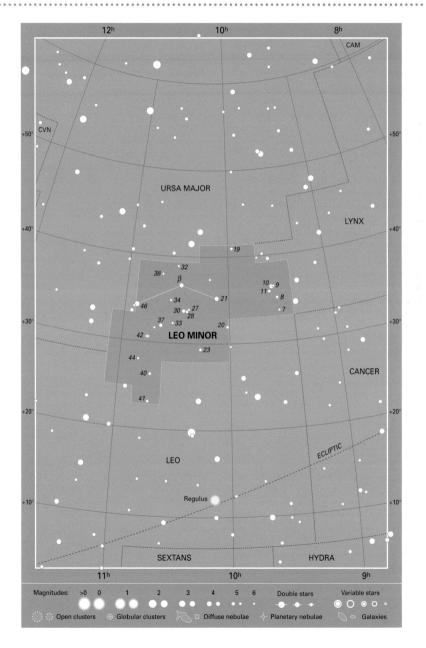

Lepus – The Hare

Lepus is a constellation known since ancient Greek times. It represents a hare, cunningly located at the feet of its hunter, Orion, and pursued endlessly across the sky by Canis Major, the hunter's dog. In one Greek legend, the inhabitants of the island of Leros began to breed hares but the population exploded and the island quickly became overrun with them. The hares destroyed the crops, reducing the inhabitants to starvation. After a concerted effort, the islanders drove the hares out of the island. They put the image of a hare among the stars as a reminder that one can easily end up with too much of a good thing.

The Hare is also associated in many legends with the Moon. For instance, the familiar figure of the man in the Moon is sometimes interpreted as a hare or rabbit, so perhaps Lepus is another incarnation of the lunar hare. Lepus is overshadowed by Orion's brilliance, but is not without interest for amateur observers.

α (alpha) Leporis, 5h 33m −17°.8, (Arneb, 'hare'), mag. 2.6, is a white supergiant star, about 1300 l.y. away.

β (beta) Lep, 5h 28m −20°.8, (Nihal), mag. 2.8, is a yellow giant star 159 l.y. away.

γ (gamma) Lep, 5h 44m −22°.5, 29 l.y. away, is an attractive binocular duo consisting of a yellow star of mag. 3.6 with an orange companion of mag. 6.2.

δ (delta) Lep, 5h 51m −20°.9, mag. 3.8, is a yellow giant 112 l.y. away.

ε (epsilon) Lep, 5h 05m −22°.4, mag. 3.2, is an orange giant 227 l.y. away.

κ (kappa) Lep, 5h 13m −12°.9, 560 l.y. away, is a mag. 4.4 blue-white star with a close mag. 7.4 companion, difficult to see in the smallest telescopes because of the magnitude contrast.

R Lep, 5h 00m −14°.8, 820 l.y. away, is an intensely red star known as Hind's Crimson Star after the English observer John Russell Hind, who described it in 1845 as 'like a drop of blood on a black field'. R Leporis is a Mira-type variable that ranges from mag. 5.5 at its brightest to as faint as 12th mag. in a period of around 430 days.

RX Lep, 5h 11m −11°.8, 447 l.y. away, is a red giant that varies semi-regularly between mags. 5.0 and 7.4 every 2 months or so.

M79 (NGC 1904), 5h 24m −24°.5, is a small but rich globular cluster 44,000 l.y. away, visible as a fuzzy 8th-mag. star in small telescopes. Nearby in the same low-power field is the multiple star Herschel 3752, consisting of a mag. 5.4 primary with two companions, a close one of mag. 6.6 and a wide one of mag. 9.1, all visible in small telescopes.

NGC 2017, 5h 39m −17°.8, is a small but remarkable star cluster, also known as the multiple star Herschel 3780. Modest amateur telescopes reveal a group of five well-spaced stars ranging from 6th to 10th mag. In addition, the brightest star has a mag. 7.9 companion that requires a telescope of at least 200 mm aperture to split, while an aperture of at least 100 mm shows that one of the 9th-mag. stars is a close double. There is also a 12th-mag. component which should be visible with 100 mm or upwards, so this is actually a group of at least eight stars. However, it is not a true cluster: the stars lie at a range of distances and are moving in different directions.

Lepus · Lep · Leporis

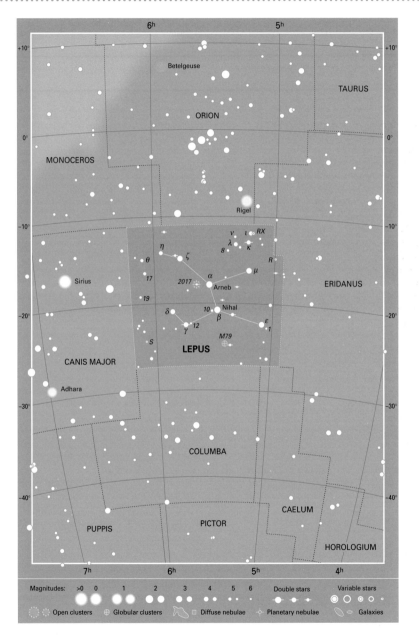

Libra – The Scales

A faint, easily overlooked constellation of the zodiac, through which the Sun passes during November. The ancient Greeks knew it as the Claws of the Scorpion, an extension of neighbouring Scorpius, and this old identification lives on in the Arabic-derived names Zubenelgenubi and Zubeneschamali ('southern claw' and 'northern claw', respectively) applied to the stars α (alpha) and β (beta) Librae.

The Romans made this region into a separate constellation during the time of Julius Caesar in the first century BC. Since then the Scales have come to be regarded as the symbol of justice, held aloft by the goddess of justice, Astraeia, in the shape of the neighbouring figure of Virgo.

Libra once contained the September equinox, the point at which the Sun passes south of the celestial equator each year. Because of precession, this point moved into neighbouring Virgo around 730 BC, but the September equinox is still sometimes referred to as the First Point of Libra. Despite its relative faintness, Libra contains several stars of interest.

α (alpha) Librae, 14h 50m −16°.0, (Zubenelgenubi, 'the southern claw'), 77 l.y. away, is a wide binocular double consisting of a blue-white star of mag. 2.7 with a white companion of mag. 5.2.

β (beta) Lib, 15h 17m −9°.4, (Zubeneschamali, 'the northern claw'), mag. 2.6, the brightest in the constellation, is celebrated as one of the few bright stars to show a distinct greenish tinge. It lies 160 l.y. away.

γ (gamma) Lib, 15h 36m −14°.8, mag. 3.9, is an orange giant 152 l.y. away.

δ (delta) Lib, 15h 01m −8°.5, 304 l.y. away, is an eclipsing variable of the Algol type. It varies between mags. 4.9 and 5.9 in 2 days 8 hours.

ι (iota) Lib, 15h 12m −19°.8, is a multiple star 377 l.y. away. Its main blue-white component, of mag. 4.5, has a wide mag. 9.4 companion that is difficult to see in the smallest telescopes because of the brightness difference. An aperture of 75 mm or above with high magnification will split this fainter companion into two stars of 10th and 11th mags. The brightest component of ι Lib is itself a binary with a 23-year period, but too close for amateur telescopes. Binoculars show a mag. 6.1 star nearby called 25 Librae, which is a foreground object, 219 l.y. away.

μ (mu) Lib, 14h 49m −14°.1, 235 l.y. away, is a close double star consisting of components of mags. 5.7 and 6.8, divisible in a telescope of 75 mm aperture.

48 Lib, 15h 58m −14°.3, mag. 4.9, is a blue giant 513 l.y. away with an abnormally high speed of rotation that causes it to throw off rings of gas from its equator, as a result of which it varies irregularly by a few tenths of a magnitude in similar fashion to γ (gamma) Cassiopeiae and Pleione in Taurus. It also bears the variable-star designation FX Lib.

NGC 5897, 15h 17m −21°.0, is a large but loosely scattered 9th-mag. globular cluster 40,000 l.y. away, unspectacular in small instruments.

Libra · Lib · Librae

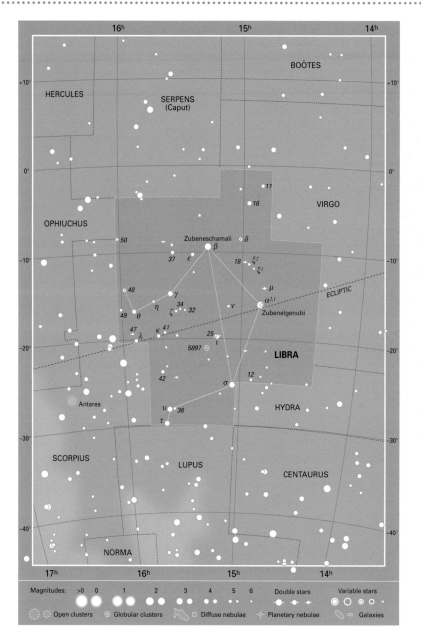

Lupus – The Wolf

Lupus is stocked with numerous interesting objects, although it is often overlooked in favour of its more spectacular neighbours Scorpius and Centaurus. The constellation was regarded by the Greeks and Romans as an unspecified wild animal, impaled on a pole by Centaurus the Centaur. Some mythologists said that Centaurus was about to sacrifice the animal on the altar, represented by nearby Ara. The constellation's identification with a wolf seems to have become common in Renaissance times. Lupus lies in the Milky Way and is rich in double stars. None of its stars is named.

α (alpha) Lupi, 14h 42m −47°.4, mag. 2.3, is a blue giant star 548 l.y. away.

β (beta) Lup, 14h 59m −43°.1, mag. 2.7, is a blue giant 524 l.y. away.

γ (gamma) Lup, 15h 35m −41°.2, mag. 2.8, is a blue-white star 570 l.y. away. It is a close binary with a 190-year orbital period, divisible only in apertures of 200 mm and above.

ε (epsilon) Lup, 15h 23m −44°.7, 504 l.y. away, is a blue-white star of mag. 3.4 with a wide mag. 8.8 companion visible in a small telescope. The primary is itself a close double, divisible only with large apertures.

η (eta) Lup, 16h 00m −38°.4, 493 l.y. away, is a double star consisting of a mag. 3.4 blue-white primary with a mag. 7.9 companion, not easy to see in a small telescope because of the magnitude contrast.

κ (kappa) Lup, 15h 12m −48°.7, 188 l.y. away, is an easy double star for small telescopes, consisting of blue-white components of mags. 3.9 and 5.7.

μ (mu) Lup, 15h 19m −47°.9, 291 l.y. away, is a multiple star. Small telescopes reveal a mag. 4.3 blue-white primary with a wide mag. 6.9 companion. But in telescopes of at least 100 mm, with high magnification, the primary itself is seen to be double, consisting of two near-identical stars of mags. 5.0 and 5.1.

ξ (xi) Lup, 15h 57m −34°.0, 200 l.y. away, is a neat pair of mag. 5.1 and 5.6 blue-white stars, well seen in a small telescope.

π (pi) Lup, 15h 05m −47°.1, 500 l.y. away, appears to the naked eye as mag. 3.9, but telescopes above 75 mm aperture show that it consists of two close blue-white stars of mags. 4.6 and 4.7.

GG Lup, 15h 19m −40°.8, 514 l.y., is an eclipsing binary of Algol type, ranging between mags. 5.6 and 6.1 with a period of 1.85 days.

NGC 5822, 15h 05m −54°.3, is a large, loose open cluster of about 150 faint stars, 2400 l.y. away, visible in binoculars or small telescopes.

NGC 5986, 15h 46m −37°.8, is an 8th-mag. globular cluster, 33,000 l.y. distant, visible as a rounded patch in a small telescope.

Lupus · Lup · Lupi

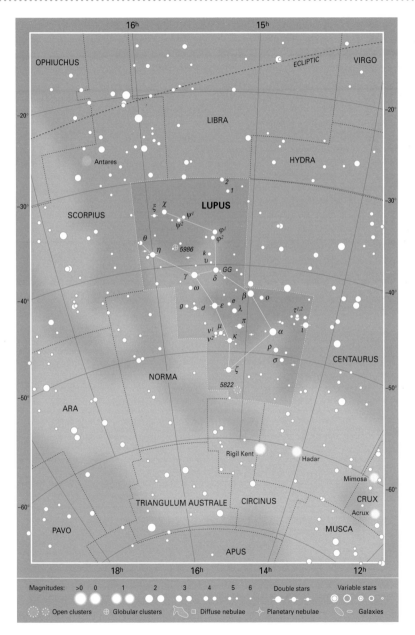

Lynx – The Lynx

A decidedly obscure constellation, despite its considerable size (larger than Gemini, for example), introduced in 1687 by the Polish astronomer Johannes Hevelius to fill the gap between the traditional figures of Ursa Major and Auriga. Hevelius named it Lynx because, he said, only the lynx-eyed would be able to see it – a reference to the fact that his own eyesight was exceptionally keen. Indeed, Hevelius continued to measure star positions with naked-eye instruments throughout his life, even though others had by then moved on to telescopes. Despite the constellation's faintness, owners of small telescopes will find many exquisite double stars within it.

α (alpha) Lyncis, 9h 21m +34°.4, mag. 3.1, is a red giant 222 l.y. away.

5 Lyn, 6h 27m +58°.4, 680 l.y. away, is an orange giant star of mag. 5.2 with a wide, unrelated mag. 7.9 companion, visible in a small telescope.

12 Lyn, 6h 46m +59°.4, 229 l.y. away, is a fascinating triple star. A small telescope will show a mag. 5.0 blue-white star with a fainter mag. 7.2 companion. Telescopes of 75 mm aperture and over reveal that the brighter component is itself a binary, of mags. 5.5 and 6.1, which orbit each other every 900 years.

15 Lyn, 6h 57m +58°.4, 170 l.y. away, is a close double star for telescopes of 150 mm aperture and above. The components are of mags. 4.7 and 5.8, the brighter star appearing a deep yellow colour.

19 Lyn, 7h 23m +55°.3, is an attractive triple star for small telescopes, with blue-white components of mags. 5.8 and 6.9. The very wide third star is of mag. 7.6. The trio lies about 500 l.y. away.

38 Lyn, 9h 19m +36°.8, 122 l.y. away, is a pair of mag. 3.9 and 6.3 blue-white stars, difficult in the smallest telescopes because of their closeness.

41 Lyn, 9h 29m +45°.6, 288 l.y. away, is actually over the border in Ursa Major (hence the designation '41 Lyncis' has now fallen out of use, but we retain it here for identification purposes). It is a yellow giant of mag. 5.4, and small telescopes reveal a wide companion of mag. 8.0. A 10th-mag. star nearby forms a triangle, making this an apparent triple.

NGC 2419, 7h 38m +38°.9, is an unusually remote globular cluster, of the type known as an intergalactic tramp. It is of 11th mag. and lies about 330,000 l.y. from the centre of our Galaxy, farther than either of the Magellanic Clouds.

Flamsteed numbers

Apart from α Lyncis, all the main stars in Lynx are referred to not by Greek letters but by so-called Flamsteed numbers. These numbers originate from a catalogue of 2935 stars, *Historia Coelestis Britannica*, compiled by the first Astronomer Royal of England, John Flamsteed (1646–1719), and published posthumously in 1725. In the catalogue, Flamsteed listed the stars in each constellation in order of right ascension. The numbers that are now known as Flamsteed numbers were not actually allocated by Flamsteed himself but were added later by the French astronomer Joseph Jérôme de Lalande (1732–1807). In a French edition of the catalogue published in 1783 Lalande inserted a column numbering the stars consecutively in each constellation in the order that Flamsteed had listed them.

Lynx · Lyn · Lyncis

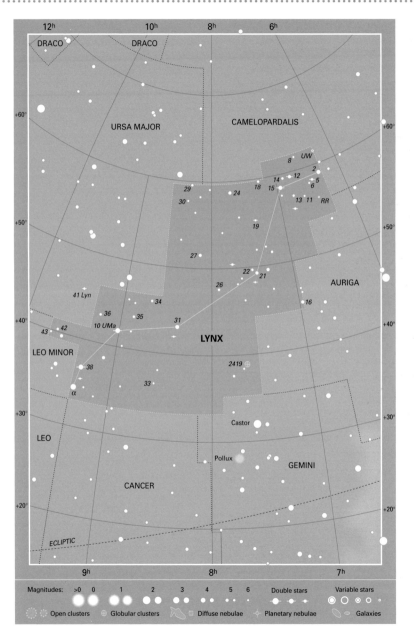

Lyra – The Lyre

A constellation dating from ancient times, representing the stringed instrument invented by Hermes and subsequently given by his half-brother Apollo to the great musician Orpheus. This constellation has also been visualized as an eagle or vulture. Although small, Lyra is bright and prominent. It contains the fifth-brightest star in the sky, Vega, which forms one corner of the northern Summer Triangle (Deneb in Cygnus and Altair in Aquila mark the other two corners). Because of precession, Vega will be the pole star between AD 13,000 and 14,000, although it will come no closer than 5°.7 to the pole. In a quite separate effect, our Sun's motion around the Galaxy is carrying us in the general direction of Vega at a velocity of 20 km/s relative to nearby stars. The Lyrid meteor shower emanates from this constellation each year, reaching a peak of about 10 per hour on April 21–22.

α (alpha) Lyrae, 18h 37m +38°.8, (Vega, 'the swooping eagle'), mag. 0.03, is a brilliant blue-white main-sequence star 25 l.y. away. It is the fifth-brightest star in the sky, and is surrounded by a disk of dust from which planets may be forming.

β (beta) Lyr, 18h 50m +33°.4, (Sheliak, 'the harp'), 882 l.y. away, is a remarkable multiple star. Small telescopes easily resolve it as a double star of cream and blue components. The fainter, blue star is of mag. 7.2, while the brighter star is an eclipsing binary that varies between mags. 3.3 and 4.4 in 12.9 days. β Lyrae is the prototype of a class of eclipsing variables in which the stars are so close together that gravity distorts them into egg-shapes, and hot gas spirals off them into space.

γ (gamma) Lyr, 18h 59m +32°.7, (Sulafat, 'the tortoise'), mag. 3.2, is a blue-white giant 635 l.y. away.

δ¹ δ² (delta¹ delta²) Lyr, 18h 54m +37°.0, is a wide naked-eye or binocular double consisting ▶

The four component stars of the Double Double, ε (epsilon) Lyrae, as seen through a telescope. See page 182 for a description. (Wil Tirion)

Lyra · Lyr · Lyrae

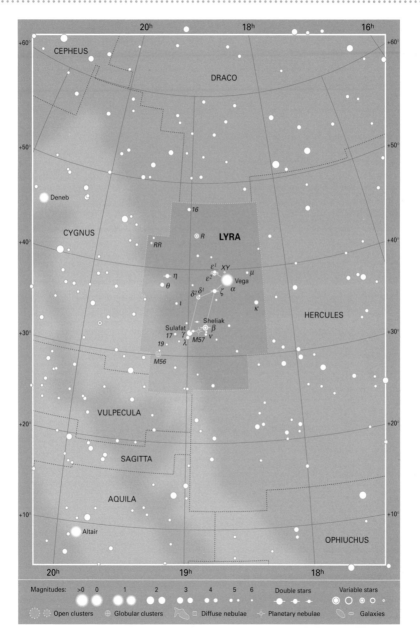

of two unrelated stars: δ^1, blue-white, mag. 5.6, 1080 l.y. away; and δ^2, a red giant 899 l.y. away, which varies erratically from mag. 4.2 to 4.3.

ϵ^1 ϵ^2 (epsilon^1 epsilon^2) Lyr, 18h 44m +39°.7, 161 l.y. away, is a celebrated quadruple star known popularly as the Double Double. Binoculars or even keen eyesight separate it into two stars, ϵ^1 and ϵ^2, of mags. 4.7 and 4.6 respectively. But a telescope of 60 to 75 mm aperture and high magnification reveals that each star is itself double, the two pairs being oriented almost at right angles to each other (see diagram on page 180). The ϵ^1 pair has mags. of 5.0 and 6.1 and an orbital period of about 1750 years; the ϵ^2 pair, slightly closer together has mags. of 5.2 and 5.5, and a period of about 720 years. Quadruple stars are rare, and this is the finest of them.

ζ (zeta) Lyr, 18h 45m +37°.6, 152 l.y. away, is a double star, easily split in small telescopes or binoculars into components of mags. 4.4 and 5.7.

η (eta) Lyr, 19h 14m +39°.1, 1040 l.y. away, is a blue-white star of mag. 4.4 with a wide mag. 9.1 companion visible in a small telescope.

R Lyr, 18h 55m +43°.9, 350 l.y. away, is a red giant that varies semi-regularly between mags. 3.9 and 5.0 every 6 or 7 weeks.

RR Lyr, 19h 25m +42°.8, 745 l.y. away, is the prototype of an important class of variable stars used as 'standard candles' for indicating distances in space. RR Lyrae variables are often found in globular clusters and are thus known as cluster-type variables. Related to Cepheid variables, they are giant stars that pulsate in size, varying by about one magnitude usually in less than a day. RR Lyrae itself varies from mag. 7.1 to 8.1 in 13.6 hours.

M57 (NGC 6720), 18h 54m +33°.0, the Ring Nebula, is a famous 9th-mag. planetary nebula 2000 l.y. away, conveniently placed between β (beta) and γ (gamma) Lyrae. On photographs taken through large telescopes it looks like a celestial smoke ring, but in fact it is a cylinder seen end-on. A small telescope shows it as a noticeably elliptical misty disk, but a larger aperture is needed to see the central hole. It is one of the brightest planetary nebulae and appears larger in the sky than the planet Jupiter. Its true diameter is about one light year. (For a photograph see page 277.)

Mensa – The Table Mountain

The faintest of all the constellations, with no star brighter than magnitude 5.0. Mensa was introduced by the Frenchman Nicolas Louis de Lacaille to commemorate Table Mountain at the Cape of Good Hope from where he surveyed the southern skies in 1751–52. Part of the Large Magellanic Cloud strays from neighbouring Dorado over the border into Mensa, possibly reminding Lacaille of the cloud that frequently caps the real Table Mountain. Unfortunately, there is little else of interest here.

α (alpha) Mensae, 6h 10m −74°.8, mag. 5.1, is a yellow star similar to the Sun, 33 l.y. away.

β (beta) Men, 5h 03m −71°.3, mag. 5.3, is a yellow giant 642 l.y. away.

γ (gamma) Men, 5h 32m −76°.3, mag. 5.2, is an orange giant 101 l.y. away.

η (eta) Men, 4h 55m −74°.9, mag. 5.5, is an orange giant 712 l.y. away.

Mensa · Men · Mensae

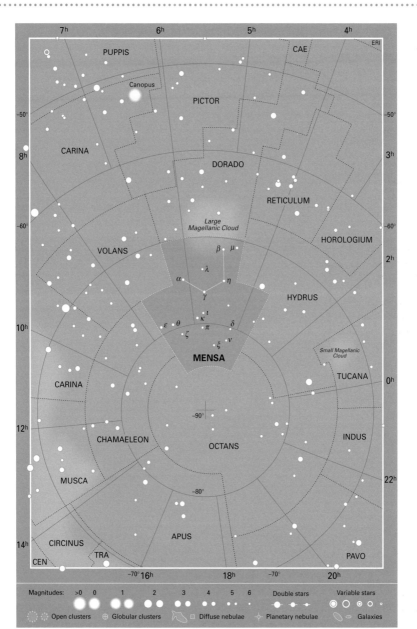

Microscopium – The Microscope

Another of the southern hemisphere constellations representing scientific instruments that were introduced in the 1750s by the Frenchman Nicolas Louis de Lacaille. The instrument in this case was an early form of compound microscope (i.e. one that uses more than one lens). As with so many of Lacaille's constellations, Microscopium is little more than a filler, encompassing a few faint stars between better-known figures.

α (alpha) Microscopii, 20h 50m −33°.8, mag. 4.9, is a yellow giant star 381 l.y. away. It has a 10th-mag. companion, visible in small telescopes.

γ (gamma) Mic, 21h 01m −32°.3, mag. 4.7, is a yellow giant 224 l.y. away.

ε (epsilon) Mic, 21h 18m −32°.2, mag. 4.7, is a blue-white main-sequence star 165 l.y. away.

The Rosette Nebula, NGC 2237, in Monoceros, perhaps the most beautiful nebula in the heavens, surrounds the star cluster NGC 2244. See page 186. (Nigel Sharp/NOAO/AURA/NSF)

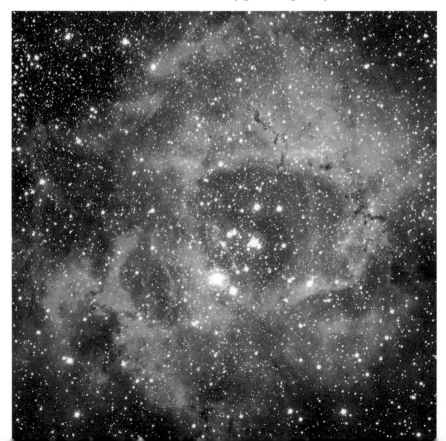

Microscopium · Mic · Microscopii

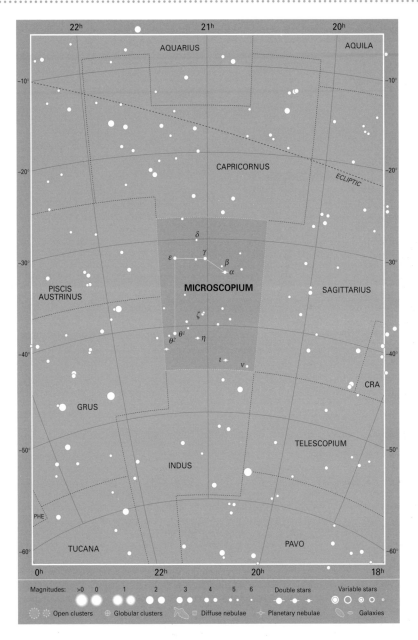

Monoceros – The Unicorn

A faint but fascinating constellation between Orion and Canis Minor, introduced by the Dutch theologian and astronomer Petrus Plancius in 1613, apparently because of several references to a unicorn in the Old Testament of the Bible. Its location in the Milky Way ensures that it is well stocked with nebulae and clusters.

Among the constellation's most celebrated features is Plaskett's Star, a mag. 6.1 spectroscopic binary named after the Canadian astronomer John Stanley Plaskett (1865–1941), who found in 1922 that it is the most massive pair of stars known; according to current data, it consists of two blue supergiants of masses 43 and 51 times that of the Sun, orbiting each other every 14.4 days. Plaskett's Star lies at 6h 37.4m, +6° 08′, near the open cluster NGC 2244, of which it may be an outlying member.

α (alpha) Monocerotis, 7h 41m −9°.6, mag. 3.9, is an orange giant star 144 l.y. away.

β (beta) Mon, 6h 29m −7°.0, 690 l.y. away, is rated as perhaps the finest triple star in the heavens. The smallest of telescopes should separate the three components, of mags. 4.6, 5.0 and 5.4, on a steady night. They form a curving arc of blue-white stars, the faintest two being the closest together.

δ (delta) Mon, 7h 12m −0°.5, mag. 4.2, is a blue-white star 375 l.y. away. It has a wide, unrelated naked-eye companion, 21 Mon, of mag. 5.5.

8 Mon, 6h 24m +4°.6, also known as ε (epsilon) Mon, is an easy double star for small telescopes, consisting of unrelated yellow and blue-white components of mags. 4.4 and 6.7, distances 128 and 79 l.y., in an attractive low-power field.

S Mon, 6h 41m +9°.9, also known as 15 Mon, is an intensely luminous blue-white star of mag. 4.7, situated 2600 l.y. away in the star cluster NGC 2264 (see page 188). It is a double star, with a close companion of mag. 7.6 visible in a small telescope. S Mon is slightly variable, fluctuating erratically by about 0.1 mag.

M50 (NGC 2323), 7h 03m −8°.3, is an open cluster of about 80 stars, half the apparent size of the full Moon, visible in binoculars and small telescopes. Apertures of 100 mm or so resolve it into a ragged patch of stars of 8th mag. and fainter, with a reddish star near its southern edge. M50 is 3300 l.y. away.

NGC 2232, 6h 27m −4°.7, is a scattered cluster of 20 stars for binoculars, containing the mag. 5.1 blue-white star 10 Mon. The cluster, which covers the same area of sky as the full Moon, lies 1200 l.y. away.

NGC 2237, NGC 2244, 6h 32m +4°.9, is a complex combination of a faint diffuse nebula, known as the Rosette Nebula, and a cluster of stars, all about 5000 l.y. away. Long-exposure photographs show the nebula as a pink loop, twice the apparent diameter of the full Moon. Visual observations with large amateur telescopes reveal only the brightest parts of the nebula. The associated cluster, NGC 2244, consists of stars that have been born from the Rosette Nebula's gas; it is just visible to the naked eye and is an easy binocular object. The six most prominent stars of the cluster form a rectangular shape, although the brightest of them, 12 Mon, mag. 5.9, is not a true member but an unrelated foreground star. The cluster is likely to be the only part of this celebrated object visible in a small telescope, but the pale outline of the nebula can just be made out in binoculars under clear, dark skies. (See picture on page 184.)

NGC 2261, 6h 39m +8°.7, Hubble's Variable Nebula, is a small, faint, fan-shaped nebula containing the remarkable variable star R Mon. Its erratic brightness fluctuations, from mag. 9.5 down to about 12th mag., may be caused by the pangs of its birth from the surrounding nebula. ▶

Monoceros · Mon · Monocerotis

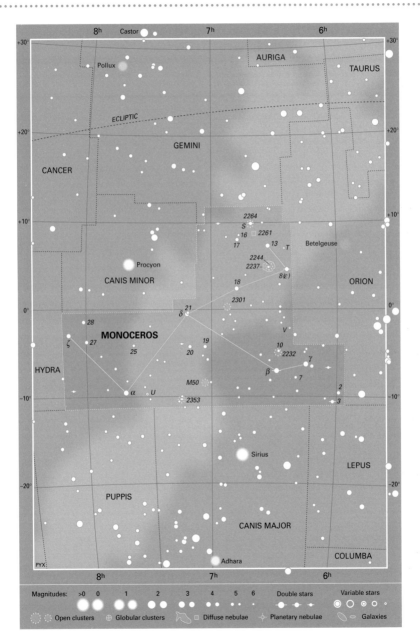

This star and its nebula are open to study only by larger amateur telescopes; the star's distance is uncertain but it may be associated with the nearby NGC 2264 complex (see below), which would place it at about 2600 l.y.

NGC 2264, 6h 41m +9°.9, is another combination of star cluster and nebula. The cluster, visible in binoculars, has about 40 members, including the 5th-mag. S Mon (see page 186). The associated nebula, known as the Cone Nebula because of its tapered shape, shows up well only on photographs such as the one at the right and is beyond the reach of visual observation with amateur telescopes. NGC 2264 is 2600 l.y. away.

NGC 2301, 6h 52m +0°.5, is a binocular cluster of 80 or so stars of 8th mag. and fainter, the brightest of them arranged in a vertical chain. It lies 2500 l.y. away.

NGC 2353, 7h 15m −10°.3, is an open cluster for small telescopes, consisting of about 30 stars of 9th mag. and fainter seemingly arranged in a spiral pattern. It lies about 3900 l.y. away.

The Cone Nebula is a dark intrusion into the southern end of the star cluster NGC 2264, but is too faint to see visually. (Carlos and Crystal Acosta/Adam Block/NOAO/AURA/NSF)

Musca – The Fly

A small southern constellation lying at the foot of the Southern Cross. It is one of the 12 constellations introduced at the end of the 16th century by the Dutch navigators Pieter Dirkszoon Keyser and Frederick de Houtman. It was also widely known under the alternative name of Apis, the Bee, which is how it was shown on several old star atlases. There is little of note in Musca other than part of the dark Coalsack Nebula, which spills over the border from Crux.

α (alpha) Muscae, 12h 37m −69°.1, mag. 2.7, is a blue-white star 306 l.y. away.

β (beta) Mus, 12h 46m −68°.1, 311 l.y. away, is a close pair of stars of mags. 3.6 and 4.0, requiring

100 mm aperture and high magnification to split. The orbital period of the pair is 400 years or so.

δ (delta) Mus, 13h 02m −71°.5, mag. 3.6, is an orange giant 91 l.y. away.

θ (theta) Mus, 13h 08m −65°.3, is a double star of mags. 5.6 and 7.6 for small telescopes. The brighter star is a blue supergiant, while its companion is a Wolf–Rayet star, a rare type of very hot star; it is the second-brightest such star in the sky, γ (gamma) Velorum being the brightest of all.

NGC 4833, 13h 00m −70°.9, is a fairly large, 7th-mag. globular cluster 18,000 l.y. away, visible in binoculars and small telescopes and resolvable into stars with an aperture of 100 mm.

Musca · Mus · Muscae

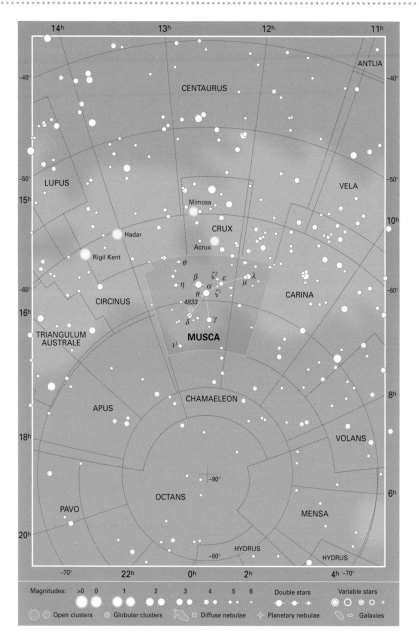

Norma – The Set Square

A superfluous constellation invented in the 1750s by Nicolas Louis de Lacaille, originally under the name Norma et Regula, the set square and ruler. Lacaille placed it next to the compasses (Circinus) and the southern triangle (Triangulum Australe), an earlier invention of Keyser and de Houtman which Lacaille visualized as a builder's level. The stars of Norma were previously part of Ara, Lupus and Scorpius, but since Lacaille's time the boundaries of Norma have been altered, so that the stars which he named α (alpha) and β (beta) Normae have been reabsorbed into Scorpius; in the chart on page 227 these two stars are labelled N and H Scorpii. Norma lies in a rich region of the Milky Way.

γ^2 (gamma^2) Normae, 16h 20m −50°.2, mag. 4.0, is a yellow giant 128 l.y. away, the brightest star in the constellation. Next to it lies the far more distant yellow-white supergiant γ^1 (gamma^1) Normae, mag. 5.0, around 1500 l.y. away.

δ (delta) Nor, 16h 06m −45°.2, mag. 4.7, is a white star 123 l.y. away.

ε (epsilon) Nor, 16h 27m −47°.6, 400 l.y. away, is a double star with components of mags. 4.5 and 6.7 visible in small telescopes.

ι^1 (iota^1) Nor, 16h 04m −57°.8, 140 l.y. away, appears in small telescopes as a double star of mags. 4.6 and 8.1. In addition, the brighter star is itself a very close binary, divisible only in very large telescopes, with a 27-year orbital period.

NGC 6087, 16h 19m −57°.9, is a loose, large binocular cluster of about 40 stars, 3000 l.y. away, with chains of stars extending from it like a spider's legs. At the centre lies its brightest star, the Cepheid variable S Nor, which ranges from mag. 6.1 to 6.8 in 9.8 days.

At 15h 51.7m −51° 31′ in Norma lies the satisfyingly symmetrical planetary nebula known variously as Shapley 1 (Sp 1), PK 329+02.1 or RCW 100. At 13th mag., it requires fairly large apertures to be seen. The central star is 14th magnitude. (Anglo–Australian Telescope Board)

Norma · Nor · Normae

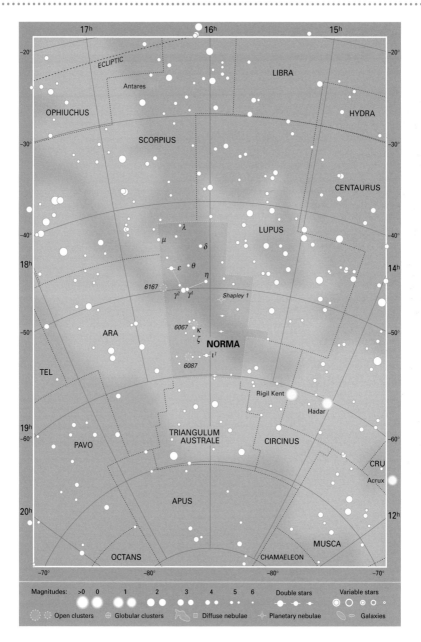

Octans – The Octant

The constellation that contains the south pole of the sky. Despite this privileged position, Octans is faint and unremarkable. The south celestial pole forms an almost exact equilateral triangle with 5th-mag. τ (tau) and χ (chi) Octantis. There is no southern equivalent of Polaris, the north pole star; the nearest naked-eye star to the southern pole is 5th-mag. σ (sigma) Octantis. Currently this star lies about 1° from the pole, but the distance is increasing due to precession (see chart below); ο Octantis was closest to the pole, under ¾°, around 1860.

Octans commemorates the navigational instrument known as the octant, a forerunner of the sextant, invented in 1730 by the English instrument maker John Hadley (1682–1744). The octant consisted of an arc of 45°, i.e. an eighth of a circle, from which came its name. The constellation itself was introduced in the 1750s by Nicolas Louis de Lacaille, and its dullness is a memorial to his lack of imagination.

α (alpha) Octantis, 21h 05m −77°.0, mag. 5.1, is a spectroscopic binary consisting of white and yellow giants 148 l.y. away.

β (beta) Oct, 22h 46m −81°.3, mag. 4.1, is a white star 140 l.y. away.

δ (delta) Oct, 14h 27m −83°.7, mag. 4.3, is an orange giant 279 l.y. away.

ε (epsilon) Oct, 22h 20m −80°.4, also known as BO Oct, is a red giant that varies semi-regularly between mags. 4.6 and 5.3 with an approximate period of 8 weeks. It lies 268 l.y. away.

λ (lambda) Oct, 21h 51m −82°.7, 435 l.y. away, is a double star with yellow and white components of mags. 5.5 and 7.2, individually visible in small telescopes.

ν (nu) Oct, 21h 41m −77°.4, mag. 3.7, the brightest star in the constellation, is an orange giant 69 l.y. away.

σ (sigma) Oct, 21h 09m −89°.0, mag. 5.4, the nearest naked-eye star to the south celestial pole, is a white giant 270 l.y. away.

The movement of the south celestial pole over 800 years as a result of precession. (Wil Tirion)

Octans · Oct · Octantis

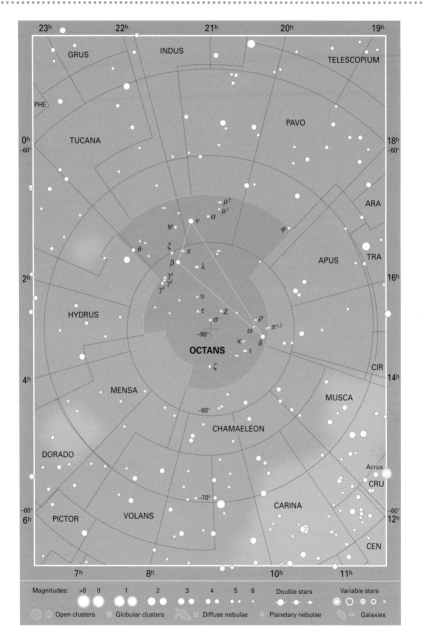

Ophiuchus – The Serpent Holder

An ancient constellation, representing a man encoiled by a serpent (the constellation Serpens). Ophiuchus is usually identified as Aesculapius, a mythical healer who was a forerunner of Hippocrates; his reputed powers included the ability to raise the dead. The serpent he holds is a symbol of this power, since snakes are seemingly reborn every year when they shed their skin. Perhaps the most celebrated star in Ophiuchus is Barnard's Star, a mag. 9.5 red dwarf 5.9 l.y. away, the second-closest star to the Sun. It lies at 17h 57.8m, +4° 42′, and is named after the American astronomer E. E. Barnard who found in 1916 that it has the greatest proper motion of any star, covering the apparent diameter of the Moon every 180 years.

The southernmost regions of Ophiuchus extend into rich starfields of the Milky Way, in the direction of the centre of the Galaxy; consequently the constellation is replete with star clusters. Ophiuchus was the site of the last supernova seen to erupt in our Galaxy, popularly called Kepler's Star, which appeared in 1604 at 17h 30.6m, −21° 29′, reaching mag. −3 at its brightest.

α (alpha) Ophiuchi, 17h 35m +12°.6, (Rasalhague, 'head of the serpent collector'), mag. 2.1, is a white main-sequence star 47 l.y. away.

β (beta) Oph, 17h 43m +4°.6, (Cebalrai, 'the shepherd's dog'), mag. 2.8, is an orange giant 82 l.y. away.

γ (gamma) Oph, 17h 48m +2°.7, mag. 3.7, is a blue-white main-sequence star 95 l.y. away.

δ (delta) Oph, 16h 14m −3°.7, (Yed Prior, 'the preceding star of the hand'), mag. 2.7, is a red giant 170 l.y. away.

ε (epsilon) Oph, 16h 18m −4°.7, (Yed Posterior, 'the following star of the hand'), mag. 3.2, is a yellow giant 108 l.y. away.

ζ (zeta) Oph, 16h 37m −10°.6, mag. 2.5, is a blue main-sequence star 458 l.y. away.

η (eta) Oph, 17h 10m −15°.7, (Sabik), mag. 2.4, is a blue-white main-sequence star 84 l.y. away.

ρ (rho) Oph, 16h 26m −23°.4, 400 l.y. away, is a striking multiple star for small instruments. The brightest component, of mag. 5.0, has a close companion of mag. 5.7 visible in a small telescope at high magnification; either side of this pair are wide binocular companions of mags. 6.7 and 7.3.

τ (tau) Oph, 18h 03m −8°.2, 170 l.y. away, is a close pair of cream-coloured stars, mags. 5.2 and 5.9, orbiting every 260 years and gradually closing; currently at least 75 mm aperture is required to separate them and 100 mm will be needed by the year 2025.

36 Oph, 17h 15m −26°.6, 20 l.y. away, is a pair of mag. 5.1 orange dwarf stars divisible with small apertures. Their calculated orbital period is nearly 500 years.

70 Oph, 18h 05m +2°.5, 16 l.y. away, is a celebrated double star, consisting of yellow and orange components of mags. 4.2 and 6.0 which orbit each other in a period of 88 years. They are easily divisible in the smallest apertures and will continue to be so throughout the first half of the 21st century, being at their widest around 2025. For a diagram showing how the positions and separations of the pair change throughout the 21st century, see page 282.

RS Oph, 17h 50m −6°.7, is a recurrent nova seen to have erupted six times, breaking the record it previously shared with T Pyxidis and the fainter U Scorpii. Normally around 12th mag., RS Oph flared up to naked-eye brightness in 1898, 1933, 1958, 1967, 1985 and 2006.

M10 (NGC 6254), 16h 57m −4°.1, is a 7th-mag. globular cluster visible in binoculars or a small ▶

Ophiuchus · Oph · Ophiuchi

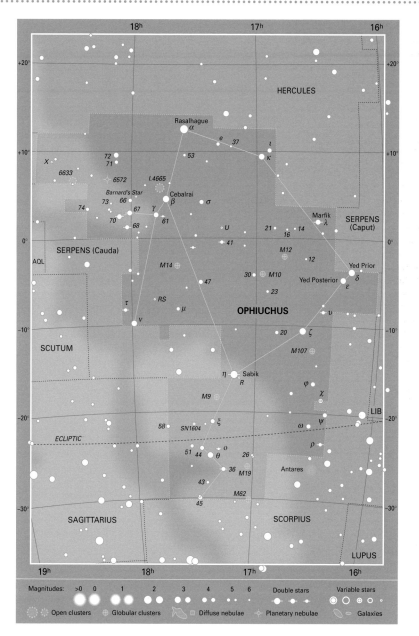

telescope. It is 14,000 l.y. away, somewhat closer than its neighbour M12 (see below). Individual stars in the cluster can be resolved with telescopes of 75 mm aperture.

M12 (NGC 6218), 16h 47m −1°.9, is a 7th-mag. globular cluster 18,000 l.y. away, visible in binoculars or a small telescope. In small apertures it appears slightly larger and less easy to resolve than its neighbour M10, and its stars are more loosely scattered. There are several other globular clusters in Ophiuchus worthy of attention, but M10 and M12 are the finest.

NGC 6572, 18h 12m +6°.8, is a planetary nebula of 9th mag., visible in apertures of 75 mm or more as a tiny blue-green ellipse. It lies 2000 l.y. away.

NGC 6633, 18h 28m +6°.6, is a scattered binocular cluster of about 30 stars of 8th mag. and fainter covering an area similar to that of the full Moon. It is 950 l.y. away.

IC 4665, 17h 46m +5°.7, is a loose and irregular open cluster of two dozen or so stars of 7th mag. and fainter, 1100 l.y. away, larger than the apparent size of the Moon and best seen in binoculars.

Orion – The Hunter

Without doubt the brightest and grandest constellation of all, crammed with objects of interest for all sizes of instrument. Orion's impressiveness stems in large measure from the fact that it contains an area of star formation in a nearby arm of our Galaxy, centred on the famous Orion Nebula. The Greek poet Homer described Orion as a giant hunter, armed with an unbreakable club of solid bronze. Legend tells that the boastful Orion was stung to death by a scorpion, and was placed in the sky so that he sets at the same time as his slayer, represented by the constellation Scorpius, rises.

In the sky, Orion is depicted brandishing his club and shield at the snorting Taurus, the Bull, while the hunter's dogs (the constellations Canis Major and Canis Minor) follow at his heels, in pursuit of the hare (Lepus). The Orion Nebula, M42, marks the hunter's sword, hanging from his belt. The belt itself is formed by a distinctive line of three bright stars, while the outline of his body is picked out by α (alpha), β (beta), γ (gamma) and κ (kappa) Orionis. Orion is one of the constellations in which the star labelled α is not the brightest; that honour goes to β Orionis, better known as Rigel. Each year the Orionid meteors, caused by dust from Halley's Comet,

radiate from a point near the border with Gemini. As many as 25 Orionid meteors per hour may be seen around October 22.

α (alpha) Orionis, 5h 55m +7°.4, (Betelgeuse, corrupted from an Arabic name referring to a hand), 427 l.y. away, is a red supergiant star about 500 times the size of the Sun, so large that it is unstable. It fluctuates erratically in size, changing in brightness as it does so from mag. 0.0 to 1.3, making it the most obviously variable of all first-magnitude stars; the average brightness is around mag. 0.5, fainter than Rigel.

β (beta) Ori, 5h 15m −8°.2, (Rigel, 'foot'), at mag. 0.2 the brightest star in Orion, is a blue-white supergiant 773 l.y. away; note its colour contrast with the ruddy hue of Betelgeuse. Rigel has a mag. 6.8 companion, difficult to see in small telescopes, particularly in poor seeing, because of glare from Rigel itself.

γ (gamma) Ori, 5h 25m +6°.3, (Bellatrix, 'the female warrior'), mag. 1.6, is a blue giant star 243 l.y. away.

δ (delta) Ori, 5h 32m −0°.3, (Mintaka, 'belt'), about 2000 l.y. away, is a complex multiple star. It is a blue giant which appears of mag. 2.2 to the   ▶

Orion · Ori · Orionis

naked eye. Binoculars or small telescopes reveal a wide companion of mag. 6.9. The brighter star is also an eclipsing binary that varies by about 0.1 mag. every 5.7 days.

ε (epsilon) Ori, 5h 36m −1°.2, (Alnilam, 'the string of pearls'), mag. 1.7, is a blue supergiant about 1300 l.y. away.

ζ (zeta) Ori, 5h 41m −1°.9, (Alnitak, 'the girdle'), 820 l.y. away, is a blue supergiant that appears to the naked eye as mag. 1.7. But telescopes of 75 mm aperture and above reveal a close companion of mag. 3.9 that is estimated to orbit it every 1500 years. There is also a much wider 10th-mag. star.

η (eta) Ori, 5h 24m −2°.4, 900 l.y. away, is a complex multiple–variable. A telescope of at least 100 mm aperture at high magnification is needed to show that it consists of two close stars, of mags. 3.8 and 4.8. The brighter star is also an eclipsing binary, varying by 0.3 mag. every 8 days.

θ^1 (theta^1) Ori, 5h 35m −5°.4, about 1500 l.y. away, is a multiple star at the heart of the Orion Nebula, from which it has recently formed and which it now illuminates. This star is popularly known as the Trapezium, because a small telescope shows four stars here; but 100 mm aperture also reveals two others, of 11th mag. The four main stars of the Trapezium are of mags. 5.1, 6.7, 6.7 and 8.0. Nearby lies θ^2 (theta^2) Orionis, a binocular double of mags. 5.1 and 6.4.

ι (iota) Ori, 5h 35m −5°.9, 1300 l.y. away, is a double star on the southern edge of the Orion Nebula, divisible in a small telescope. Its components are of mags. 2.8 and 6.9. Also visible in the same field of view is a wider double of blue-white stars, Σ 747 (Struve 747), of mags. 4.8 and 5.7.

κ (kappa) Ori, 5h 48m −9°.7, (Saiph, 'sword'), mag. 2.1, is a blue supergiant 722 l.y. away.

λ (lambda) Ori, 5h 35m +9°.9, is a blue giant of mag. 3.5 with a mag. 5.6 companion visible in small telescopes under high magnification. The stars lie about 1060 l.y. away.

σ (sigma) Ori, 5h 39m −2°.6, 1150 l.y. away, is perhaps the most impressive of all Orion's stellar treasures. To the naked eye it appears as a blue-white star of mag. 3.8, but small telescopes reveal much more. On one side of the star are blue-white companions of mags. 6.8 and 6.6, the wider of which can be glimpsed in binoculars; it is an eclipsing binary with a range of about 0.1 mag. On the opposite side is a closer 9th-mag. companion, which is more difficult to see because of glare from the primary. The effect is like a planet with moons. To complete the picture, in the same telescopic field of view is a faint triple star called Σ 761 (Struve 761), consisting of a narrow triangle of 8th- and 9th-mag. stars. This is an extraordinarily rich and unexpected sight, to be returned to again and again.

U Ori, 5h 56m +20°.2, is a huge red giant variable of Mira type, many hundreds of times the diameter of the Sun, whose brightness ranges between 5th and 13th mags. in about 1 year.

M42, M43 (NGC 1976, NGC 1982), 5h 35m −5°.4, the Orion Nebula, is one of the greatest deep-sky wonders – a gigantic cloud of gas and dust, 1500 l.y. away and 20 l.y. in diameter, from which a star cluster is being born. Behind the visible part of the nebula, illuminated by the stars of the Trapezium (θ^1 Ori), radio and infrared astronomers have detected an even larger dark cloud in which more stars are forming. M42 covers an area greater than 1° × 1°, and is indisputably the finest diffuse nebula in the sky, clearly visible to the naked eye as a hazy cloud. Binoculars and small telescopes reveal some of the more prominent wreaths and swirls of gas, which become more complex and breathtaking with increasing aperture. Although colour photographs such as the one on page 268 depict the nebula as red and blue, to the eye it appears distinctly greenish; this is because the human eye has a different colour sensitivity to that of photographic film. A dark lane of dust separates M42 from M43, a smaller and rounder patch to the north that is centred on a 7th-mag. star. Although M43 has a separate catalogue number from M42, it is really part of the same huge gas cloud.

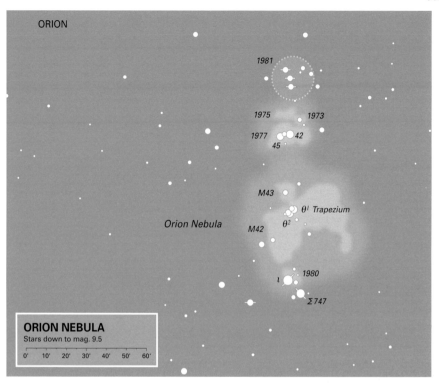

ORION

1981

1975   1973

1977   42
45

M43

θ¹ *Trapezium*

Orion Nebula   θ²

M42

ι   1980

Σ747

ORION NEBULA
Stars down to mag. 9.5

0'   10'   20'   30'   40'   50'   60'

The region south of Orion's Belt known as the Sword of Orion stretches from the large open cluster NGC 1981 at the top to the double star ι (iota) Orionis at its southern tip. In between is the misty mass of the Orion Nebula. (Wil Tirion)

M78 (NGC 2068), 5h 47m +0°.0, is a small, elongated reflection nebula lying almost exactly on the celestial equator. Through small to moderate apertures it looks like a short-tailed comet, with a 10th-mag. double star at its head.

NGC 1977, 5h 36m −4°.9, 1500 l.y. away, is an elongated nebulosity just north of the Orion Nebula, centred on the mag. 4.6 blue giant star 42 Orionis, also known as c Orionis. This object would be more celebrated were it not so overshadowed by M42.

NGC 1981, 5h 35m −4°.4, is a scattered cluster of about 20 stars of 6th mag. and fainter, 1300 l.y.

away, to the north of the nebulosity NGC 1977. Included in this cluster is the double star Σ 750 (Struve 750), a neat pair of 7th- and 8th-mag. components.

NGC 2024, 5h 41m −2°.4, is a mushroom-shaped cloud of gas about ½° wide surrounding the star ζ (zeta) Orionis. Running south from ζ Ori is a strip of nebulosity, IC 434, into which is indented the celebrated Horsehead Nebula, a dark cloud of obscuring dust shaped like a horse's head (see the photograph on page 270). Although long-exposure photographs show NGC 2024 and the Horsehead Nebula well, they are notoriously difficult to detect with amateur telescopes.

Pavo – The Peacock

A constellation introduced at the end of the 16th century by the Dutch navigators Pieter Dirkszoon Keyser and Frederick de Houtman. It is one of several celestial birds in this region, including Apus, Tucana, Grus and Phoenix. In Greek mythology the peacock was sacred to Hera, goddess of the heavens, from whose breast the Milky Way sprang. Hera set a creature with a hundred eyes called Argos to watch over a white heifer, into which she guessed her husband Zeus had turned one of his illicit lovers, the nymph Io. At the request of Zeus, Hermes slew the watchful Argos and released the heifer. Hera placed the hundred eyes of Argos on the peacock's tail.

α (alpha) Pavonis, 20h 26m −56°.7, (Peacock), mag. 1.9, is a blue-white star 183 l.y. away.

β (beta) Pav, 20h 45m −66°.2, mag. 3.4, is a white star 138 l.y. away.

δ (delta) Pav, 20h 09m −66°.2, mag. 3.6, is a yellow star 20 l.y. away.

η (eta) Pav, 17h 46m −64°.7, mag. 3.6, is an orange giant 371 l.y. away.

κ (kappa) Pav, 18h 57m −67°.2, 544 l.y. away, is one of the brightest Cepheid variables. It is a yellow-white supergiant, varying between mags. 3.9 and 4.8 in 9.1 days.

ξ (xi) Pav, 18h 23m −61°.5, 420 l.y. away, is a red giant of mag. 4.4 with a close companion of mag. 8.6, lost in the primary's glare in small telescopes.

SX Pav, 21h 29m −69°.5, 396 l.y. away, is a red giant that varies semi-regularly between mags. 5.3 and 6.0 every 7 weeks or so.

NGC 6744, 19h 10m −63°.9, is a 9th-mag. spiral galaxy for small telescopes, with a short central bar and widespread arms, as seen in the photograph below. It is one of the largest barred spirals, but its outer regions are faint and in modest apertures only the brightest central portion will be seen. It is about 30 million l.y. away.

NGC 6752, 19h 11m −60°.0, is a large 6th-mag. globular cluster, visible in binoculars and resolvable in 75-mm telescopes, covering half the apparent diameter of the Moon. A double star of 8th and 9th mags. at one edge is a foreground object. The cluster lies 15,000 l.y. away.

NGC 6744, a beautiful face-on spiral galaxy with a short central bar. Our own Milky Way, which is also thought to have a small central bar, might look much like this if seen from outside (M. Bessell, R. Sutherland and M. Buxton, RSAA – Australian National University)

Pavo · Pav · Pavonis

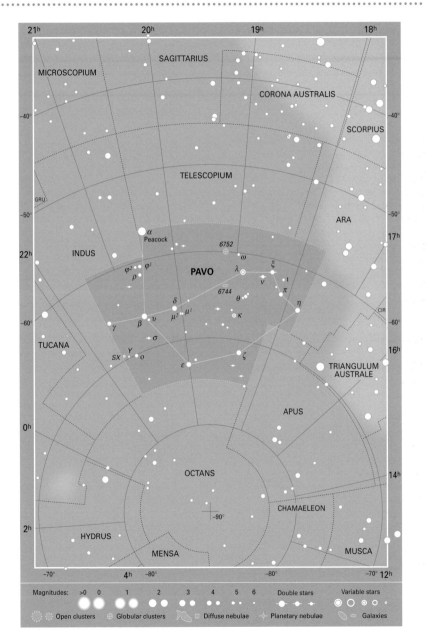

Pegasus

An ancient constellation representing the winged horse of Greek mythology, born from the blood of Medusa after she was slain by the hero Perseus, who lies nearby in the sky. Only the front half of the horse is shown. Its body is represented by the constellation's most famous feature, the Great Square, the corners of which are marked by four stars. One of these stars, once known as δ (delta) Pegasi, is now assigned to Andromeda. The Great Square of Pegasus is over 15° wide and 13° high, yet contains surprisingly few naked-eye stars for such a large area. The brightest of them are: υ (upsilon), mag. 4.4; τ (tau), mag. 4.6; ψ (psi), mag. 4.6; 56, mag. 4.8; φ (phi), mag. 5.1; 71, mag. 5.3; and 75, mag. 5.5.

α (alpha) Pegasi, 23h 05m +15°.2, (Markab, 'shoulder'), mag. 2.5, is a blue-white giant star 140 l.y. away.

β (beta) Peg, 23h 04m +28°.1, (Scheat, 'shin'), 199 l.y. away, is a red giant that varies from mag. 2.3 to 2.7 with no definite period.

γ (gamma) Peg, 0h 13m +15°.2, (Algenib, 'the side'), mag. 2.8, is a blue-white star 333 l.y. away. It is a pulsating variable of the β Cephei type, but its fluctuations every 3 hours 40 minutes are only 0.1 mag., too slight to be discernible with the naked eye.

ε (epsilon) Peg, 21h 44m +9°.9, (Enif, 'nose'), 670 l.y. away, is an orange supergiant of mag. 2.4. A small telescope, or even good binoculars, reveals a wide bluish mag. 8.4 companion star. Larger telescopes also show an 11th-mag. companion closer to ε Peg, making this an apparent triple system. ε Peg is catalogued as variable because of two

unexplained fluctuations: one night in November 1847 it was seen about a magnitude fainter than normal, while on a night in September 1972 it was seen at about magnitude 0.7 for four minutes before fading back to normal.

ζ (zeta) Peg, 22h 41m +10°.8, (Homam), mag. 3.4, is a blue-white star 209 l.y. away.

η (eta) Peg, 22h 43m +30°.2, (Matar), mag. 2.9, is a yellow giant 215 l.y. away.

π (pi) Peg, 22h 10m +33°.2, is a very wide binocular duo of white and yellow stars, both giants. They are of mags. 4.3 and 5.6, and lie 252 and 283 l.y. away respectively.

51 Peg, 22h 57m +20°.8, mag. 5.5, is a yellow main-sequence star 50 l.y. away, similar to the Sun. In 1995 it became the first star other than the Sun known to have a planet orbiting it; the planet's estimated mass is about half that of Jupiter.

M15 (NGC 7078), 21h 30m +12°.2, is an out-standing 6th-mag. globular cluster, 33,000 l.y. distant, at the limit of naked-eye visibility but easily seen in binoculars; a 6th-mag. star nearby is a sure guide to its location. A small telescope shows it as a glorious misty sight in an attractive field. Apertures of 150 mm or so resolve its outer regions into a mottled ground of sparkling stars, and larger telescopes show stars all the way to the bright and condensed core.

NGC 7331, 22h 37m +34°.4, is a 10th-mag. spiral galaxy seen nearly edge-on, visible under good conditions in apertures of 100 mm or so as an elongated smudge. It lies about 50 million l.y. away.

Pegasus · Peg · Pegasi

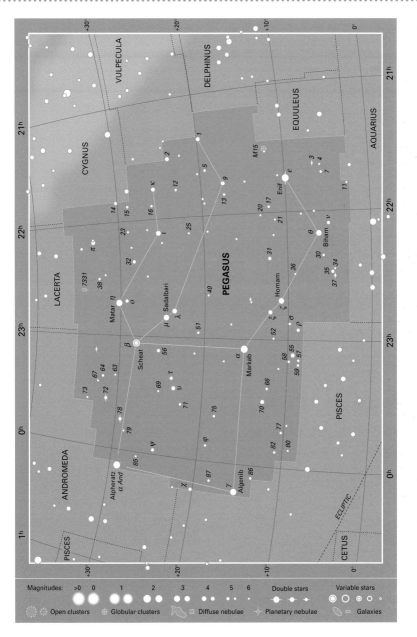

Perseus

Perseus was the hero of Greek mythology who rescued the chained maiden Andromeda from the clutches of the sea monster Cetus. Prior to that, Perseus had slain Medusa the Gorgon, whose head he is pictured holding in one hand. The Gorgon's head is marked by the winking star Algol, or β (beta) Persei, sometimes imagined as Medusa's evil eye.

Perseus lies in a rich part of the Milky Way and is well worth sweeping with binoculars; note, in particular, the Double Cluster, a sparkling pair of open clusters near the border with Cassiopeia. In 1901 Nova Persei flared up to mag. 0.2 at 3h 31.2m, +43° 54', throwing off a shell of gas that is now visible in large telescopes. NGC 1499, the California Nebula, so named because its shape resembles California, spans the apparent width of five full Moons north of 4th-mag. ξ (xi) Persei, an exceedingly hot blue giant or supergiant which illuminates it. Despite its considerable size the California Nebula is elusive visually, but shows up well on long-exposure photographs.

At 3h 19.8m, +41° 31', lies the radio source Perseus A, associated with the 12th-mag. peculiar galaxy NGC 1275, which is at the centre of the Perseus cluster of galaxies, 250 million l.y. away. Near γ (gamma) Persei lies the radiant of the Perseid meteors, the most glorious meteor shower of the year: around August 12–13 as many as 75 bright meteors can be seen flashing from Perseus each hour.

α (alpha) Persei, 3h 24m +49°.9, (Mirphak, 'elbow', or Algenib, 'the side'), mag. 1.8, is a yellow-white supergiant 592 l.y. away. Binoculars reveal a brilliant scattering of stars covering 3° in this region forming a loose cluster, Melotte 20. α Per itself seems to form the head of a snaking chain of stars within this cluster.

β (beta) Per, 3h 08m +41°.0, (Algol, 'the demon'), 93 l.y. away, is one of the most celebrated variable stars in the sky. It is the prototype of the eclipsing binary class of variables, in which two close stars periodically eclipse each other as they orbit their common centre of gravity. Algol's eclipses occur every 2.87 days, when the star's apparent brightness sinks from mag. 2.1 to 3.4 before returning to normal about 10 hours later.

γ (gamma) Per, 3h 05m +53°.5, mag. 2.9, is a yellow giant 256 l.y. away. It is an eclipsing binary with the unusually long period of 14.6 years, dimming by 0.3 mag. for 10 days. This star's variability was first detected in 1990.

δ (delta) Per, 3h 43m +47°.8, mag. 3.0, is a blue giant 528 l.y. away.

ε (epsilon) Per, 3h 58m +40°.0, 538 l.y. away, is a blue-white star of mag. 2.9 with an unrelated mag. 7.6 companion which is difficult to see through the smallest telescopes because of the brightness contrast.

ζ (zeta) Per, 3h 54m +31°.9, 980 l.y. away, is a blue supergiant of mag. 2.9 with a mag. 9.5 companion visible in a small telescope. ▶

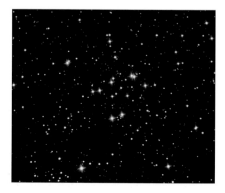

M34 is an attractive open cluster for binoculars and small telescopes. Its brightest stars are of 8th magnitude. (Peter Wienerroither)

Perseus · Per · Persei

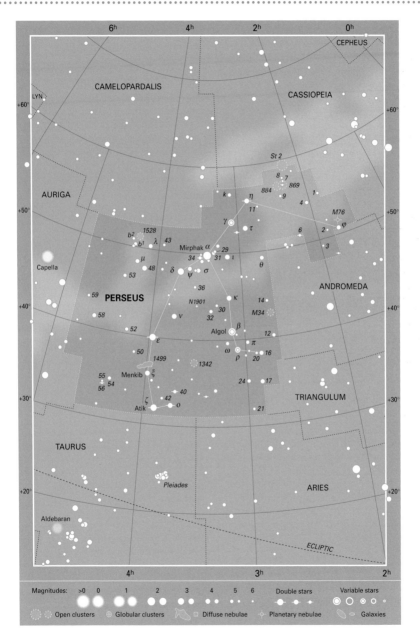

η (eta) Per, 2h 51m +55°.9, about 1300 l.y. away, is an orange supergiant of mag. 3.8 with a mag. 8.5 blue companion that forms an attractive double for small telescopes. The field of view contains a sprinkling of background stars.

ρ (rho) Per, 3h 05m +38°.8, 325 l.y. away, is a red giant that varies between mags. 3.3 and 4.0 in semi-regular fashion every 7 weeks or so.

M34 (NGC 1039), 2h 42m +42°.8, is a bright open cluster at the limit of naked-eye visibility, containing about 60 stars splashed over an area larger than the apparent size of the full Moon. Binoculars resolve it into stars, and it is well seen in a small telescope, many of the stars seeming to form pairs. M34 lies about 1500 l.y. away.

M76 (NGC 650–1), 1h 42m +51°.6, the Little Dumbbell, is a planetary nebula, the faintest object on Charles Messier's famous list of deep-sky objects. At about magnitude 10 it is difficult to see, although it can be picked up in 100 mm apertures on a dark night. It is relatively large for a planetary nebula, similar in size to the Ring Nebula in Lyra, but smaller than its namesake the Dumbbell Nebula in Vulpecula, the elongated shape of which it resembles. Each end of the Little Dumbbell has a separate NGC number. It lies about 3500 l.y. away.

NGC 869, NGC 884, 2h 19m +57°.2, 2h 22m +57°.1, the famous Double Cluster in Perseus, also known as h and χ (chi) Persei. They are two open star clusters visible to the naked eye and superb in binoculars, each covering an area equal to the full Moon. NGC 869 is the brighter and richer of the pair, containing an estimated 200 stars, while NGC 884 appears more scattered. They both lie about 7500 l.y. away and are relatively young, only a few million years old. Small telescopes have an advantage when observing these objects, for at low powers both clusters fit into the same field of view, which is not always the case with larger and more powerful telescopes. Most of the stars in the clusters are blue-white, but there are several red giants to be spotted in and around NGC 884. In binoculars, note a curving chain of stars leading northwards towards a large starry patch known as the cluster Stock 2 (abbreviated St 2). This whole area is a breathtaking sight in all apertures. (For a photograph of the Double Cluster, see page 274.)

Phoenix – The Phoenix

An inconspicuous constellation near the southern end of Eridanus, representing the mythical bird that was regularly reborn from its own ashes. It was introduced at the end of the 16th century by the Dutch navigators Pieter Dirkszoon Keyser and Frederick de Houtman in an area that had been known by the Arabs as the Boat, moored on the bank of the river Eridanus.

α (alpha) Phoenicis, 0h 26m −42°.3, (Ankaa), mag. 2.4, is an orange giant star 77 l.y. away.

β (beta) Phe, 1h 06m −46°.7, 185 l.y. away, appears to the naked eye as a yellow star of mag. 3.3. In fact, it is a close double with well-matched components of mags. 4.0 and 4.2, currently too close for resolution in small telescopes.

γ (gamma) Phe, 1h 28m −43°.6, mag. 3.4, is an orange giant 234 l.y. away.

ζ (zeta) Phe, 1h 08m −55°.2, 280 l.y. away, is a complex variable and multiple star. The main star is a blue-white eclipsing binary of the same type as Algol that rises and falls between mags. 3.9 and 4.4 every 1.67 days. It has a companion of 8th mag. visible in a small telescope. There is also a much closer 7th-mag. companion visible only in large apertures.

Phoenix · Phe · Phoenicis

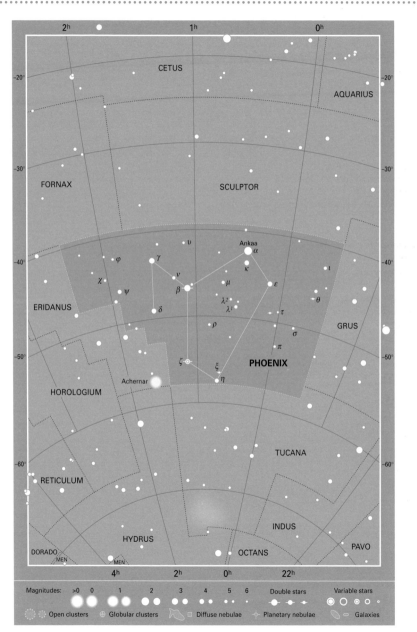

Pictor – The Painter's Easel

A faint constellation overshadowed by the neighbouring brilliant star Canopus in Carina on one side and the Large Magellanic Cloud in Dorado on the other. The constellation was invented in the 1750s by Nicolas Louis de Lacaille, who originally called it Equuleus Pictorius, which has since been shortened. At 5h 11.7m, −45° 01′ lies the mag. 8.9 red dwarf known as Kapteyn's Star, 12.8 l.y. away, named after the Dutch astronomer who discovered in 1897 that it has the second-largest proper motion of any known star (the record is held by Barnard's Star in Ophiuchus). Kapteyn's Star moves 1° in 415 years; see illustration below.

α (alpha) Pictoris, 6h 48m −61°.9, mag. 3.2, is a white star 99 l.y. away.

β (beta) Pic, 5h 47m −51°.1, mag. 3.9, is a blue-white main-sequence star 63 l.y. away. This star became famous in 1984 when astronomers photographed a disk of dust and gas around it, thought to be a planetary system in the process of formation. (See also page 307.)

γ (gamma) Pic, 5h 50m −56°.2, mag. 4.5, is an orange giant 174 l.y. away.

δ (delta) Pic, 6h 10m −55°.0, is a blue-white star about 1700 l.y. away. It is an eclipsing binary of the β (beta) Lyrae type, varying from mag. 4.7 to 4.9 every 1.67 days.

ι (iota) Pic, 4h 51m −53°.5, 120 l.y. away, is an easy double of mags. 5.6 and 6.4 for small telescopes.

Chart showing the proper motion of Kapteyn's Star over a period of 200 years, during which it crosses almost one Moon's width of sky. (Wil Tirion)

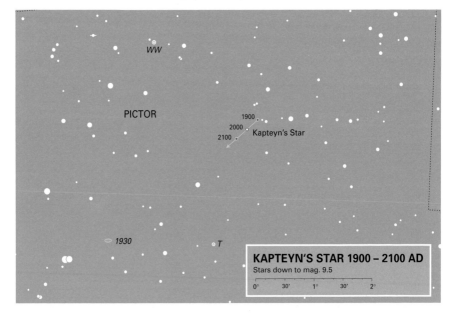

Pictor · Pic · Pictoris

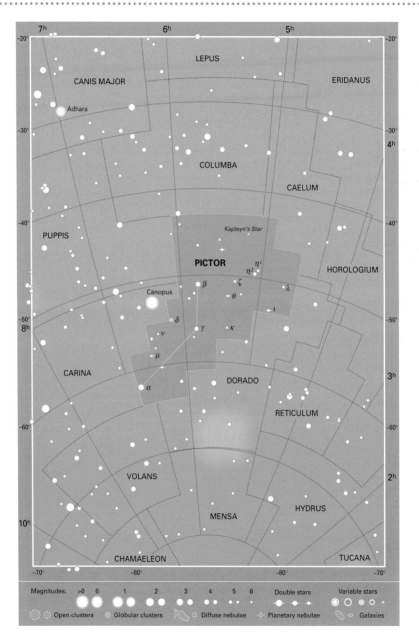

Pisces – The Fishes

An ancient constellation representing two fishes with their tails tied together by cords; the knot where the two cords join is marked by the star α (alpha) Piscium. One legend identifies the constellation with Aphrodite and her son Eros, who swam away from the attack of the monster Typhon in the guise of fishes.

The Sun is in Pisces from mid-March to late April, so the constellation contains the March (or *vernal*) equinox – the point at which the Sun crosses the celestial equator into the northern celestial hemisphere each year. This point originally lay in neighbouring Aries, but the effect of precession has now carried it into Pisces; eventually, in AD 2597, it will pass on into Aquarius. The constellation's most distinctive feature is a ring of seven stars called the Circlet, marking the body of one of the fish; clockwise from north the seven stars are θ (theta), 7, γ (gamma), κ (kappa), λ (lambda), TX (or 19) and ι (iota) Piscium.

α (alpha) Piscium, 2h 02m +2°.8, (Alrescha, 'the cord'), 139 l.y. away, appears to the naked eye as a star of mag. 3.8, but in fact is a challenging double with an orbital period of over 900 years. Its components, of mags. 4.2 and 5.2, are gradually closing but will remain within range of 100-mm aperture telescopes for at least the first half of the 21st century. The brighter star is a spectroscopic binary, and the fainter one may be also. Their colour is bluish-white, although some observers see the brighter star as greenish. ▶

M74 in Pisces is a beautiful face-on spiral galaxy with loosely wound arms. See page 212. (Paul Vreeswijk and Nik Szymanek)

Pisces · Psc · Piscium

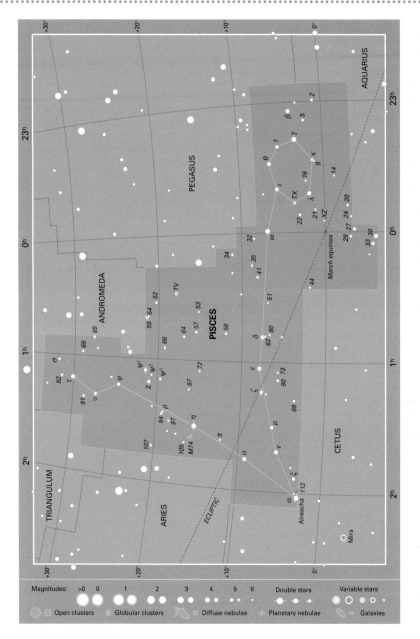

β (beta) Psc, 23h 04m +3°.6, mag. 4.5, is a blue-white star 493 l.y. away.

γ (gamma) Psc, 23h 17m +3°.3, mag. 3.7, is a yellow giant 131 l.y. away.

ζ (zeta) Psc, 1h 14m +7°.6, 148 l.y. away, is a wide double of mags. 5.2 and 6.4, divisible in the smallest telescopes.

η (eta) Psc, 1h 31m +15°.3, mag. 3.6, is the brightest star in the constellation. It is a yellow giant, 294 l.y. away.

κ (kappa) Psc, 23h 27m ⌐1°.3, mag. 4.9, is a blue-white star 162 l.y. away. It forms a wide binocular double with 9 Piscium, mag. 6.3, actually a back-ground star.

ρ (rho) Psc, 1h 26m +19°.2, mag. 5.3, is a white star 85 l.y. away that forms an easy binocular duo with the unrelated orange giant 94 Piscium, mag. 5.5, 307 l.y. away.

ψ¹ (psi¹) Psc, 1h 06m +21°.5, is a wide pair of blue-white stars of mags. 5.3 and 5.6 about 230 l.y. away, visible in small telescopes or even good binoculars.

TV Psc, 0h 28m +17°.9, 490 l.y. away, is a red giant that varies semi-regularly between mags. 4.7 and 5.4 every 7 weeks.

TX Psc, 23h 46m +3°.5, also known as 19 Psc, 760 l.y. away, is a deep-red irregular variable star in the Circlet of Pisces, visible with the naked eye or binoculars. It fluctuates between about mags. 4.8 and 5.2.

M74 (NGC 628), 1h 37m +15°.8, is a spiral galaxy of 9th mag., presented face-on to us. In dark conditions it can be glimpsed through a small telescope as a pale disk with a starlike nucleus and faint foreground stars dotted across it, but it needs an aperture of at least 150 mm to be well seen. M74 lies some 30 million l.y. away. (See the photograph on page 210.)

Piscis Austrinus – The Southern Fish

This constellation has been known since ancient times and is visualized as a fish lying on its back, drinking the stream of water flowing southwards from the urn of neigh-bouring Aquarius. It mouth is marked by the bright star Fomalhaut. This fish was said to be the parent of the two smaller zodiacal fishes, represented by Pisces.

α (alpha) Piscis Austrini, 22h 58m −29°.6, (Fomalhaut, 'the fish's mouth'), mag. 1.2, is a blue-white main-sequence star 25 l.y. away. It is sur-rounded by a disk of cool dust from which a planetary system may be forming.

β (beta) PsA, 22h 32m −32°.3, 148 l.y. away, is a wide double star consisting of a mag. 4.3 blue-white primary and a mag. 7.7 companion visible in small telescopes.

γ (gamma) PsA, 22h 53m −32°.9, 222 l.y. away, is a double star of mags. 4.5 and 8.0, made difficult to split in small telescopes by the magnitude contrast.

η (eta) PsA, 22h 01m −28°.5, 1000 l.y. away, is a close pair of blue-white stars of mags. 5.8 and 6.8, divisible with an aperture of 100 mm and high magnification.

Piscis Austrinus · PsA · Piscis Austrini

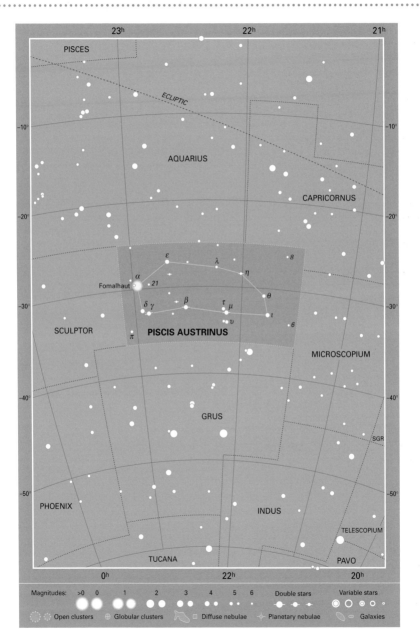

Puppis – The Stern

This is the largest of the three sections into which the ancient constellation of Argo Navis, the ship of the Argonauts, was dismembered in 1763 by Nicolas Louis de Lacaille; the other sections are Carina and Vela. When Argo Navis was split up its stars retained their existing Greek letters, and it so happens that the labelling of the stars assigned to Puppis begins with ζ (zeta). Puppis lies in the Milky Way and contains rich starfields for sweeping with binoculars.

ζ (zeta) Puppis, 8h 04m −40°.0, (Naos, 'ship'), mag. 2.2, is a brilliant blue supergiant about 1400 l.y. away, the hottest (and hence bluest) of all naked-eye stars, with a surface temperature of about 40,000°C.

ξ (xi) Pup, 7h 49m −24°.9, mag. 3.3, is a yellow supergiant about 1350 l.y. away. Binoculars reveal a wide, unrelated mag. 5.3 yellow companion, 321 l.y. away.

π (pi) Pup, 7h 17m −37°.1, mag. 2.7, is an orange supergiant 1100 l.y. away.

ρ (rho) Pup, 8h 08m −24°.3, mag. 2.8, is a yellow-white star 63 l.y. away. It is a variable of the δ (delta) Scuti type, fluctuating by 0.2 mag. with a period of 3 hours 23 minutes.

k Pup, 7h 39m −26°.8, 454 l.y. away, is a striking double star with blue-white components of mags. 4.5 and 4.6 easily divisible in small telescopes.

L¹ L² Pup, 7h 13m −45°.2, is an optical double consisting of two unrelated and contrasting stars. L¹ is mag. 4.9, a blue-white star 182 l.y. away. L² is a red giant semi-regular variable, 198 l.y. away, that fluctuates between 3rd and 6th mags. every 140 days or so.

V Pup, 7h 58m −49°.2, is an eclipsing binary of the β (beta) Lyrae type, about 1200 l.y. away. It varies from mag. 4.4 to 4.9 with a period of 35 hours.

M46 (NGC 2437), 7h 42m −14°.8, is a 6th-mag. open cluster of about 100 faint stars of remarkably uniform brightness, most of them around 10th mag. With its neighbour M47 it is visible to the naked eye as a brighter knot in the Milky Way. In binoculars it appears as a smudgy patch two-thirds the size of the full Moon, while a small telescope shows it as a sprinkling of stardust. M46 lies 5200 l.y. away. On its northern edge lies the 10th-mag. planetary nebula NGC 2438. This is not a member of the cluster, but is a foreground object about 3000 l.y. from us.

M47 (NGC 2422), 7h 37m −14°.5, is a scattered naked-eye cluster of similar apparent size to the full Moon. It contains three dozen or so stars, the brightest being of mag. 5.7. M47 lies about 1500 l.y. away, less than one-third the distance of its richer neighbour M46.

M93 (NGC 2447), 7h 45m −23°.9, is a 6th-mag. binocular cluster, 3600 l.y. away, consisting of 80 stars of 8th mag. and fainter, arranged in a wedge-shape.

NGC 2451, 7h 45m −38°.0, is a large and bright open cluster of 40 stars 1050 l.y. away, well seen in binoculars, centred on the mag. 3.6 orange giant c Puppis.

NGC 2477, 7h 52m −38°.5, is a large 6th-mag. open cluster consisting of a swarm of faint stars, looking in binoculars like a loose globular, seemingly with arms. This excellent object would undoubtedly have featured on Messier's list had he lived farther south. It lies 4000 l.y. away.

Puppis · Pup · Puppis

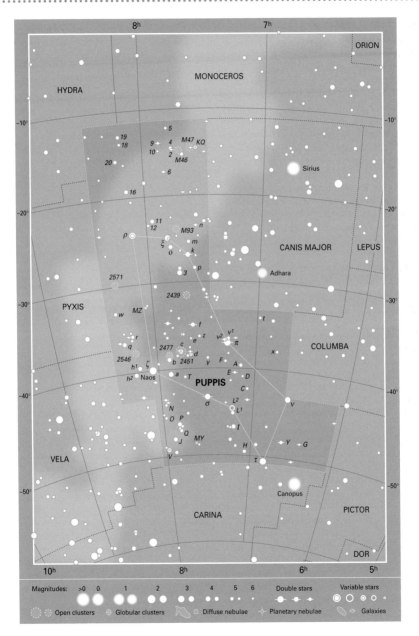

Pyxis – The Compass

A small, faint constellation invented by Nicolas Louis de Lacaille in the 1750s, representing a magnetic compass as used by sailors. It lies near Puppis, the stern of the Argonauts' ship, and was formed from stars that Ptolemy had catalogued as being part of the mast of Argo; but that ship, of course, would not have had a magnetic compass, so Pyxis cannot really be considered a genuine part of the old Argo. This area has also been known as Malus, the mast of Argo, a suggestion made in 1844 by the English astronomer John Herschel, although it was not widely adopted. The German astronomer Johann Bode introduced an additional constellation here called Lochium Funis, the log and line, representing a device used for measuring distance travelled at sea, which he visualized as curving around the compass.

Pyxis contains no objects of particular interest to users of small telescopes, despite the fact that it lies in the Milky Way, and its brightest stars are only of 4th magnitude.

α (alpha) Pyxidis, 8h 44m −33°.2, mag. 3.7, is a blue giant star 845 l.y. away.

β (beta) Pyx, 8h 40m −35°.3, mag. 4.0, is a yellow giant 388 l.y. away.

γ (gamma) Pyx, 8h 51m −27°.7, mag. 4.0, is an orange giant 209 l.y. away.

T Pyx, 9h 05m −32°.4, is a recurrent nova that has undergone five recorded eruptions, in 1890, 1902, 1920, 1944 and 1966. Normally it is of mag. 14, but brightens to 6th or 7th magnitude. Further outbursts may be expected.

Nicolas Louis de Lacaille (1713–1762)

Lacaille, a French astronomer, was the first person to map the southern skies comprehensively, as a result of which he became known as the Father of Southern Astronomy. He directed an expedition of the French Academy of Sciences to the Cape of Good Hope in 1750–54, where he systematically surveyed the southern celestial hemisphere, listing nearly 10,000 stars. The accurate positions of 2000 of these, along with a star map, were published posthumously in 1763 under the title *Coelum Australe Stelliferum*. Lacaille is usually best remembered for the 14 new constellations he introduced, representing instruments used in science and the fine arts: Antlia, Caelum, Circinus, Fornax, Horologium, Mensa, Microscopium, Norma, Octans, Pictor, Pyxis, Reticulum, Sculptor and Telescopium. Lacaille also dismantled one constellation, Robur Carolinum, Charles's Oak; this had been formed in 1678 by the English astronomer Edmond Halley from some of the stars of Argo Navis to commemorate the oak tree in which his patron, King Charles II, hid after his defeat by Oliver Cromwell at the Battle of Worcester. It is said that this inspired piece of flattery earned Halley his master's degree from Oxford by the king's express command. Lacaille, less impressed, uprooted the oak and returned its stars to Argo Navis.

Pyxis · Pyx · Pyxidis

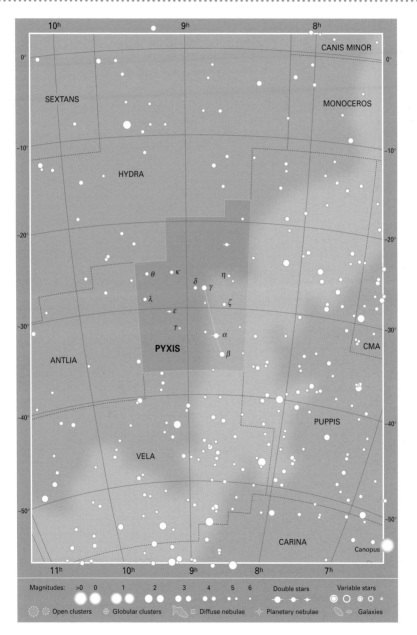

Reticulum – The Net

A constellation introduced in the 1750s by the French astronomer Nicolas Louis de Lacaille to commemorate a grid-like device in his telescope's eyepiece known as a reticle, which assisted him in measuring star positions during his surveys of the southern sky. The reticle was formed by silk threads and was diamond-shaped, like the constellation. Reticulum lies near the Large Magellanic Cloud, but is not prominent.

α (alpha) Reticuli, 4h 14m −62°.5, mag. 3.3, is a yellow giant star 163 l.y. away.

β (beta) Ret, 3h 44m −64°.8, mag. 3.8, is an orange star 100 l.y. away.

ζ^1 ζ^2 (zeta^1 zeta^2) Ret, 3h 18m −62°.5, 39 l.y. away, is a wide naked-eye or binocular double of near-identical yellow main-sequence stars similar to the Sun, of mags. 5.5 and 5.2 respectively.

Across the border of Reticulum in neighbouring Dorado lies this attractive 10th-mag. spiral galaxy, NGC 1566. It is a type known as a Seyfert galaxy, with a bright, variable centre. It is the brightest member of a small cluster of over two dozen galaxies nearly 70 million l.y. away. (SAAO)

Reticulum · Ret · Reticuli

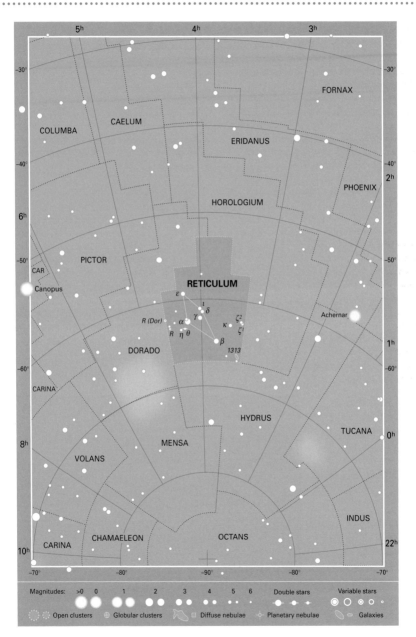

Sagitta – The Arrow

Despite its diminutive size – it is the third-smallest constellation – this arrow-shaped group is distinctive enough to have been recognized by the ancient Greeks. In the sky, the Arrow seems to be flying between Cygnus the Swan and Aquila the Eagle; in one legend, the arrow was shot by Hercules, although it has also been seen as the arrow of Apollo or Eros, the god of love. Like its northerly neighbour Vulpecula, Sagitta lies in a rich part of the Milky Way.

α (alpha) Sagittae, 19h 40m +18°.0, mag. 4.4, is a yellow giant 473 l.y. away.

β (beta) Sge, 19h 41m +17°.5, mag. 4.4, is a yellow giant 467 l.y. away.

γ (gamma) Sge, 19h 59m +19°.5, mag. 3.5, the brightest star in the constellation, is an orange giant 274 l.y. away.

δ (delta) Sge, 19h 47m +18°.5, mag. 3.7, is a red giant 448 l.y. away.

ζ (zeta) Sge, 19h 49m +19°.1, 326 l.y. away, is a blue-white star of mag. 5.0 with a 9th-mag. companion, divisible in a small telescope.

S Sge, 19h 56m +16°.6, is a yellow supergiant Cepheid variable, ranging between mag. 5.2 and 6.0 in 8.4 days.

VZ Sge, 20h 00m +17°.5, 746 l.y. away, is a pulsating red giant that varies irregularly between mags. 5.3 and 5.6.

WZ Sge, 20h 08m +17°.7, is a so-called dwarf nova that flared up from 15th mag. to 7th or 8th mag. in 1913, 1946, 1978 and 2001; its location is worth checking in case of another outburst.

M71 (NGC 6838), 19h 54m +18°.6, is an 8th-mag. globular cluster 13,000 l.y. away, visible as a small, somewhat elongated misty patch in binoculars or a small telescope and resolvable with 100 mm aperture. Its stars are scattered and it lacks a central condensation, so it looks more like a dense open cluster than a typical globular.

Just north of Sagitta's border, in neighbouring Vulpecula, lies the Dumbbell Nebula, M27, a shell of gas thrown off by a dying star. It is visible in binoculars and small telescopes. See page 262. (From The IAC Morphological Catalog of Northern Galactic Planetary Nebulae, by A. Manchado, M.A. Guerrero, L. Stanghellini and M. Serra-Ricart/IAC)

Sagitta · Sge · Sagittae

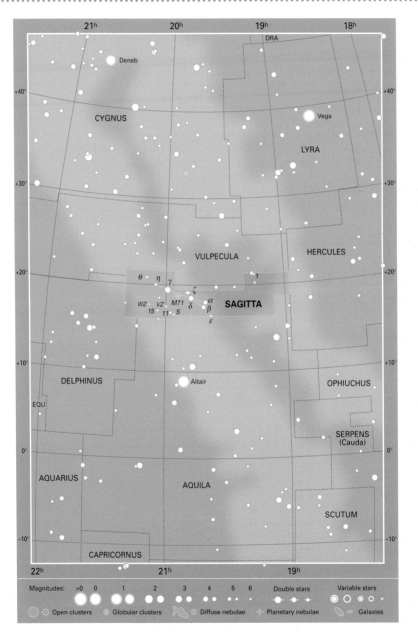

Sagittarius – The Archer

An ancient constellation depicting a centaur, a creature half man, half beast, with a raised bow and arrow. It is an older constellation than the other celestial centaur, Centaurus, and is different in character. Whereas Centaurus is identified as a scholarly, beneficent creature, Sagittarius is depicted with a threatening look, aiming his arrow at the heart of Scorpius, the Scorpion. The bow is marked by the stars λ (lambda), δ (delta) and ε (epsilon) Sagittarii, while γ (gamma) is the tip of the arrow. The shape outlined by the eight main stars of Sagittarius is popularly referred to as the Teapot, while λ (lambda), φ (phi), σ (sigma), τ (tau) and ζ (zeta) Sagittarii form a ladle shape known as the Milk Dipper – a suitable implement to dip into this rich region of the Milky Way.

The centre of our Galaxy lies in Sagittarius, so the Milky Way starfields are particularly rich here, as well as in neighbouring Scutum and Scorpius. The actual centre of the Galaxy is marked by a radio and infrared source known as Sagittarius A, at 17h 46.1m, −28° 51′. The main attraction of Sagittarius is its clusters and nebulae. Charles Messier catalogued a total of 15 objects in Sagittarius, more than in any other constellation; only a selection of them can be mentioned here. The Sun passes through the constellation from mid-December to mid-January, and thus lies in Sagittarius at the December solstice, its farthest point south of the equator.

α (alpha) Sagittarii, 19h 24m −40°.6, (Rukbat, 'knee', or Alrami, 'the archer'), mag. 4.0, is one of several instances in which the star labelled α in a constellation is not the brightest. It is a blue-white main-sequence star 170 l.y. away.

β¹ β² (beta¹ beta²) Sgr, 19h 23m −44°.5, (Arkab, from the Arabic for 'Achilles tendon'), is a pair of unrelated naked-eye stars. β¹ Sgr, a blue-white star of mag. 4.0, 378 l.y. away, has a mag. 7.2 companion visible in a small telescope. β² Sgr is a white star of mag. 4.3, 139 l.y. away. All three stars appear in the same line of sight by chance.

γ (gamma) Sgr, 18h 06m −30°.4, (Alnasl, 'the point', i.e. of the arrow), mag. 3.0, is an orange giant 96 l.y. away.

δ (delta) Sgr, 18h 21m −29°.8, (Kaus Media, 'middle of the bow'), mag. 2.7, is an orange giant 306 l.y. away.

ε (epsilon) Sgr, 18h 24m −34°.4, (Kaus Australis, 'southern part of the bow'), mag. 1.8, is the brightest star in Sagittarius. It is a blue-white giant 145 l.y. away.

λ (lambda) Sgr, 18h 28m −25°.4, (Kaus Borealis, 'northern part of the bow'), mag. 2.8, is an orange giant 77 l.y. away.

σ (sigma) Sgr, 18h 55m −26°.3, (Nunki), mag. 2.1, is a blue-white star 224 l.y. away.

W Sgr, 18h 05m −29°.6, is a yellow supergiant Cepheid variable ranging between mags. 4.3 and 5.1 with a period of 7.6 days. It lies about 2100 l.y. away.

X Sgr, 17h 48m −27°.8, is a yellow-white giant Cepheid variable ranging between mags. 4.2 and 4.9 in 7 days. It lies about 1100 l.y. away.

Y Sgr, 18h 21m −18°.9, is a yellow-white giant Cepheid that varies between mags. 5.3 and 6.2 in 5.8 days. It lies about 1300 l.y. away.

RR Sgr, 19h 56m −29°.2, is a red giant variable of Mira type, with a range from mag. 5.4 to 14.0 and a period of about 11 months. Its distance is too great to measure accurately.

RY Sgr, 19h 17m −33°.5, is the southern equivalent of R Coronae Borealis – a star like a reverse nova that normally shines at around 6th ▶

Sagittarius · Sgr · Sagittarii

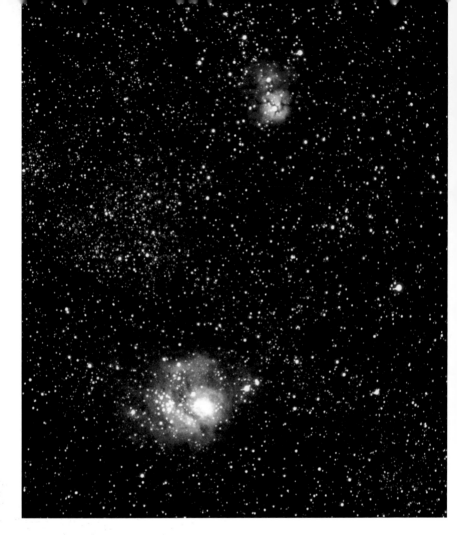

Among the Milky Way starfields of Sagittarius are two outstanding nebulae: M20, the Trifid Nebula (top), and M8, the Lagoon Nebula. The Trifid consists of a pink emission nebula and a blue reflection nebula. (Nik Szymanek)

mag., but which can suddenly and unpredictably drop to 14th mag.

M8 (NGC 6523), 18h 04m −24°.4, the Lagoon Nebula, is a famous gaseous nebula, visible to the naked eye, elongated in shape and encompassing the star cluster NGC 6530. M8 is a fine object for binoculars or telescopes, covering the area of three full Moons, with a dark rift down its centre. In the eastern half of the nebula is NGC 6530, a cluster of about 25 stars of 7th mag. and fainter, formed recently from the surrounding gas. The other (western) side of the nebula is dominated by two main stars, the brighter of which is the blue super-giant 9 Sgr, mag. 5.9. Photographs show the nebula as an intense red, but visually it appears milky-white. M8 is about 5000 l.y. away.

M17 (NGC 6618), 18h 21m −16°.2, the Omega, Horseshoe or Swan Nebula, is another gaseous nebula. Binoculars show it as a wedge-shaped object of similar apparent width to the full Moon, while in larger instruments it appears arch-shaped, variously likened to a Greek capital omega (Ω), a horseshoe or a swan, which accounts for its range of popular names. M17 lies 5000 l.y. away. About 1° south of it is M18 (NGC 6613), 4000 l.y. distant, a small, loose cluster of 20 stars of 9th mag. and fainter, unimpressive in binoculars.

M20 (NGC 6514), 18h 03m −23°.0, the Trifid Nebula, is a cloud of glowing gas far less impressive visually than photographically. Moderate-sized telescopes show it as only a diffuse patch of light centred on the double star HN 40, of 8th and 9th mags., which was evidently born from it and now illuminates it. The Trifid Nebula gets its name from three dark lanes of dust that trisect it, well shown on photographs but elusive in small apertures. It lies 5000 l.y. away, the same as M8.

M21 (NGC 6531), 18h 05m −22°.5, is a spider-like open cluster in the same low-power field of view as M20, containing about 70 stars of 7th mag. and fainter, 4000 l.y. away.

M22 (NGC 6656), 18h 36m −23°.9, is a large, rich 5th-mag. globular cluster, one of the finest in the entire heavens and ranked third only to ω (omega) Centauri and 47 Tucanae. Visible to the naked eye, M22 is an excellent binocular object and a fine sight in small telescopes, which reveal its noticeably elliptical outline. A telescope of 75 mm aperture will begin to resolve its outer regions, and larger apertures show its brightest stars to be reddish. Its nucleus is not as condensed as that of many other globulars. M22 lies 10,000 l.y. away. At the edge of the same binocular field is the smaller 7th-mag. globular M28 (NGC 6626).

M23 (NGC 6494), 17h 57m −19°.0, is a widely spread open cluster of fairly uniform appearance, just at the limit of resolution in binoculars. It is elongated in shape, consisting of a field of 10th- and 11th-mag. stars, some arranged in arcs. It lies 2100 l.y. away.

M24, 18h 18m −18°.5, is a rich and extensive Milky Way starfield south of M17 and M18, grainy and shimmering in binoculars. Some observers restrict the name M24 to a small cluster of faint stars in its northern half known also as NGC 6603, but this is not what Messier meant. The whole Milky Way star cloud that makes up M24 measures about 2° × 1° and is one of the most prominent parts of the Milky Way to the naked eye.

M25 (IC 4725), 18h 32m −19°.2, 2300 l.y. away, is a scattered cluster of about 30 stars, well seen in binoculars. The most prominent members form two bars across the cluster's centre. Its brightest star is U Sgr, a yellow supergiant Cepheid which varies from mag. 6.3 to 7.1 with a period of 6 days 18 hours.

M55 (NGC 6809), 19h 40m −31°.0, is a 6th-mag. globular cluster, nebulous-looking in binoculars with little central condensation. Small telescopes resolve individual stars and show a dark notch on one side. M55 lies 19,000 l.y. away.

M22: Easily visible as a misty patch through binoculars and small telescopes, the true nature of this huge globular cluster becomes apparent when viewed through larger apertures. (Nigel Sharp, REU program/NOAO/AURA/NSF)

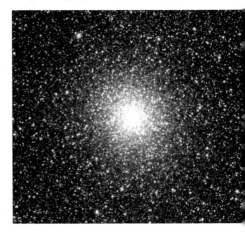

Scorpius – The Scorpion

A resplendent constellation, lying in a rich area of the Milky Way and packed with exciting objects for users of small telescopes. In mythology, Scorpius was the scorpion whose sting killed Orion. In the sky Orion still flees from the scorpion, for Orion sets below the horizon as Scorpius rises.

Originally, in ancient Greek times and earlier, Scorpius was a much larger constellation, but in the first century BC the stars that comprised its claws were made into the separate constellation of Libra by the Romans. Despite the loss of its claws, Scorpius still bears a clear resemblance to the creature after which it is named, with a distinctive curve of stars forming its stinging tail. Its heart is marked by the bright star Antares, a name that can be translated either as 'rival of Mars' or 'like Mars', in reference to its strong red colour. North of β (beta) Scorpii, at 16h 19.9m, −15° 38′, lies the strongest X-ray source in the sky, Scorpius X-1. This has been identified with a 13th-magnitude spectroscopic binary some 9000 l.y. away consisting of a neutron star accreting matter from a companion.

A scattering of several hundred bright stars centred between about 400 and 500 l.y. from us extends from Scorpius via Lupus into Centaurus and Crux; this is known as the Scorpius–Centaurus (or Sco–Cen) association. Antares is a prominent member. The Sun passes briefly through Scorpius during the last week of November.

α (alpha) Scorpii, 16h 29m −26°.4, (Antares), 604 l.y. away, is a red supergiant 400 times the diameter of the Sun. It is a semi-regular variable, fluctuating between mags. 0.9 and 1.2 with an approximate period of 5 years. Antares has a mag. 5.4 blue companion that requires at least 75 mm aperture and the steadiest atmospheric conditions to be visible against the primary's glare. The orbital period of the companion around Antares is estimated to be over 1200 years.

β (beta) Sco, 16h 05m −19°.8, (Graffias, 'claws', or Acrab, 'scorpion') is a striking double star divisible in the smallest telescopes, consisting of unrelated blue-white main-sequence stars of mags. 2.6 and 4.9, distances 530 and about 1100 l.y.

δ (delta) Sco, 16h 00m −22°.6, (Dschubba, 'forehead'), is a blue-white star 402 l.y. away. Normally of mag. 2.3, in 2000 it began to brighten, apparently as a result of throwing off a shell of gas, reaching a maximum of around mag. 1.6 in 2002. It subsequently subsided to near-normal brightness but may surge again in future.

ε (epsilon) Sco, 16h 50m −34°.3, mag. 2.3, is an orange giant 65 l.y. away.

ζ¹ ζ² (zeta¹ zeta²) Sco, 16h 54m −42°.4, is a naked-eye double of unrelated stars, ζ² being a mag. 3.6 orange giant 151 l.y. away, and ζ¹ a blue supergiant that varies erratically between mags. 4.7 and 4.9; ζ¹ is probably an outlying member of the open cluster NGC 6231 (see page 229).

θ (theta) Sco, 17h 37m −43°.0, mag. 1.9, is a white giant 272 l.y. away with a mag. 5.3 companion visible in small telescopes.

λ (lambda) Sco, 17h 34m −37°.1, (Shaula, 'sting'), mag. 1.6, is a blue-white star 703 l.y. away.

μ¹ μ² (mu¹ mu²) Sco, 16h 52m −38°.0, is a naked-eye double of unrelated stars. μ¹, 822 l.y. away, is an eclipsing binary that varies from mag. 2.9 to 3.2 in 34 hours 43 minutes; μ² is a blue-white star of mag. 3.6, 517 l.y. away.

ν (nu) Sco, 16h 12m −19°.5, 437 l.y. away, is a quadruple star similar to the famous Double Double in Lyra. A small telescope, or even powerful binoculars, shows ν Sco as a wide double, with blue-white components of mags. 4.0 and 6.3. Telescopes of 75 mm and above reveal at high magnification that the fainter star is itself a close double of mags. 6.7 and 7.7. The brighter star is

Scorpius · Sco · Scorpii

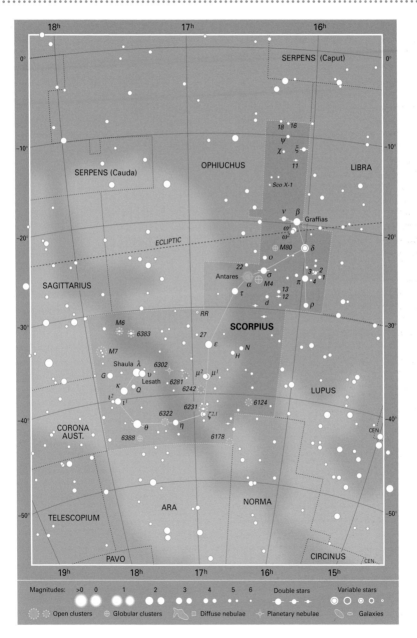

M6 and M7 are two glorious open clusters. Right: M6 is popularly known as the Butterfly Cluster because of its shape, although there is also a strong resemblance to the outline of a bird. Most of its stars are blue, with the exception of its brightest member, the orange giant BM Scorpii, seen at the left of the cluster. (Nigel Sharp, Mark Hanna, REU Program/NOAO/AURA/NSF)

Left: M7 is a large naked-eye cluster, seen against a backdrop of shimmering Milky Way starfields, unlike M6 to the north which has a much darker background. M7 is only just over half the distance of M6, so its individual members appear brighter than those of M6, and are scattered over a much wider area. (Nigel Sharp, REU program/NOAO/AURA/NSF)

an even closer double of mags. 4.3 and 5.4, requiring an aperture of 100 mm to divide.

ξ (xi) Sco, 16h 04m −11°.4, about 95 l.y. away, is a celebrated multiple star. A small telescope shows it as a white star of mag. 4.2 with a mag. 7.3 orange companion; also visible in the same field is a fainter and wider pair, Σ 1999 (Struve 1999), composed of mag. 7.4 and 8.0 stars that are gravitationally connected to ξ Sco. Therefore, at first sight, ξ Sco looks like another Double Double. But the brighter star is itself a close pair, consisting of yellow-white stars of mags. 4.8 and 5.1 orbiting each other with a period of 46 years. They were closest in 1996 and are currently moving apart; by 2007 they were divisible with 150 mm aperture and 100 mm should separate them by 2015.

ω^1 ω^2 (omega^1 omega^2) Sco, 16h 07m −20°.7, is a pair of unrelated stars distinguishable with the naked eye: ω^1, a blue-white main-sequence star of mag. 4.0, 424 l.y. away, and ω^2, a yellow giant of mag. 4.3, 265 l.y. away.

RR Sco, 16h 57m −30°.6, is a red giant variable of the Mira type, ranging between 5th and 12th mag. every 9 months or so. It lies about 1150 l.y. away.

M4 (NGC 6121), 16h 24m −26°.5, is a 6th-mag. globular cluster appearing almost as large as the full Moon. It looks like a woolly ball in binoculars, but is not as easy to spot as its magnitude would suggest because its light is spread over a large area. In 100-mm telescopes individual stars are resolved and there is a noticeable bar of stars running north–south across its centre. M4 is more loosely scattered than many globulars, and does not have a strong central condensation. It is the closest globular to us, 5600 l.y. away.

M6 (NGC 6405), 17h 40m −32°.2, is an impressive 4th-mag. open cluster of about 80 stars arranged in radiating chains, popularly called the Butterfly Cluster. Binoculars and small telescopes resolve the main stars which form the butterfly shape. The brightest of them, BM Sco, on one of the 'wings', is an orange giant semi-regular variable which ranges between 5th and 7th mags. every 27 months or so. M6 lies 1600 l.y. away. See the top photograph on the facing page.

M7 (NGC 6475), 17h 54m −34°.8, the most southerly object in the Messier catalogue, is a huge, scattered 3rd-mag. open cluster of 80 or so stars individually of 6th mag. and fainter, visible to the naked eye as a brighter knot in the Milky Way (see the photograph on the facing page). The apparent diameter is over twice that of the full Moon, and it is easily resolved in binoculars. With M6 at the edge of the same binocular field, this is an exceptionally rich sight. The central group of stars is arranged in an X-shape with scattered outliers that form triangular surroundings like a Christmas tree, set against the backdrop of a dense star cloud. An outstanding cluster and a classic for small apertures. M7 lies 950 l.y. away, just over half the distance of M6; the two clusters are not related.

M80 (NGC 6093), 16h 17m −23°.0, is a small, 7th-mag. globular cluster visible in binoculars or a small telescope, appearing like the fuzzy head of a comet. It lies 27,000 l.y. away.

NGC 6231, 16h 54m −41°.8, is a naked-eye open cluster of over 100 stars in a rich area of the Milky Way that is well worth sweeping with binoculars. The brightest stars of the group are of 6th mag. and give the impression of a mini-Pleiades in binoculars and small telescopes. The 5th-mag. blue supergiant ζ^1 (zeta^1) Sco (see page 226) is probably an outlying member of this cluster, which lies 6500 l.y. away. NGC 6231 is connected to a larger, scattered cluster of fainter stars visible in binoculars, known both as Trumpler 24 and Harvard 12, which lies 1° to the north. The chain of stars linking the two clusters delineates one of the spiral arms of our Galaxy.

NGC 6242, 16h 56m −39°.5, is a wedge-shaped open cluster for small telescopes containing about two dozen stars of 7th mag. and fainter. It lies about 3500 l.y. away.

NGC 6281, 17h 05m −37°.9, is an open cluster visible in small telescopes, consisting of over 50 stars of 8th mag. and fainter arranged in a rectangular shape, 1800 l.y. away.

Sculptor – The Sculptor

One of the faint and half-forgotten constellations introduced in the 1750s by Nicolas Louis de Lacaille to fill in the southern skies. It represents a sculptor's studio. Sculptor contains the south pole of our Galaxy, 90° from the plane of the Milky Way. In this direction we can look out into deep space, unobscured by stars or dust, and see many faint galaxies. Among these is a very faint member of our Local Group, the Sculptor Dwarf, detectable only on long-exposure photographs taken through large telescopes.

α (alpha) Sculptoris, 0h 59m −29°.4, mag. 4.3, is a blue-white giant 672 l.y. away.

β (beta) Scl, 23h 33m −37°.8, mag. 4.4, is a blue-white star 178 l.y. away.

γ (gamma) Scl, 23h 19m −32°.5, mag. 4.4, is an orange giant 179 l.y. away.

δ (delta) Scl, 23h 49m −28°.1, mag. 4.6, is a blue-white main-sequence star 143 l.y. away.

ε (epsilon) Scl, 1h 46m −25°.1, 89 l.y. away, is a binary of mags. 5.3 and 8.6, visible in small telescopes. The estimated orbital period is 1200 years.

κ¹ (kappa¹) Scl, 0h 09m −28°.0, 224 l.y. away, is a tight pair of white stars, of mags. 6.1 and 6.2, at the limit of resolution of a 75-mm telescope.

R Scl, 1h 27m −32°.5, is a deep-red semi-regular variable star that ranges from about mag. 5.8 to 7.7 with a period of about a year. It lies about 1500 l.y. away.

S Scl, 0h 15m −32°.0, is a red giant variable of Mira type, ranging from mag. 5.5 to 13.6 in around a year. It lies about 1500 l.y. away.

NGC 55, 0h 15m −39°.2, is an 8th-mag. spiral galaxy seen nearly edge-on, so it appears elongated. One half is more prominent that the other. It is similar in size and shape to NGC 253 (see below), although not quite as bright. Its distance is 6 million l.y.

NGC 253, 0h 48m −25°.3, is a 7th-mag. spiral galaxy seen nearly edge-on which gives it a cigar-shaped appearance, as seen in the photograph below. Nearly ½° long, it can be picked up in binoculars under dark skies but at least 100 mm aperture is required to distinguish the mottling caused by dust clouds in its spiral arms. NGC 253 lies 13 million l.y. away.

NGC 253, a spiral galaxy seen nearly edge-on and hence appearing elliptical. It has no central bulge, but its arms are a seething mass of stars and dust.
(Todd Boroson/ NOAO/AURA/NSF)

Sculptor · Scl · Sculptoris

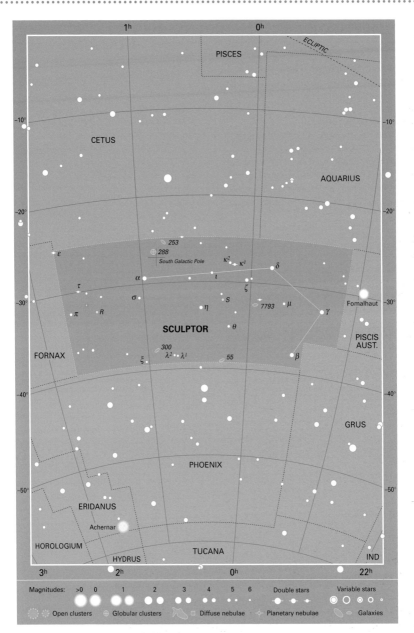

Scutum – The Shield

A faint constellation between Aquila and Serpens, introduced in 1684 by the Polish astronomer Johannes Hevelius under the title Scutum Sobiescianum, Sobieski's Shield, in honour of his patron, King John III Sobieski of Poland. Rich Milky Way star fields are the main attraction here, notably the Scutum star cloud in the northern half of the constellation, about 6° across and generally considered to be the brightest part of the Milky Way outside Sagittarius. The prominent open cluster M11 lies near a notch of dark nebulosity at the northern edge of the Scutum star cloud.

α (alpha) Scuti, 18h 35m −8°.2, mag. 3.8, is an orange giant 174 l.y. away.

δ (delta) Sct, 18h 42m −9°.1, 187 l.y. distant, is the prototype of a rare class of variable stars that pulsate in size every few hours, producing small-amplitude brightness changes. δ Scuti itself is a white giant that varies from mag. 4.6 to 4.8 with a period of 4 hours 39 minutes.

R Sct, 18h 48m −5°.7, is a pulsating orange super-giant about 1400 l.y. away that varies between mags. 4.2 and 8.6 every 5 months or so.

M11 (NGC 6705), 18h 51m −6°.3, the Wild Duck Cluster, is a showpiece open cluster of about 200 stars, half the apparent width of the full Moon. At 6th mag. it is at the limit of naked-eye visibility, but binoculars show it as a misty patch. In a telescope with a magnification of around ×100 it breaks up into a sparkling field of faint stardust. The cluster gets its popular name from the fact that its brightest members form a distinct fan-shape, resembling a flight of ducks, an effect that is noticeable visually through a telescope but which becomes lost on long-exposure photographs. An 8th-mag. star, slightly brighter than the rest, lies at the fan's apex, with a double star nearby. M11 is 6500 l.y. away.

M26 (NGC 6694), 18h 45m −9°.4, is an open cluster for small telescopes, of similar size to M11 but containing only about two dozen stars and hence much fainter. It lies about 5000 l.y. away.

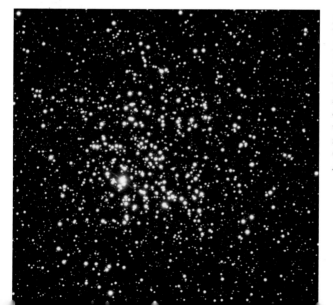

Among the rich Milky Way starfields of Scutum lies M11, the Wild Duck cluster. In small instruments it appears to be shaped like a fan or arch, because of a lower density of bright stars towards one side, the right in this image. (Adam Block/NOAO/AURA/NSF)

Scutum · Sct · Scuti

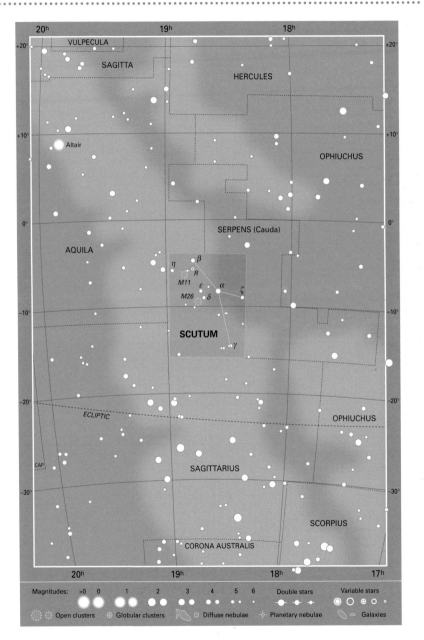

Serpens – The Serpent

An ancient constellation, representing a snake wound around the body of Ophiuchus. Serpens is actually split into two halves, one either side of Ophiuchus: Serpens Caput, the head, which is the larger and more prominent half; and Serpens Cauda, the tail. It is the only constellation to be split in two, but both halves count as one constellation.

α (alpha) Serpentis, 15h 44m +6°.4, (Unukalhai, 'the serpent's neck'), mag. 2.6, is an orange giant star 73 l.y. away.

β (beta) Ser, 15h 46m +15°.4, 153 l.y. distant, is a blue-white main-sequence star in the serpent's head of mag. 3.7, with a 10th-mag. companion visible in small telescopes. An unrelated mag. 6.7 background star, 29 Ser, is visible just north of it in binoculars.

γ (gamma) Ser, 15h 56m +15°.7, mag. 3.8, is a white main-sequence star 36 l.y. away.

δ (delta) Ser, 15h 35m +10°.5, 210 l.y. away, is a white star of mag. 4.2 with a close mag. 5.2 companion visible in a small telescope at high magnification.

η (eta) Ser, 18h 21m −2°.9, mag. 3.2, is an orange giant star 62 l.y. away.

θ (theta) Ser, 18h 56m +4°.2, (Alya), 132 l.y. distant, is an elegant pair of white stars of mags. 4.6 and 5.0, easily split in the smallest telescopes. ▶

M16 and the surrounding Eagle Nebula in Serpens is a spectacular combination of star cluster and gas cloud. At the nebula's centre are dark fingers of dust and cooler gas that were made famous by the "pillars of creation" image taken by the Hubble Space Telescope. This photograph was taken with the 4-m telescope at Kitt Peak, Arizona. (Bill Schoening/NOAO/AURA/NSF)

Serpens · Ser · Serpentis

ν (nu) Ser, 17h 21m −12°.8, 193 l.y. away, is a blue-white star of mag. 4.3 with a wide mag. 8.3 companion visible in binoculars or small telescopes.

τ¹ (tau¹) Ser, 15h 26m +15°.4, 920 l.y. away, a red giant of mag. 5.2, is the brightest member of a loose scattering of eight stars of 6th mag. near β Ser, all visible in binoculars.

R Ser, 15h 51m +15°.1, is a red giant variable of Mira type, ranging between mags. 5.2 and 14.4 in approximately a year. It lies about 900 l.y. away.

M5 (NGC 5904), 15h 19m +2°.1, is a 6th-mag. globular cluster 26,000 l.y. away, visible in binoculars or small telescopes. It is rated as one of the finest globulars in the northern sky, second only to the famous M13 in Hercules. A telescope of 100 mm aperture reveals its brilliant, condensed centre and mottled outer regions with curving chains of stars. Close to M5 (but actually in the foreground) lies 5 Ser, a mag. 5.0 yellow-white star with a mag. 10 companion.

M16 (NGC 6611), 18h 19m −13°.8, is a hazy-looking open star cluster 8500 l.y. distant, of similar apparent size to the full Moon, embedded in the larger Eagle Nebula. The cluster is a grouping of about 60 stars of 8th mag. and fainter at the limit of resolution in binoculars. Small telescopes show that most of its members congregate in a V-shape in the northern half. The surrounding Eagle Nebula adds a touch of haziness to the cluster when seen in binoculars. The nebula is too faint to be seen well in amateur telescopes, but shows up beautifully on long-exposure photographs. It was the subject of a famous Hubble Space Telescope image taken in 1995 (see page 271).

IC 4756, 18h 39m +5°.4, is a scattered open cluster of 8th-mag. stars and fainter, about two Moon diameters wide, visible in binoculars. It lies 1300 l.y. away.

Sextans – The Sextant

A faint and insignificant constellation south of Leo, introduced in 1687 by the Polish astronomer Johannes Hevelius. It commemorates the instrument he used for measuring star positions. Hevelius continued to make naked-eye sightings of star positions with his sextant long after telescopes were available, and it is perhaps to demonstrate how good his eyesight was that he formed Sextans out of stars no brighter than mag. 4.5.

α (alpha) Sextantis, 10h 08m −0°.4, mag. 4.5, is a blue-white giant star 287 l.y. away.

β (beta) Sex, 10h 30m −0°.6, mag. 5.1, is a blue-white main-sequence star 345 l.y. away.

γ (gamma) Sex, 9h 53m −8°.1, mag. 5.1, is a blue-white main-sequence star 262 l.y. away.

δ (delta) Sex, 10h 30m −2°.7, mag. 5.2, is a blue-white main-sequence star 300 l.y. away.

17, 18 Sex, 10h 10m −8°.4, is a neat pairing of unrelated stars, mags. 5.9 and 5.6, distances 527 and 473 l.y., easily divisible in binoculars.

NGC 3115, 10h 05m −7°.7, is a 9th-mag. galaxy known as the Spindle Galaxy, some 30 million l.y. away. It is of the type known as a lenticular galaxy, with a disk of stars surrounding a central bulge but no spiral arms. Moderate-sized amateur telescopes show its elongated outline and brighter centre.

Sextans · Sex · Sextantis

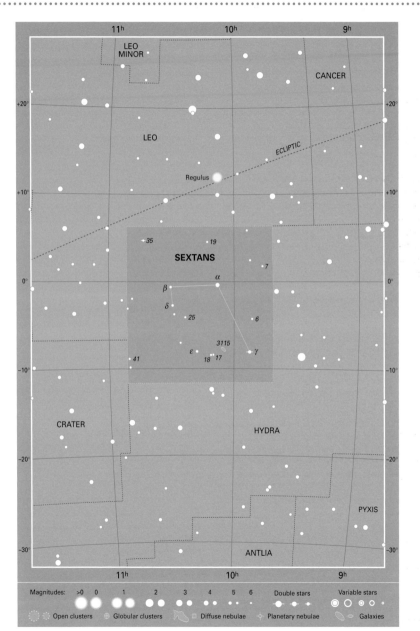

Magnitudes: >0 0 1 2 3 4 5 6   Double stars   Variable stars

Open clusters · Globular clusters · Diffuse nebulae · Planetary nebulae · Galaxies

Taurus – The Bull

One of the most ancient constellations, recognized since the dawn of civilization. In Greek mythology, Taurus represents the animal disguise adopted by Zeus to carry off Princess Europa to Crete. Only the front half of the bull is depicted in the sky, its face being formed by the V-shaped cluster of stars known as the Hyades. Its glinting red eye is marked by the star Aldebaran, and its long horns are tipped by the stars β (beta) and ζ (zeta) Tauri. As well as the Hyades, Taurus contains another celebrated star cluster, the Pleiades or Seven Sisters. Taurus was the site of a supernova seen from Earth in AD 1054 that gave rise to the Crab Nebula, M1. At 4h 22.0m, +19° 32′, lies the faint Hind's Variable Nebula, NGC 1554–5, discovered in the 19th century by the English astronomer John Russell Hind; within this nebula lies the 10th-mag. star T Tauri, 576 l.y. away, prototype of a class of irregular variables that are stars still in the process of formation.

Each year, the Taurid meteors radiate from south of the Pleiades, reaching a maximum of about 10 per hour around November 4. The Sun passes through the constellation from mid-May to late June, and is in Taurus at the June solstice; precession carried the position of the June solstice into Taurus from Gemini at the end of 1989.

α (alpha) Tauri, 4h 36m +16°.5, (Aldebaran, 'the follower', i.e. of the Pleiades), is an orange giant irregular variable that fluctuates between about mags. 0.75 and 0.95. Although it appears to be part of the Hyades cluster, it is in fact an unrelated foreground star, 65 l.y. away.

β (beta) Tau, 5h 26m +28°.6, (Alnath or Elnath, 'the butting one'), mag. 1.7, is a blue-white giant 131 l.y. away.

ζ (zeta) Tau, 5h 38m +21°.1, 417 l.y. away, is a blue giant that is slightly variable, ranging erratically between about mags. 2.9 and 3.2.

θ¹ θ² (theta¹ theta²) Tau, 4h 29m +15°.9, is a naked-eye or binocular double in the Hyades cluster, consisting of yellow and white giants of mags. 3.8 and 3.4 respectively, distances 158 and 149 l.y. θ² is the brightest member of the Hyades.

κ¹ κ² (kappa¹ kappa²) Tau, 4h 25m +22°.3, are a pair of white stars of mags. 4.2 and 5.3 that form a naked-eye or binocular duo, 153 and 144 l.y. away respectively. Both are outlying members of the Hyades.

λ (lambda) Tau, 4h 01m +12°.5, 370 l.y. away, is an eclipsing binary of the Algol type, varying between mags. 3.4 and 3.9 with a period of 4 days.

σ¹ σ² (sigma¹ sigma²) Tau, 4h 39m +15°.8, is a wide binocular double of blue-white stars in the Hyades, mags. 5.1 and 4.7 and distances 152 and 159 l.y. respectively.

φ (phi) Tau, 4h 20m +27°.4, 342 l.y. away, is an optical double divisible in small telescopes, consisting of a mag. 5.0 orange giant and a white star of mag. 8.4.

χ (chi) Tau, 4h 23m +25°.6, 268 l.y. away, is a double star for small telescopes, with blue and gold components of mags. 5.4 and 7.6.

The Hyades, 4h 27m +16°, is a large and bright open cluster of about 200 stars covering over 5° of sky. The brightest members form a distinctive V-shape, easily visible to the naked eye. In Greek mythology the Hyades were the daughters of Atlas and Aethra, and half-sisters of the Pleiades. Because of its considerable size, the cluster is best studied with binoculars rather than a telescope. The bright star Aldebaran is not a member of the Hyades, but is superimposed on it by chance; the brightest true member is actually θ² (theta²) Tauri (see separate entry above). The centre of the cluster lies 150 l.y. away; the distance of the Hyades is important, for it marks the first step in our distance scale of the Galaxy. ▶

Taurus · Tau · Tauri

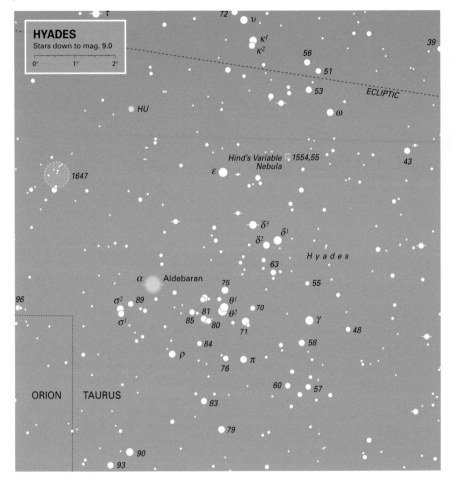

Detail chart of the Hyades cluster. (Wil Tirion)

M1 (NGC 1952), 5h 35m +22°.0, is the Crab Nebula, the remains of a star that exploded as a supernova. It can be glimpsed through binoculars on clear, dark nights. Despite its fame, the Crab Nebula is a disappointing object for small telescopes, appearing as an elliptical 8th-mag. wisp of nebulosity several tines the apparent diameter of the disk of Jupiter – in fact the Crab Nebula can be missed because it is larger than expected. (For a size comparison, see the picture on page 280.) At

the heart of the nebula, beyond the reach of amateur telescopes, is a 16th-mag. object, the remains of the star that exploded. This faint object is now known to be a pulsar that spins 30 times a second, emitting flashes at radio, optical and other wavelengths as it does so. The Crab Nebula and pulsar lie about 6500 l.y. away.

M45, 3h 47m +24°, the Pleiades, is the brightest and most famous star cluster in the sky; it is popularly termed the Seven Sisters, after a group of mythological nymphs, the daughters of Atlas and Pleione. Approximately seven stars are visible to the naked eye, arranged into a mini-dipper shape covering three full-Moon widths of sky; binoculars bring dozens more into view. About 100 stars belong to the cluster, which is centred 378 l.y. away. Unlike the stars of the Hyades, which are older and more evolved, the Pleiades formed within the last 50 million years and include many young blue giants. The brightest member is η (eta) Tauri (Alcyone), mag. 2.9. Other prominent members are 16 Tau (Celaeno), mag. 5.5; 17 Tau (Electra), mag. 3.7; 19 Tau (Taygeta), mag. 4.3; 20 Tau (Maia), mag. 3.9; 21 Tau (Asterope), mag. 5.8; 23 Tau (Merope), mag. 4.1; 27 Tau (Atlas), mag. 3.6; and BU Tau (Pleione), a so-called shell star that throws off rings of gas at irregular intervals, causing it to fluctuate unpredictably between mags. 4.8 and 5.5. The whole of the Pleiades cluster is embedded in a faint nebulosity which reflects the light of the hot blue stars. This nebula is noticeable on long-exposure photographs such as the one on the following page, and under very clear conditions its brightest part, around Merope, may be glimpsed in binoculars or small telescopes. This nebulosity was long thought to be the remains of the cloud from which the cluster formed, but now it seems more likely that it is an entirely separate cloud into which the stars have since drifted by chance.   ▶

Detail chart of the Pleiades cluster and its surrounding nebulosity. (Wil Tirion)

M45, the Pleiades in Taurus, is the most glorious star cluster in the entire sky. Light from the hot, young stars is reflected from surrounding dust, producing a blue nebulosity that is brightest near Merope, bottom centre. The Moon has here been superimposed to the same scale for a size comparison. (Pleiades image Kevin Bays/NOAO/AURA/NSF.)

Telescopium – The Telescope

A constellation invented in the 1750s by the Frenchman Nicolas Louis de Lacaille to honour the most important of astronomical instruments, the telescope. Lacaille had in mind a specific instrument, the great refractor used by J. D. Cassini at Paris Observatory which, in common with other large refractors of the day, had an exceptionally long tube to combat chromatic aberration (false colour) of the image and was suspended from a mast. Lacaille's version of Telescopium originally extended northwards, including stars from Corona Australis, Sagittarius, Scorpius and Ophiuchus, but modern astronomers have cut off the top of the telescope's tube and supporting mast. As with so many of Lacaille's constellations, it is faint and contrived, containing little to interest owners of small telescopes.

α (alpha) Telescopii, 18h 27m −46°.0, mag. 3.5, is a blue-white star 249 l.y. away.

δ¹ δ² (delta¹ delta²) Tel, 18h 32m −45°.9, is a pair of blue-white stars of mags. 4.9 and 5.1, visible separately in binoculars. They are unrelated to each other, being 800 and 1120 l.y. from us respectively.

ε (epsilon) Tel, 18h 11m −46°.0, mag. 4.5, is a yellow giant 409 l.y. away.

ζ (zeta) Tel, 18h 29m −49°.1, mag. 4.1, is a yellow giant 127 l.y. away.

Telescopium · Tel · Telescopii

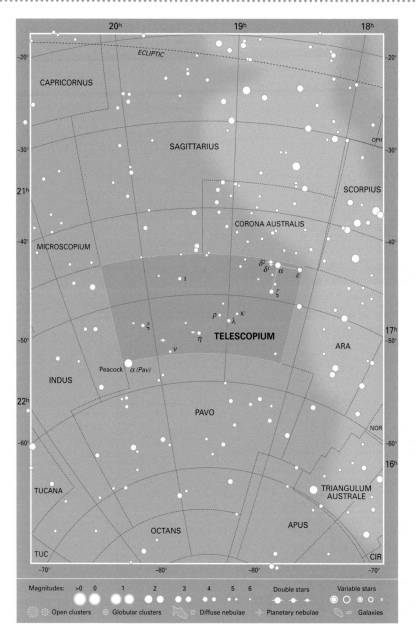

Triangulum – The Triangle

A small but distinctive constellation lying between Andromeda and Aries, consisting of three main stars that form a shape like a thin capital delta; as a result, the Greeks referred to it as Deltoton. It has also been visualized as the Nile river delta and the island of Sicily. Its most important feature is the spiral galaxy M33, the third-largest member of our Local Group of galaxies, after the Andromeda Galaxy and our own Milky Way.

α (alpha) Trianguli, 1h 53m +29°.6, mag. 3.4, is a yellow-white star 64 l.y. away.

β (beta) Tri, 2h 10m +35°.0, mag. 3.0, the constellation's brightest member, is a white star 124 l.y. away.

γ (gamma) Tri, 2h 17m +33°.8, mag. 4.0, is a blue-white star 118 l.y. away.

6 Tri, 2h 12m +30°.3, 305 l.y. away, is a mag. 5.2 yellow giant with a close mag. 6.6 companion visible in a small telescope.

R Tri, 2h 37m +34°.3, is a red giant variable of Mira type, ranging from mag. 5.4 to 12.6 every 9 months or so. It lies about 1300 l.y. away.

M33 (NGC 598), 1h 34m +30°.7, is a spiral galaxy 2.7 million l.y. away in our Local Group, sometimes popularly termed the Pinwheel Galaxy. Presented almost face-on, it covers a larger area of sky than the full Moon. Despite its size and proximity M33 is not prominent visually because its light is spread over such a large area. M33 is best picked up on a dark night in binoculars or a small telescope using low power to enhance the contrast. Unlike most galaxies, it does not have a noticeably stellar nucleus. Quite large amateur telescopes are needed to trace the spiral arms.

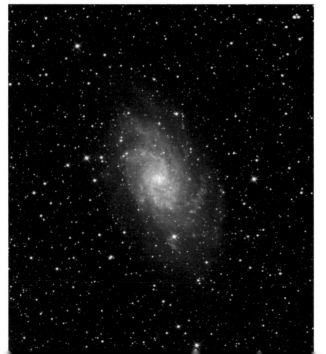

M33, the Triangulum spiral galaxy, is a member of the Local Group and hence one of the closest galaxies to us, lying around 2.7 million light years away.
(Adam Block/NOAO/AURA/NSF)

Triangulum · Tri · Trianguli

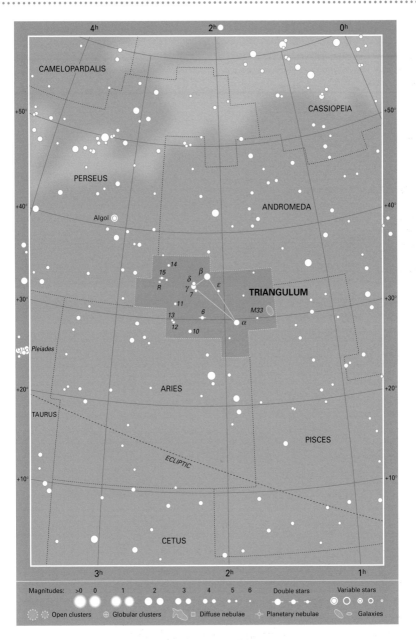

Triangulum Australe – The Southern Triangle

A small but readily distinguishable constellation near α (alpha) Centauri, introduced at the end of the 16th century by the Dutch navigators Pieter Dirkszoon Keyser and Frederick de Houtman. The French astronomer Nicolas Louis de Lacaille visualized it as a surveyor's level with an attached plumbline and placed two of his own related inventions next to it, Circinus (the compasses) and Norma (the ruler and set square). Its three main stars are brighter than those of its northern equivalent, Triangulum, although the constellation itself is smaller.

α (alpha) Trianguli Australis, 16h 49m −69°.0, (Atria), mag. 1.9, is an orange giant star 415 l.y. away.

β (beta) TrA, 15h 55m −63°.4, mag. 2.8, is a white star 40 l.y. away.

γ (gamma) TrA, 15h 19m −68°.7, mag. 2.9, is a blue-white star 183 l.y. away.

NGC 6025, 16h 04m −60°.5, is a binocular cluster, elongated in shape, containing about 60 stars of 7th mag. and fainter, 2500 l.y. away.

Johann Bayer (1572–1625)

Johann Bayer was a German lawyer in Augsburg and an amateur astronomer who, in 1603, published the first star atlas that covered the entire sky, *Uranometria*. He based his maps of the northern heavens mainly on the observations made by the great Danish astronomer Tycho Brahe, while his information on the southern skies came from the work of the Dutch navigator Pieter Dirkszoon Keyser.

In addition to the 48 constellations known since ancient times, Bayer showed the 12 new constellations around the southern celestial pole invented by Keyser and his countryman Frederick de Houtman: Apus, Chamaeleon, Dorado, Grus, Hydrus, Indus, Musca, Pavo, Phoenix, Triangulum Australe, Tucana and Volans. Bayer's most important legacy to astronomers was his system of identifying stars by Greek letters. In each constellation the brightest stars were assigned letters of the Greek alphabet, usually, but not always, in approximate order of brightness (Gemini, Orion and Sagittarius are notable examples of constellations in which the star labelled α (alpha) is not the brightest).

Before Bayer's time, stars without proper names were identified by the cumbersome descriptions of the Greek astronomer Ptolemy such as 'in the left forearm of the advance twin', a reference to the 4th-magnitude star we now term θ (theta) Geminorum. Clearly, to make sense of such descriptions astronomers had to know their constellation figures well, and even then there was considerable room for confusion. The system of so-called Bayer letters was a vast improvement, and remains in use today.

Triangulum Australe · TrA · Trianguli Australis

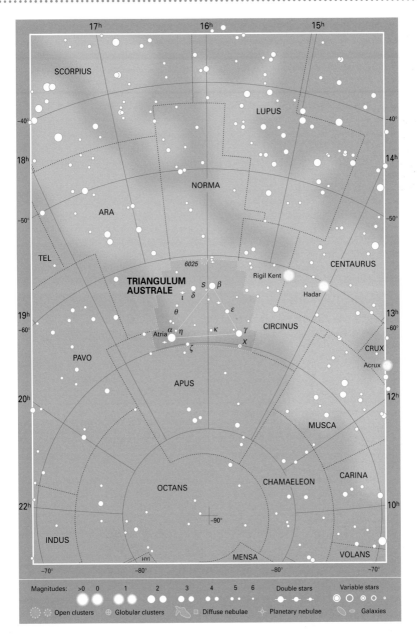

Tucana – The Toucan

A constellation near the south pole of the sky, introduced in the late 16th century by the Dutch navigators Pieter Dirkszoon Keyser and Frederick de Houtman. It represents a toucan, the South American bird with the large beak. Its most notable features are the Small Magellanic Cloud, actually a neighbouring mini-galaxy, and the globular cluster known as 47 Tucanae.

α (alpha) Tucanae, 22h 19m −60°.3, mag. 2.9, is an orange giant star 199 l.y. away.

β (beta) Tuc, 0h 32m −63°.0, is a complex multiple star. Binoculars or small telescopes show that it consists of two almost identical blue-white stars, $β^1$ and $β^2$, mags. 4.4 and 4.5; $β^2$ is itself a close binary with a period of 44 years, requiring an aperture larger than 200 mm to divide. Nearby lies a mag. 5.1 white star which is unlettered, although on some older charts and catalogues it was labelled $β^3$. All three stars share the same proper motion through space, but their distances of 140, 172 and 152 l.y. suggest they are not truly connected.

γ (gamma) Tuc, 23h 17m −58°.2, mag. 4.0, is a white star 72 l.y. away.

δ (delta) Tuc, 22h 27m −65°.0, mag. 4.5, is a blue-white star 267 l.y. away with a 9th-mag. companion visible in small telescopes.

κ (kappa) Tuc, 1h 16m −68°.9, 67 l.y. away, is a double star consisting of components of mags. 5.1 and 7.3 visible in small telescopes. This pair moves through space with a wide mag. 7.2 star, itself a close binary with an orbital period of 85 years, just divisible in 150 mm aperture.

47 Tuc (NGC 104), 0h 24m −72°.1, is a prominent globular cluster the same apparent size as the full Moon, visible to the naked eye as a fuzzy 4th-mag. star; on early charts, it was actually catalogued as a star and given a stellar designation. Among globular clusters it is second only to

ω (omega) Centauri in size and brightness. Telescopes of 100 mm aperture begin to resolve 47 Tuc, and even binoculars show the brilliant blaze of its star-packed core. Its diameter is about 200 light years and it is among the closest globulars to us, 16,000 l.y. away.

NGC 362, 1h 03m −70°.8, is a 6th-mag. globular cluster visible in binoculars near the northern edge of the Small Magellanic Cloud, but not associated with it. NGC 362 actually lies 29,000 l.y. away, in our own Galaxy.

Small Magellanic Cloud (SMC), 0h 53m −73°, is a satellite galaxy of the Milky Way, as is its larger sibling, the Large Magellanic Cloud in Dorado. The Small Magellanic Cloud appears to the naked eye as a nebulous, tadpole-shaped patch 3½° across. Binoculars and small telescopes resolve star clusters and glowing gas clouds within it, although they are smaller and less impressive than those in the Large Magellanic Cloud. It lies about 200,000 l.y. away but is elongated along our line of sight. (See the photograph on page 164.)

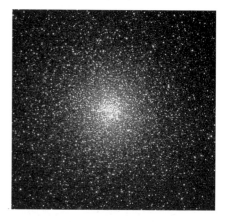

47 Tucanae (NGC 104), one of the nearest and brightest globular clusters, is an excellent object for small instruments. (ESO)

Tucana · Tuc · Tucanae

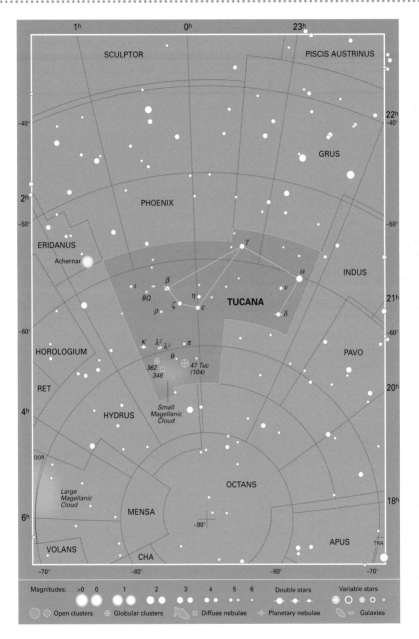

Ursa Major – The Great Bear

The third-largest constellation in the sky. Its central feature is the seven stars that make up the familiar shape variously called the Plough or the Big Dipper, the best-known of all star patterns, although why so many people, including the North American Indians, visualized this group as a bear remains a mystery. In Europe the pattern was seen as a wagon or chariot. Others, notably the Arabs, viewed the dipper shape not as a bear, but as a bier or coffin. In Greek mythology the bear represented Callisto, who was turned into a bear in punishment for her illicit love affair with Zeus.

The two end stars in the dipper's bowl, Merak and Dubhe, are known as the Pointers, since they indicate the direction of Polaris, the north pole star in neighbouring Ursa Minor. The curving handle of the Big Dipper points to the bright star Arcturus in Boötes. At 11h 03.3m, +35° 58', lies the mag. 7.5 red dwarf Lalande 21185, which is the Sun's fourth-closest stellar neighbour, 8.3 l.y. away. Its designation comes from its number in a catalogue drawn up by the 18th-century French astronomer Joseph Lalande.

With the exception of Alkaid and Dubhe, the stars of the Big Dipper are travelling together through space, along with a number of other stars in the region; together, these stars make up the so-called Ursa Major moving cluster. Ursa Major contains numerous galaxies, but only a few of them are easily visible in amateur telescopes.

α (alpha) Ursae Majoris, 11h 04m +61°.8, (Dubhe, 'the bear'), mag. 1.8, is a yellow-orange giant 124 l.y. away. It has a close mag. 4.8 companion that orbits it in 44 years. Usually the pair are too close to split in amateur telescopes but could be divisible with 250 mm when at their widest, in 2023–2026. ▶

Mizar and Alcor are a well-known naked-eye or binocular double in the handle of the Plough, but form an attractive multiple star when seen through a small telescope. (Wil Tirion)

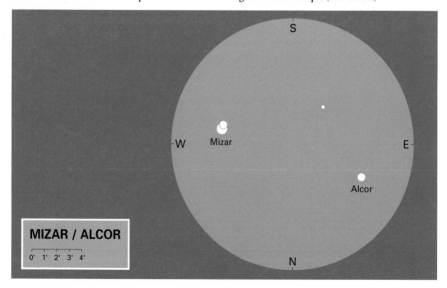

Ursa Major · UMa · Ursae Majoris

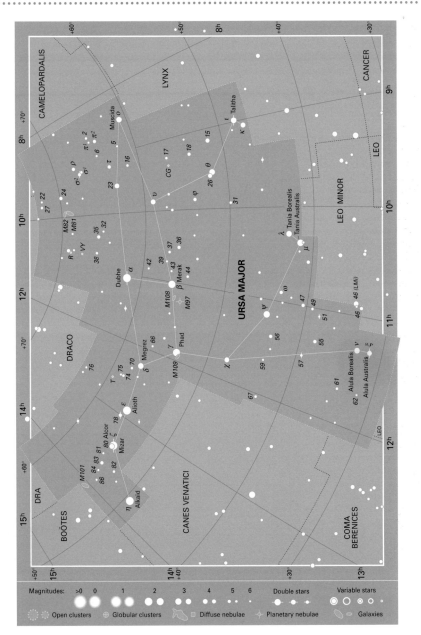

Magnitudes: >0 0 1 2 3 4 5 6   Double stars   Variable stars

Open clusters   Globular clusters   Diffuse nebulae   Planetary nebulae   Galaxies

β (beta) UMa, 11h 02m +56°.4, (Merak, 'flank'), mag. 2.3, is a blue-white star 79 l.y. away.

γ (gamma) UMa, 11h 54m +53°.7, (Phad or Phecda, 'thigh'), mag. 2.4, is a blue-white star 84 l.y. away.

δ (delta) UMa, 12h 15m +57°.0, (Megrez, 'root of the tail'), mag. 3.3, is a blue-white star 81 l.y. away.

ε (epsilon) UMa, 12h 54m +56°.0, (Alioth), mag. 1.8, is a blue-white star with a peculiar spectrum, 81 l.y. away.

ζ (zeta) UMa, 13h 24m +54°.9, (Mizar), mag. 2.2, is a celebrated multiple star. Good eyesight, or binoculars, reveals a mag. 4.0 companion, Alcor. Mizar is 78 l.y. from Earth and Alcor 81 l.y. away, too far apart to be a genuine binary. However, a small telescope reveals that Mizar has another mag. 4.0 companion closer to it, which definitely is related (see the eyepiece view on page 250). This companion was first seen by the Italian astronomer Giovanni Riccioli in 1650, making Mizar the first double star to be discovered telescopically. Mizar

itself was also the first star discovered to be a spectroscopic binary, by the American astronomer Edward C. Pickering in 1889. The companion of Mizar is another spectroscopic binary, as is Alcor, making this a highly complex group.

η (eta) UMa, 13h 48m +49°.3, (Alkaid or Benetnasch, both from the Arabic for 'leader of the mourners'), mag. 1.9, is a blue-white main-sequence star 101 l.y. away.

ξ (xi) UMa, 11h 18m +31°.5, 26 l.y. away, was the first binary star to have its orbit computed. Its two yellow components (both also spectroscopic binaries), of mags. 4.3 and 4.8, orbit each other with a period of 60 years. They are currently divisible in a 75-mm telescope and after 2015 will come within range of the smallest apertures, continuing to open out until 2035. For a diagram of the orbit, see page 282.

M81 (NGC 3031), 9h 56m +69°.1, is a beautiful 7th-mag. spiral galaxy, one of the brightest in the sky, visible in binoculars. A small telescope shows it as a roundish, softly glowing patch noticeably

M97, the Owl Nebula, gets its popular name from the two dark patches, like an owl's eyes, either side of the 16th-magnitude central star. This star, a white dwarf, originally ejected the gas that forms the nebula and the star's energy now makes the nebula glow. An aperture of at least 100 mm is needed to see the "eyes". Visually, the nebula appears grey rather than green.
(NOAO/AURA/NSF)

brighter towards the centre. Being tilted at an angle to us it appears somewhat elliptical in outline, covering over half a Moon's width at its longest. In the same telescopic field of view ½° to the north is M82 (see separate entry below); the two galaxies are about 12 million l.y. from us. For a photograph of M81 see page 288.

M82 (NGC 3034), 9h 56m +69°.7, is a neighbour galaxy of M81, a quarter as bright and less than half its size but still visible in binoculars. In a small telescope it appears as an elongated blur and can actually seem more prominent than M81 because of its higher surface brightness. Detailed studies by professional astronomers have shown that M82 is an edge-on spiral galaxy mottled by dust clouds and experiencing a burst of star formation as a result of a recent interaction with M81.

M97 (NGC 3587), 11h 15m +55°.0, is an elusive 11th-mag. planetary nebula known as the Owl Nebula because of two dark patches like eyes that give it the appearance of an owl's face when seen

M101, a face-on spiral galaxy, displays an asymmetric shape in photographs. In amateur instruments only the core and traces of the spiral arms will be seen. (George Jacoby, Bruce Bohannan, Mark Hanna/NOAO/AURA/NSF)

through a large telescope. In moderate apertures it appears as only a pale disk over three times the size of Jupiter, and will probably need an aperture of at least 75 mm to be seen at all. The Owl lies about 1300 l.y. away.

M101 (NGC 5457), 14h 03m +54°.3, is a spiral galaxy visible in binoculars as a pale, rounded smudge about half the apparent size of the full Moon; because of its large size it is less prominent than its quoted brightness of 8th mag. would suggest. Long-exposure photographs show it as a face-on galaxy with widely spread spiral arms, but these are not apparent in small telescopes which tend to show only the elliptical central region. M101 lies 22 million l.y. away.

Ursa Minor – The Little Bear

A constellation said to have been introduced in about 600 BC by the Greek astronomer Thales. Ursa Minor currently contains the north celestial pole, which lies within 1° of the conveniently placed 2nd-magnitude star popularly known as Polaris, α (alpha) Ursae Minoris. Precession will carry the pole closest to Polaris around AD 2100, when the two will be separated by just under ½°, after which it will start to move away again towards Cepheus, crossing the border in AD 2234 (see chart below).

Ursa Minor is also termed the Little Dipper because its seven brightest stars outline a shape like a smaller version of the Big Dipper in Ursa Major. The stars β (beta) and γ (gamma) Ursae Minoris (Kochab and Pherkad) in the Little Dipper's bowl are called the Guardians of the Pole.

α (alpha) Ursae Minoris, 2h 32m +89°.3, (Polaris), mag. 2.0, is a yellow-white supergiant 431 l.y. away. It is a Cepheid variable – in fact, the nearest Cepheid to us. Its pulsations diminished during the 20th century, stabilizing by the 1990s at a few hundredths of a magnitude. Polaris is also a double star, with a mag. 8.2 companion visible in a small telescope. Binoculars and small telescopes show a ring of 8th- to 11th-mag. stars some ¾° across, on which Polaris appears to be strung like some lustrous pearl on a necklace.

β (beta) UMi, 14h 51m +74°.1, (Kochab), mag. 2.1, is an orange giant 126 l.y. away.

γ (gamma) UMi, 15h 21m +71°.8, (Pherkad), mag. 3.0, is a blue-white giant star 480 l.y. away. The mag. 5.0 orange giant 11 UMi that appears near it, as seen by the naked eye or in binoculars, is unrelated, lying 390 l.y. away.

ε (epsilon) UMi, 16h 46m +82°.0, mag. 4.2, 347 l.y. away, is a yellow giant eclipsing binary that varies in a period of 39.5 days by under 0.1 mag., not discernible to the naked eye.

η (eta) UMi, 16h 18m +75°.8, mag. 5.0, is a white main-sequence star 97 l.y. away. A wide mag. 5.5 companion, 19 UMi, is an unrelated background star.

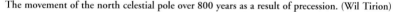

The movement of the north celestial pole over 800 years as a result of precession. (Wil Tirion)

Ursa Minor · UMi · Ursae Minoris

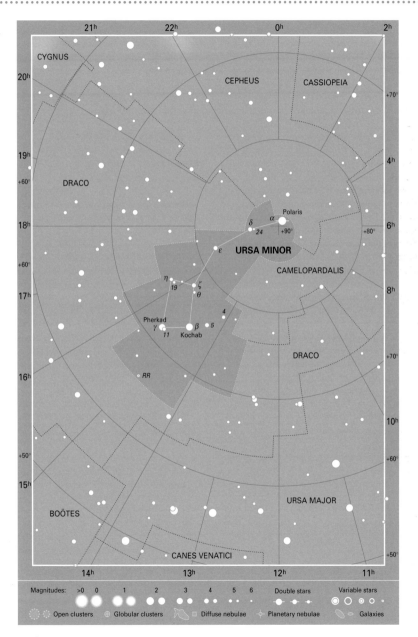

Vela – The Sails

Formerly part of the ancient constellation Argo Navis, the ship of Jason and the Argonauts, Vela was made into a separate constellation (along with Carina and Puppis) by the French astronomer Nicolas Louis de Lacaille in 1763. Vela represents the ship's sails while Carina is the keel and Puppis the stern. When Argo Navis was subdivided into these three parts, its stars were not given new Greek letters; as a result, the labelling of the stars in Vela begins with γ (gamma). In fact, Argo Navis was so large that astronomers ran out of Greek letters to label its stars and turned to Roman letters; many of these Roman-lettered stars have ended up in Vela.

The stars κ (kappa) and δ (delta) Velorum, in conjunction with ι (iota) and ε (epsilon) Carinae, form a shape known as the False Cross, sometimes mistaken for the real Southern Cross. Vela lies in a part of the Milky Way rich in faint nebulosity visible on long-exposure photographs. This nebulosity, named the Gum Nebula after the Australian astronomer Colin S. Gum who drew attention to it in 1952, is believed to be the remains of one or more supernovae that occurred long ago. Another remnant of a supernova in the constellation is the Vela Pulsar, which flashes 11 times per second, one of the few pulsars that can be seen flashing optically as well as at radio wavelengths.

γ (gamma) Velorum, 8h 10m −47°.3, is an interesting multiple star. Binoculars or small telescopes divide it into two blue-white components of mags. 1.8 and 4.3. The brighter of these is the brightest known example of a Wolf–Rayet type, a rare class of stars with very hot surfaces which seem to be ejecting gas. It is 840 l.y. away, while the fainter component is estimated to lie 1600 l.y. away and hence is unrelated. There are also two wider companions of 8th and 9th mags., distances unknown.

δ (delta) Vel, 8h 45m −54°.7, mag. 1.9, is a blue-white main-sequence star 80 l.y. away with a mag.

5.1 companion requiring apertures of 100 mm to be detected. The brighter star is an eclipsing binary which dips to mag. 2.3 for a few hours every 45 days.

κ (kappa) Vel, 9h 22m −55°.0, mag. 2.5, is a blue-white star 539 l.y. away.

λ (lambda) Vel, 9h 08m −43°.4, mag. 2.2, is an orange supergiant, irregularly variable (by less than 0.2 mag.), 573 l.y. away.

H Vel, 8h 56m −52°.7, 376 l.y. away, is a neat double of mags. 4.8 and 7.4, difficult in the smallest telescopes because of the magnitude contrast.

NGC 2547, 8h 11m −49°.3, 1400 l.y. away, is an open cluster of about 80 stars of mag. 6.5 and fainter, just visible to the naked eye and best seen in binoculars.

NGC 3132, 10h 08m −40°.5, is a relatively large and bright planetary nebula of 8th mag., known as the Eight-Burst Nebula. Through a small telescope it appears larger than Jupiter, with a central star of 10th mag. It lies 2600 l.y. away.

NGC 3228, 10h 22m −51°.7, is an open cluster of about 15 faint stars for binoculars and small telescopes, 1600 l.y. away.

IC 2391, 8h 40m −53°.1, is a large naked-eye cluster of 50 stars, 500 l.y. away, scattered around the mag. 3.6 blue-white star o (omicron) Vel, a small-amplitude variable of β Cephei type. About 1° from it is the binocular cluster NGC 2669.

IC 2395, 8h 41m −48°.2, is a binocular cluster of 40 stars, 3100 l.y. away. A mag. 5.5 star which seems to be the brightest member is probably a foreground star. Also visible ½° to the south is the 8th-mag. open cluster NGC 2670.

Vela · Vel · Velorum

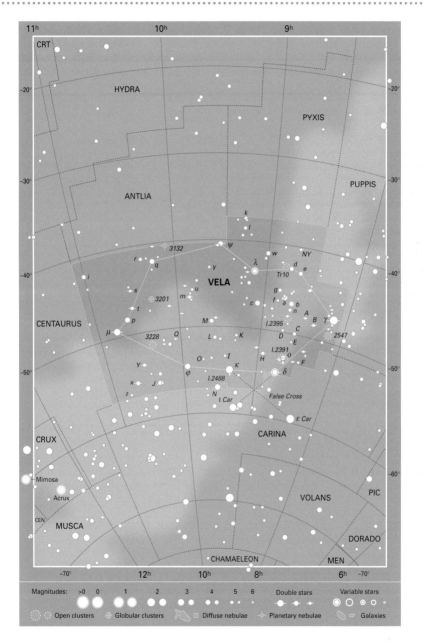

Virgo – The Virgin

The second-largest constellation in the sky (only Hydra is bigger), and the largest in the zodiac. Virgo is usually identified as Dike, the goddess of justice, the scales of justice being represented by neighbouring Libra. But another legend sees her as Demeter, the corn goddess, and in the sky she is pictured holding an ear of wheat (the star Spica). The Sun passes through the constellation from mid-September to early November, and thus is within Virgo's boundaries at the time of the September equinox each year, when the Sun moves south of the celestial equator.

Virgo contains the nearest major cluster of galaxies to us, which spills over into neighbouring Coma Berenices; the area is sometimes known as 'the realm of the galaxies'. The Virgo Cluster lies 55 million l.y. away and contains about 3000 members, several dozen of which are visible in apertures of 150 mm or so, although they appear as little more than hazy patches; some of the brightest members are mentioned below. Another famous feature in Virgo, unrelated to the Virgo Cluster of galaxies, is the optically brightest quasar, 3C 273, located at 12h 29.1m, +2° 03′. It appears as a 13th-mag. blue star and is estimated to lie about 3000 million l.y. away.

α (alpha) Virginis, 13h 25m −11°.2, (Spica, 'ear of wheat'), mag. 1.0, is a blue-white main-sequence star 262 l.y. away. It is a spectroscopic binary, tidally distorted by its companion so that it varies slightly with a period of 4 days as it rotates, although only by 0.1 mag.

β (beta) Vir, 11h 51m +1°.8, (Zavijava), mag. 3.6, is a yellow-white main-sequence star 36 l.y. away.

γ (gamma) Vir, 12h 42m −1°.4, (Porrima), 39 l.y. away, is a celebrated double star. Together, the stars shine as mag. 2.7. But a small telescope reveals γ Vir to consist of a matching pair of white stars, each of mag. 3.5. They orbit each other every 169

years and were last closest in 2005, when a 250-mm telescope was needed to split them. Now moving apart rapidly, they become divisible in 100 mm by 2010 and in 60 mm after 2012, remaining within range of small telescopes for the rest of the 21st century. See diagram page 282.

δ (delta) Vir, 12h 56m +3°.4, mag. 3.4, is a red giant 202 l.y. away.

ε (epsilon) Vir, 13h 02m +11°.0, (Vindemiatrix, 'grape gatherer'), mag. 2.8, is a yellow giant 102 l.y. away.

θ (theta) Vir, 13h 10m −5°.5, mag. 4.4, is a blue-white star 415 l.y. away with a 9th-mag. companion visible in a small telescope.

τ (tau) Vir, 14h 02m +1°.5, mag. 4.3, a blue-white star 218 l.y. away, forms a wide optical double for small telescopes with a 9th-mag. companion.

φ (phi) Vir, 14h 28m −2°.2, 135 l.y. away, is a yellow giant of mag. 4.8 with a 9th-mag. companion, difficult to see in the smallest telescopes because of the brightness contrast.

M49 (NGC 4472), 12h 30m +8°.0, is an 8th-mag. elliptical galaxy visible as a rounded glow in a 75-mm telescope under low power. It is one of the largest and brightest members of the Virgo Cluster of galaxies.

M58 (NGC 4579), 12h 38m +11°.8, is a 10th-mag. barred spiral galaxy with a noticeably brighter core.

M59 (NGC 4621), 12h 42m +11°.6, is a 10th-mag. elliptical galaxy with a starlike nucleus, lying about a quarter of the way from M60 to M58.

M60 (NGC 4649), 12h 44m +11°.5, is a 9th-mag. elliptical galaxy, one of the most prominent members of the Virgo Cluster, detectable in 75 mm aperture.

▶

Virgo · Vir · Virginis

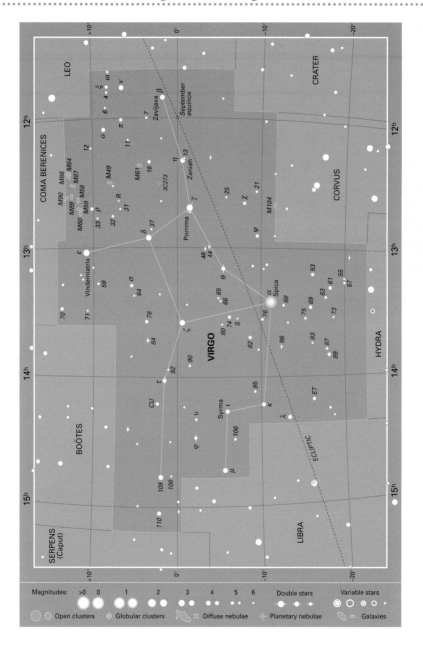

M84 (NGC 4374), 12h 25m +12°.9, and M86 (NGC 4406), 12h 26m +12°.9, are a pair of 9th-mag. elliptical galaxies appearing in the same telescopic field as fuzzy patches with noticeably brighter cores. M86 is slightly the larger of the two and noticeably elongated, whereas M84 appears round.

M87 (NGC 4486), 12h 31m +12°.4, is a celebrated giant elliptical galaxy. It is also a strong radio source, known as Virgo A, and an X-ray source. Photographs taken through large telescopes show a jet of matter emerging from M87 (see right for an example). In amateur telescopes, M87 appears as a rounded, 9th-mag. glow with a noticeable nucleus.

M90 (NGC 4569), 12h 37m +13°.2, is a large 9th-mag. spiral galaxy, tilted at an angle to us so that it appears elongated.

M104 (NGC 4594), 12h 40m −11°.6, is an 8th-mag. spiral galaxy seen edge-on so that it appears elongated. It is popularly known as the Sombrero because its appearance on long-exposure photographs has been likened to that of a Mexican hat

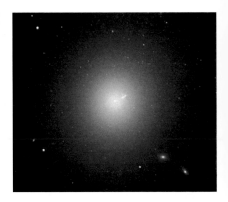

M87, a giant elliptical galaxy in the Virgo cluster, has a jet of gas extending from its central core. (Adam Block/NOAO/AURA/NSF)

(see page 130). However, with its bulging nucleus ringed by tightly coiled spiral arms, its appearance is more reminiscent of Saturn. Apertures of 150 mm reveal a dark lane of dust along its rim, silhouetted against the brighter arms and nucleus. The Sombrero is not in the Virgo Cluster but somewhat closer, about 30 million l.y. away.

Volans – The Flying Fish

This figure was invented at the end of the 16th century by the Dutch navigators Pieter Dirkszoon Keyser and Frederick de Houtman. It represents the type of fish found in tropical waters that can leap out of the water and glide through the air on wings. None of its stars is particularly bright, but there are two fine double stars for small telescopes.

α (alpha) Volantis, 9h 02m −66°.4, mag. 4.0, is a blue-white star 124 l.y. away.

β (beta) Vol, 8h 26m −66°.1, mag. 3.8, is an orange giant 108 l.y. away.

γ (gamma) Vol, 7h 09m −70°.5, 142 l.y. away, is a pair of gold and cream stars of mags. 3.8 and 5.7, beautiful in a small telescope.

δ (delta) Vol, 7h 17m −68°.0, mag. 4.0, is a yellow-white giant 660 l.y. away.

ε (epsilon) Vol, 8h 08m −68°.6, 642 l.y. away, is a mag. 4.4 blue-white star with an 8th-mag. companion visible in a small telescope.

Volans · Vol · Volantis

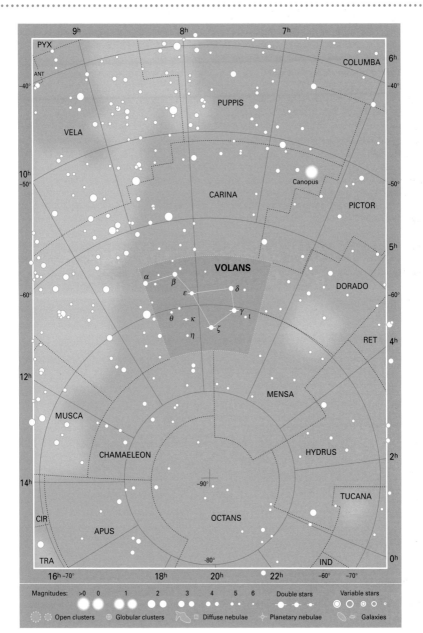

Vulpecula – The Fox

A faint constellation at the head of Cygnus. It was originated in 1687 by the Polish astronomer Johannes Hevelius, who called it Vulpecula cum Anser, the Fox and Goose. Since then the goose has fled, leaving the fox on its own. The first pulsar, or flashing radio source, was discovered in Vulpecula in 1967 by radio astronomers at Cambridge, England. It lies about 1½° north of the notable grouping called Brocchi's Cluster.

α (alpha) Vulpeculae, 19h 29m +24°.7, mag. 4.4, is a red giant 297 l.y. away. An unrelated mag. 5.8 orange giant, 8 Vul, 484 l.y. away, is seen nearby in binoculars.

T Vul, 20h 51m +28°.3, is a yellow-white supergiant Cepheid variable that ranges between mags. 5.4 and 6.1 every 4.4 days. It lies about 1700 l.y. away.

M27 (NGC 6853), 20h 00m +22°.7, the Dumbbell Nebula, is a large and bright planetary nebula, reputedly most conspicuous of its kind, visible in binoculars as an elliptical misty glow. Its dumbbell shape is better seen through telescopes, which also reveal its greenish tinge. M27 is of 8th mag. and spans about one-quarter the apparent diameter of the full Moon. M27 is about 1400 l.y. away. (For a photograph see page 220.)

Brocchi's Cluster (Collinder 399), 19h 25m +20°.2, is a striking binocular group of stars on the border with Sagitta, popularly known as the Coathanger because of its distinctive shape. An almost straight line of six stars extends for three Moon diameters; from the centre of this line emerges a curve of four more stars, forming the Coathanger's hook. The brightest member of the group, in the hook, is 4 Vul, mag. 5.1. Distances of the individual stars range from just over 200 l.y. to more than 1000 l.y., and all have different motions through space, so this is not a true cluster but a chance alignment.

The stars of Brocchi's Cluster form the shape of a coathanger. (Wil Tirion)

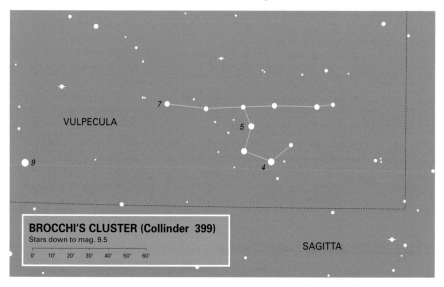

VULPECULA

BROCCHI'S CLUSTER (Collinder 399)
Stars down to mag. 9.5

0'   10'   20'   30'   40'   50'   60'

SAGITTA

Vulpecula · Vul · Vulpeculae

SECTION II

The Pleiades star cluster in Taurus hovers above a winter landscape in this composite image. (Robin Scagell)

Stars and nebulae

Stars are balls of gas made incandescent by energy from nuclear reactions deep in their interiors. They come in a wide range of sizes and brightnesses, from faint dwarfs a hundredth of the Sun's diameter to dazzling supergiants hundreds of times larger than the Sun. They range in temperature from intensely hot blue-white specimens with surface temperatures of more than 20,000°C to cool red stars with surfaces at around 3000°C. The Sun, which is a medium-temperature yellow star (surface temperature 5800°C) turns out to be pretty average in all respects.

Stars are born from massive clouds of gas and dust within our Galaxy. An interstellar gas cloud is termed a *nebula* (plural: *nebulae*), from the Latin for 'cloud'. A nebula is not uniformly distributed in space, but contains denser knots – the seeds of future stars. If the knot is dense enough it begins to contract under the inward pull of its own gravity.

As it gets smaller and denser it heats up, until the temperature and pressure at the centre of the shrinking blob become so great that nuclear reactions begin. The gas blob has switched on to become a true star, generating its own heat and light for millions or, more usually, billions of years.

Several star-spawning clouds are well within reach of observation by amateurs. Most famous is the Orion Nebula, marking the sword in the constellation of Orion the Hunter. This nebula is visible as a hazy green glow to the naked eye; binoculars show it more clearly, while on photographs it appears complex and colourful (see next page).

Embedded within the Orion Nebula is a star called θ¹ (theta¹) Orionis, recently born from the surrounding gas. Small telescopes divide this star into four components which form a group known as the Trapezium. Energy emitted by the brightest of these four stars makes the surrounding nebula shine.

Behind the bright, visible part of the cloud is an even larger, still-dark area where stars are being born at this moment. The Orion Nebula is estimated to contain enough matter to produce hundreds of stars: it is a star cluster in the making.

The Orion Nebula is by no means unique in our own Galaxy, and there are plenty of similar nebulae in other galaxies, too. One such example is the Tarantula Nebula in the southern constellation Dorado, which dwarfs the Orion Nebula. The Tarantula is part of a small nearby galaxy called the Large Magellanic Cloud.

Star clusters and associations

One celebrated group of stars that has come into being recently is the Pleiades cluster, popularly known as the Seven Sisters, in the constellation Taurus, the Bull. At least five members of the Pleiades can be distinguished by normal eyesight; binoculars and small telescopes bring dozens more members into view. The whole cluster is estimated to contain about a hundred stars. The brightest and youngest of these formed no more than 2 million years ago, making them extremely youthful by astronomical standards.

The Pleiades is an example of a type of cluster referred to as an *open cluster* or *galactic cluster*. About a thousand such clusters are known to astronomers, and the most prominent of them are listed in this book. Not far from the Pleiades in Taurus is a larger and older open cluster, the Hyades, which is estimated to be about 500 million years old. Being older than the Pleiades, the stars have had more time to drift apart. Eventually, most open clusters disperse completely. The Sun was probably a member of such a cluster when it was born 4600 million years ago.

Much larger than open clusters are *stellar associations*, vast scatterings of young stars hundreds of light years across. It is no coinci-

dence that most of the bright stars in Orion lie at similar distances from us, for they are members of such an association, centred on the Orion Nebula about 1500 light years away. Three times closer to us is the extensive Scorpius–Centaurus association, which extends across 60° or more of sky from Scorpius via Lupus into Centaurus and Crux. Antares is the brightest member; other prominent members are β (beta) Centauri, α (alpha) and β (beta) Crucis, and the open cluster IC 2602 in Carina. Associations arise from particularly large clouds of gas and dust in the spiral arms of our Galaxy.

An altogether different type of cluster, containing much older stars, is a *globular cluster*; these are described on page 291.

Star sizes and lifetimes

Nebulae are composed of a 10:1 mixture of hydrogen and helium, the primary constituents of the Universe, so, naturally enough, stars have this same composition. Stars get their energy from nuclear reactions which transform hydrogen into helium. In the reactions, four hydrogen atoms are crushed together to make one atom of helium; an uncontrolled version of the same reaction occurs in a hydrogen bomb.

There are certain limits on the size of a star. A gas blob with less than about 8 per

Ghostly, glowing swirls of the Orion Nebula, M42, a cradle of star formation. At its heart is the multiple star θ¹ (theta¹) Orionis, also known as the Trapezium (see inset); this lies near the tip of the dark intrusion known as the Fish's Mouth. North of M42 is the smaller, rounded knot of M43, actually part of the same great gas cloud. Farther north is the fainter smudge of NGC 1977 and, at top, the scattered cluster NGC 1981. South of M42 is ι (iota) Orionis and the wide double Struve 747. (Ryan Steinberg and Family/Adam Block/NOAO/AURA/NSF. Inset: Jim Rada/Adam Block/NOAO/AURA/NSF)

cent of the Sun's mass (equivalent to about 80 times the mass of Jupiter, the largest planet in the Solar System) cannot become a true star, because conditions in its interior will never reach the extremes necessary for nuclear reactions to begin.

Below this threshold, the gas blob would become a *brown dwarf*, a not-quite star that shines feebly from energy released during its contraction. Smaller still, below about 1 per cent the Sun's mass, the gas blob would be regarded as a planet. In other words, if gaseous Jupiter had possessed at least 10 times its current mass it would have been not a planet but a brown dwarf, and with a mass over 80 times its current value it would have been a small star, in which case our Sun would have had a faint companion.

At the other end of the scale, the largest stars have masses of about a hundred times that of the Sun. It was once thought that stars more massive than this would produce so much energy that they would literally disintegrate, but this may not be true in all cases. A few stars are known that seem to have masses greater than a hundred Suns, one example being η (eta) Carinae in the southern constellation Carina; this star has varied erratically in brightness in the past, as though on the verge of instability.

A star's most vital statistic is its mass, for this factor affects everything else about it: its temperature, its brightness and its lifetime. The stars with the least mass are, not surprisingly, the coolest; they are known as *red dwarfs*. A typical red dwarf such as Barnard's Star, the second-closest star to the Sun, has a mass about a tenth that of the Sun and glows a dull red with a surface temperature of about 3000°C. Even though Barnard's Star is only six light years away, it is too faint to be seen with the naked eye.

Surprisingly enough, stars with the lowest mass live the longest. Their nuclear fires burn so slowly that they can survive for as much as a million million years, a hundred times as long as the Sun. The Sun itself, which by definition is of one solar mass, has

The Horsehead Nebula in Orion, a dark cloud of cooler gas and dust seen in silhouette against a bright background of glowing hydrogen, looks like an interstellar chess piece. (Nigel Sharp/ NOAO/AURA/NSF)

a surface temperature of 5800°C, and is expected to live for about 10,000 million years. It is currently in the prime of its life.

Moving up the weight scale, a star such as Sirius, which is twice the Sun's mass, can live for only about 1000 million years, a tenth of the Sun's age. The surface temperature of Sirius is a blue-white 11,000°C. Larger and hotter still, the star Spica in the constellation Virgo has a mass of about 11 Suns and a surface temperature of around 24,000°C. The lifetime of this intensely hot, highly luminous star is less than 1 per cent of the lifetime of the Sun.

Star temperatures and colours

A star's colour is a direct indicator of its temperature. The most precise way to measure a star's temperature is to study the spectrum of its light, which is done by splitting up the light in a device called a spectroscope.

Stars are classified into a sequence of so-called *spectral types* according to their temperature, as shown in the table on page 273. The bluest and hottest stars are categorized as spectral types O and B. Stars of type O are rare and none is close enough to us to appear really bright. However, B-type stars are more common; prominent examples are α (alpha) and β (beta) Crucis, β (beta) Centauri and Spica; these are in fact the bluest of the first-magnitude stars.

Then come the cooler blue-white A-type stars, which include Sirius, followed by stars of F type, which appear white or yellow-white; Procyon is an example. G-type stars are yellow; they include the Sun, α (alpha) Centauri and τ (tau) Ceti. Cooler still are K-type stars, such as ε (epsilon) Eridani, which have orange hues. Coolest of all are stars with an M-type spectrum, examples being Antares and Betelgeuse which are the reddest first-magnitude stars. Red dwarfs also have M-type spectra but none of them are bright enough to be visible to the naked eye.

Each spectral type is subdivided into ten steps from 0 to 9; on this more precise scale, the Sun ranks as a G2 star. The seemingly haphazard lettering sequence used for spectral types is the result of a previous classification scheme which was rearranged and shortened to produce the present system. The sequence of stellar spectral types can be remembered by the mnemonic: 'Oh Be A Fine Girl, Kiss Me'.

Star colours are necessarily subjective, given the individual characteristics of different people's eyes and the varying conditions under which stars are viewed. For example, astronomers define Vega, of spectral type A0, as pure white, yet to most eyes it has a distinctly bluish cast, as do the components of Castor, also of type A. At the other end of

Columns of cooler hydrogen gas and dust up to one light year long protrude into the Eagle Nebula, M16, in Serpens. Ultraviolet light flooding from hot, young stars off the top of the picture evaporates gas from the surfaces of the columns, leaving behind denser knots where new stars are forming. In this image from the Hubble Space Telescope, hydrogen gas is depicted as green rather than the usual pink. (Jeff Hester and Paul Scowen, Arizona State University/NASA)

the scale, few so-called red giant or super-giant stars appear truly red, but more a deep orange or tan colour. The first-magnitude stars on the charts in this book are coloured according to their spectral types rather than how they actually appear to the eye.

An apparent paradox is that our Sun, whose light is usually thought of as white, is classified as a yellow star. In fact, what makes the daytime Sun appear white is its over-powering brilliance. If we could see it from a distance, so it was much fainter, it would indeed appear yellowish. It turns out that a star of truly white appearance to the eye has a spectral type of around F0, similar to that of Canopus, 2000°C hotter than the Sun.

The star colours described in the constellation notes in this book are an indication of how stars appear to an observer, although the real colours are subtle and your own visual impressions may well disagree. Indeed, you will find that the apparent intensity of the coloration varies from night to night with

NGC 2070 in the Large Magellanic Cloud is popularly known as the Tarantula Nebula because of its spidery shape. At its heart is an open cluster, R 136, containing some of the hottest and most massive stars known. Such massive stars have short lifetimes and are likely to die in supernova explosions. (ESO)

STELLAR SPECTRAL TYPES

Type	Assigned colour	Temperature range (°C)	Examples
O	Blue	40,000–25,000	ζ Puppis (supergiant)
B	Blue	25,000–11,000	Spica (main sequence) Regulus (main sequence) Rigel (supergiant)
A	Blue-white	11,000–7500	Vega (main sequence) Sirius (main sequence) Deneb (supergiant)
F	White	7500–6000	Canopus (supergiant) Procyon (subgiant) Polaris (supergiant)
G	Yellow	6000–5000	Sun (main sequence) α Centauri (main sequence) τ Ceti (main sequence) Capella (giant)
K	Orange	5000–3500	ε Eridani (main sequence) Arcturus (giant) Aldebaran (giant)
M	Red	3500–3000	Barnard's Star (main sequence) Antares (supergiant) Betelgeuse (supergiant)

changing atmospheric conditions. For faint stars, you will be unlikely to see any colour at all without optical aid.

When the spectral type of stars is plotted against their actual luminosity (absolute magnitude), all stars that are in stable, hydrogen-burning middle age are found to lie in a well-defined band across the graph known as the *main sequence*. Such a plot of stellar brightness against spectral type is known as a Hertzsprung–Russell diagram, after the Danish astronomer Ejnar Hertzsprung and the American Henry Norris Russell who jointly devised it in 1911–13. An example is shown on page 275.

A star's position along the main sequence is fixed by its mass, with the least massive stars at the bottom and the most massive stars at the top. The Sun, as befits its middle-of-the-road nature, lies about halfway along the main sequence.

Although most stars lie on the main sequence, a number of particularly bright stars lie above and to the right of it, while a few faint stars lie below and to the left of it. These stars are all in late stages of evolution. We can best understand what is happening to them by following the future evolution predicted for the Sun.

Evolving stars

As mentioned earlier, the Sun formed about 4600 million years ago and is now about halfway through its expected lifespan. In a few thousand million years, though, it will start to run out of hydrogen at its core. In search of more hydrogen to use as fuel, the nuclear reactions inside the Sun will start to move outwards, releasing more energy. Eventually, when they are surrounded by a

The Double Cluster in Perseus, NGC 869 and 884, is a pair of open clusters in a spiral arm of our Galaxy. NGC 884, at left, contains several red giants, suggesting that it is the more evolved of the two. (Leland Wehland/Flynn Haase/ NOAO/AURA/NSF)

shell of burning hydrogen, even the helium atoms in the Sun's core will enter into nuclear reactions of their own, fusing together to form carbon atoms.

With all this extra energy being given off, the Sun will become much brighter than it is today and will start to swell alarmingly in size. But as the Sun's outer layers expand they will also cool, becoming redder in colour, and the Sun will turn into a *red giant*, similar to the bright stars Aldebaran and Arcturus. At its largest, the red giant Sun will grow to at least a hundred times its present diameter, engulfing Mercury, Venus and perhaps even the Earth within its outer layers. Needless to say, all life on our planet will long since have become extinct.

On the Hertzsprung–Russell diagram, the Sun's increase in brightness will move it upwards, off the main sequence, and its change in spectral type will in addition move it to the right. Stars at the top end of the main sequence, which are far more massive

In the Hertzsprung–Russell diagram, right, the true brightness of stars (their absolute magnitude) is plotted against their temperature (spectral type). Stars of different types fall in different areas of the diagram. The band running from top left to bottom right is the main sequence; the Sun falls roughly halfway along it. Giants and supergiants lie above the main sequence, while white dwarfs are below and to the left of it. (Wil Tirion)

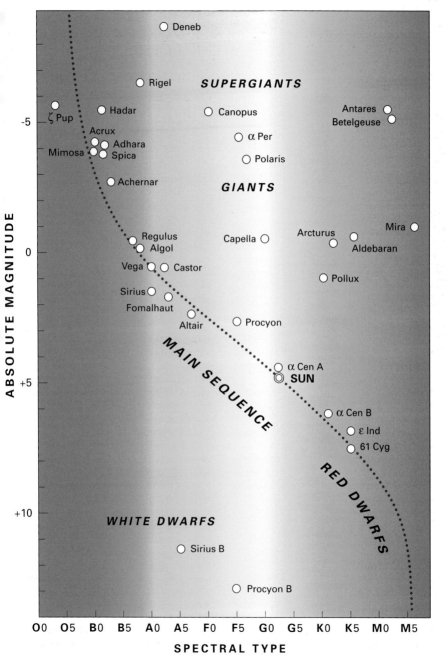

STELLAR LUMINOSITY CLASSES	
Ia0	Exceptionally bright supergiant
Ia	Bright supergiant
Iab	Less bright supergiant
Ib	Supergiant
II	Bright giant
III	Giant
IV	Subgiant
V	Main sequence
VI or sd	Subdwarf

than the Sun, become so big and bright at this stage of their evolution that they are referred to not as mere giants but as *supergiants*. Prominent examples of red supergiants are Betelgeuse and Antares, both of which are hundreds of times larger than the Sun. Other stars which have not yet evolved enough to become red in colour but which are nevertheless firmly in the supergiant bracket are Rigel, Deneb and Polaris.

To distinguish whether a star is, say, a giant or supergiant, or lies on the main sequence, astronomers assign stars a luminosity class (see the table above) in addition to the spectral type. For example, a G-type main-sequence star such as the Sun is classified GV, whereas a G-type giant such as Capella is GIII and the G-type supergiant Sadalmelik (Alpha Aquarii) is GIb.

As an aside, it should be noted that main-sequence stars are commonly referred to as dwarfs, even though the most massive of them may be several times larger in diameter than the Sun. Hence, although the Sun is in most ways an average star, in astronomical terminology it is classified as a dwarf.

Taken together, the spectral type and luminosity class define the main properties of a star as it exists at present. But those properties change as the star ages. Stars spend only a few per cent of their total lifetime in the red giant phase, which in the case of stars like the Sun amounts to no more than a few hundred million years. A red giant is a star that has grown old and is about to die.

Star death

Once a red giant has swollen beyond a certain size its distended outer layers drift off into space, forming a stellar smoke ring known rather confusingly as a *planetary nebula*, even though it has nothing to do with planets. The name was first used in 1785 by William Herschel, because they looked like the small, rounded disks of planets as seen through his telescope.

Probably the best-known of all planetary nebulae is the Ring Nebula in Lyra (M57), although it is not the easiest to see. Much larger is the Dumbbell Nebula, M27, in Vulpecula, which can be picked up in binoculars on a clear, dark night. Two small but bright planetary nebulae for amateur telescopes are NGC 6826 in Cygnus and NGC 7662 in Andromeda.

At the centre of a planetary nebula, the core of the former red giant is exposed as a small, intensely hot star. Once the surrounding gases of the planetary nebula have dispersed, usually after a few thousand years, the central star remains as a so-called *white dwarf*.

A white dwarf is only about the diameter of the Earth, but contains most of the matter of the original star; only about 10 per cent of the star's mass is lost in the planetary nebula stage. White dwarfs are therefore exceptionally dense bodies. A teaspoonful of white dwarf material would have a mass of thousands of kilograms. Over thousands of millions of years, white dwarfs slowly cool off and fade into oblivion.

NGC 6543, the Cat's Eye Nebula in Draco, consists of overlapping loops of ejected gas. In this image from the Hubble Space Telescope, different colours represent gas at different temperatures. (Bruce Balick, U. of Washington/NASA)

Below: M57, the Ring Nebula in Lyra, is actually a cylinder of gas ejected by the central star; it appears ring-like because we are looking at the cylinder end-on. (Adam Block/NOAO/ AURA/NSF)

Being so small, white dwarfs are very faint. Not one is visible to the naked eye. Both the nearby bright stars Sirius and Procyon have white dwarf companions, but Procyon's companion is too close to its parent to be distinguishable in amateur telescopes, and the companion of Sirius can be glimpsed only under the most favourable conditions. The easiest white dwarf to see is a companion of the star o^2 (omicron^2) Eridani (also known as 40 Eridani); a small telescope will show it. Of added interest is a fainter third member of this system, a red dwarf, which is also visible in amateur telescopes.

Our Sun, it seems, is destined to go through the stage of being a planetary nebula before fading away as a white dwarf. But stars with several times the Sun's mass, towards the top end of the main sequence, suffer a far more spectacular end. As we have seen, they first become dazzling supergiants rather

Main picture, right: The η (eta) Carinae Nebula, NGC 3372, a large cloud of glowing hydrogen gas visible to the naked eye in the southern hemisphere, containing clusters of hot young stars. The nebula is named after η Carinae itself, a massive star that lies in the brightest central part, just above the V-shaped dark lane. (NOAO/AURA/NSF)

Inset, right: The area around η (eta) Carinae, left of centre, which is cocooned in a reddish cloud of gas thrown off in past outbursts. To the right of Eta Carinae is the bulbous Keyhole Nebula, shaped like a figure of eight, which appears slightly darker than the background nebulosity. η Carinae is a massive, unstable star that could erupt as a brilliant supernova within the next 10,000 years. (M. Bessell, R. Sutherland and M. Buxton, RSAA – Australian National University)

The Eskimo Nebula, NGC 2392 in Gemini, photographed with unprecedented clarity by the Hubble Space Telescope, gets its name from its resemblance to a face surrounded by a fur parka. In reality, the 'parka' is a disk of material thrown off the central star during its red giant phase, while the Eskimo's 'face' is a bubble of gas being blown into space by the intensely hot central star. (Andrew Fruchter, STScI/NASA/ESA)

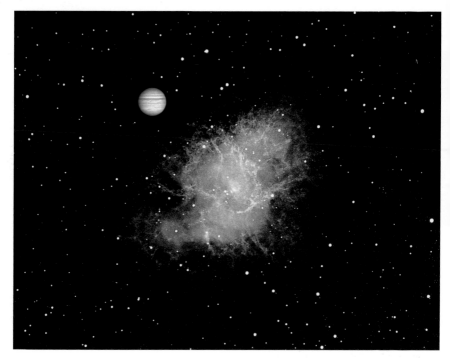

than mere giants. They do not get a chance to reach the planetary nebula stage. So massive are they that the nuclear reactions at their centres continue in runaway fashion until the star becomes unstable and explodes. Such an explosion is known as a *supernova*.

Supernovae and their remnants

In a supernova eruption a star's brightness increases millions of times, so that for a few days the star can rival the brilliance of an entire galaxy. The shattered outer layers of the star are thrown off into space at speeds of around 5000 km per second. In AD 1054 astronomers on Earth saw a star erupt as a supernova in the constellation Taurus. The star became brighter than Venus, and was visible in daylight for three weeks. It finally faded from naked-eye view more than a year after it had first appeared.

The Crab Nebula, M1, the remains of a star that exploded as a supernova, with the planet Jupiter superimposed for size comparison. The blue colour of the nebula comes from electrons spiralling in its magnetic field, while the red flecks are hydrogen gas. The Crab Pulsar is the lowermost of the two stars near the nebula's centre; the other star is an unrelated foreground object. (Crab Nebula image: Laird Thompson, University of Hawaii/CFHT)

At the site of that explosion lies one of the most famous objects in the heavens: the Crab Nebula, the shattered remains of the star that erupted. The Crab Nebula is visible as a smudgy patch in amateur telescopes, but is best seen on long-exposure photographs taken with large instruments. Over the next 50,000 years or so the gases of the Crab Nebula will disperse into space, forming delicate traceries like those of the Veil Nebula in Cygnus (see page 138), itself the remains of a former supernova.

The last supernova observed in our Galaxy was in 1604. This star, in the constellation Ophiuchus, reached a maximum magnitude of around −3, brighter than Jupiter. It was studied by the German astronomer Johannes Kepler, and is often known as Kepler's Star.

Hundreds of supernovae in other galaxies have been seen telescopically since then, but only one has become bright enough to be visible to the naked eye. That was Supernova 1987A, which erupted in our neighbouring galaxy the Large Magellanic Cloud. It was first spotted in 1987 on February 24 and rose to a maximum magnitude of 2.9 in late May, eventually fading from naked-eye view by the end of the year.

Another supernova in our Galaxy is long overdue. When it comes, it should be a spectacular sight. Many astronomers dream of seeing it outshining the other stars in the sky, dazzling the eye and casting shadows.

Supernova 1987A (the very bright star at lower right) exploded in 1987 in the Large Magellanic Cloud, near the Tarantula Nebula (left). The supernova was visible to the naked eye from the southern hemisphere for several months. (ESO)

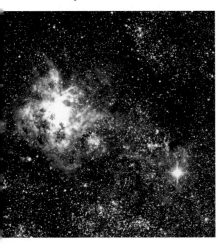

A star might not be entirely destroyed in a supernova explosion. Sometimes the central core of the exploded star is left as an object even smaller and denser than a white dwarf, termed a *neutron star*. In such an object, the protons and electrons of the star's atoms have been crushed by the tremendous forces of the supernova so that they combine to form the particles known as neutrons. A typical neutron star is a mere 20 km in diameter, but contains as much mass as one or two Suns.

Being so minute, neutron stars can spin very rapidly without flying apart. Each time they spin we receive a flash of radiation like a lighthouse beam. Astronomers have detected radio pulses from well over a thousand such sources, which they term *pulsars*; one lies at the centre of the Crab Nebula. The Crab Pulsar flashes 30 times per second; others pulse hundreds of times a second, while the slowest takes over 10 seconds. Most neutron stars are too faint to be seen optically, but the pulsar in the Crab Nebula has been seen flashing in step with the radio pulses.

Black holes

If the core of the exploded star has a mass of more than three Suns, then even a neutron star is not the end for it. Instead, it becomes something still more bizarre: a *black hole*. No force can shore up a dead star weighing more than three solar masses against the inward pull of its own gravity. It continues to shrink, becoming ever smaller and denser until its gravity becomes so great that nothing can escape from it, not even its own light. It has dug its own grave – a black hole.

Since a black hole is by definition invisible, it is of only academic interest to amateur observers. However, professional astronomers have detected X-ray emissions from various sources which they believe are caused by hot gas plunging into the bottomless pits of black holes. The best-known candidate for a black hole, Cygnus X-1, orbits a 9th-magnitude star in the neck of Cygnus, the Swan.

Double and multiple stars

To the naked eye, stars appear as solitary, isolated objects. But the great majority of them actually have one or more companion stars, too faint or too close to be seen separately with the naked eye but which can be distinguished telescopically. Many attractive double and multiple stars are within the range of binoculars and small telescopes, and the best are described in the constellation notes in this book.

There are two sorts of double star. In one type, the two stars are not truly associated but happen to lie in the same line of sight by chance; in this arrangement, termed an *optical double*, one star may be many times farther from us than the other. Such doubles are comparatively rare. In most cases, the two stars are physically linked by gravity, forming a genuine *binary* system. The two stars in a binary orbit their mutual centre of gravity, which can take decades, centuries, or even millennia, depending how far apart they are. There are also triple, quadruple and even larger families of stars, and in such cases the orbits can become very complicated.

The closest star to the Sun, α (alpha) Centauri, is a triple system – a bright binary accompanied by a much fainter red dwarf known as Proxima. Another celebrated multiple star is the so-called Double Double, ε (epsilon) Lyrae. Binoculars show it to be a wide double, but modest-sized amateur telescopes reveal that each of these stars is itself double, making a quadruple system.

Orbits of four fast-moving binaries: 70 Ophiuchi (period 88 years), Zeta Herculis (34.5 years), Xi Ursae Majoris (60 years) and Gamma Virginis (169 years). (Ian Ridpath/Wil Tirion)

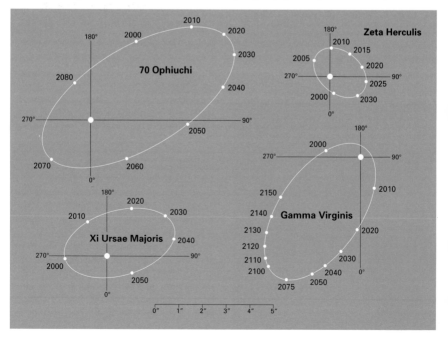

SEPARATION AND POSITION ANGLE

Three parameters describe the appearance of a double star: the brightnesses of the individual components, their separation and their relative orientation, known as position angle (PA). Separation is usually given in seconds (″) of arc, although minutes (′) may be used for wider pairs. Position angle is more tricky to understand. It is the direction of the fainter star relative to the brighter one, measured in degrees counterclockwise from north (0°) via east (90°), south (180°) and west (270°). The east–west line (90°–270°) can be established by allowing objects to drift through the field of view as the Earth rotates. The diagram below shows the view through the eyepiece of an inverting telescope, with north at the bottom and east at the right. A double star with a position angle of 150° is illustrated.

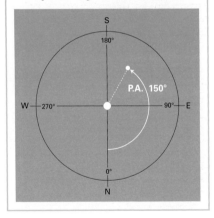

Even more remarkable is Castor, a system of six stars all linked by gravity. Amateur telescopes show that Castor consists of two bright blue-white stars close together, with a fainter third star some way off. Professional observations have revealed that each of these three stars is itself a *spectroscopic binary*. In a spectroscopic binary the two stars are too close together to be seen individually in any telescope, but analysis of the light from the star reveals the companion's presence. The two brightest stars of Castor take nearly 500

years to orbit each other, whereas their spectroscopic companions whirl around them in only a few days. Sometimes the members of a binary system eclipse each other as seen from Earth. Such *eclipsing binaries* are dealt with in the section on variable stars.

Observing double stars

For amateurs, the attraction of observing double stars lies in separating (or 'splitting') close doubles and in comparing their brightnesses and colours. Some pairs display a beautiful colour contrast, as with the yellow and blue-green components of Albireo, β (beta) Cygni. A similar pair with contrasting colours but smaller separation is γ (gamma) Andromedae. For a pair with almost identical colours look at 61 Cygni. Star colours are subtle, but can become more noticeable if you slightly defocus the telescope, or tap it gently to make the image vibrate.

The closer together two stars are, the larger the aperture of telescope needed to split them. A table giving the resolutions of various apertures can be found on page 392, but this not the whole story since the relative brightness of the two components also affects their observability. If one star is much fainter than the other it will be swamped by the primary star's light, and so will be more difficult to see than a companion that is of similar brightness to the main star.

Another factor to take into account is that the separation of binary pairs with relatively short orbital periods can change markedly over a few years as the stars orbit each other. The changing positions of four fast-moving binaries is shown in the diagram on page 282.

The notes in this book on objects of interest give guidance as to the likely aperture needed to split various double stars. But there can be no hard-and-fast rules. So much depends on the observer's eyesight, the quality of telescope and the observing conditions. The only way to find out for sure is to look yourself.

Variable stars

Certain stars change in brightness over time, and are known as *variable stars*. Amateur astronomers can make valuable observations of such stars. The observer estimates the brightness of the variable by comparing it with nearby stars of known constant magnitude, and the observations are plotted on a graph to form what is known as a *light curve*, illustrating how the star's brightness varies with time. Such a graph can reveal much about the nature of the star under study. Light curves of three famous variables are shown on the facing page.

As might be expected, the usual reason that a star appears to vary is because of actual changes in its light output, but that is not the only possible explanation. In some cases the star is a member of a binary system in which one star periodically eclipses the other. For this to happen, the stars' orbits must be oriented virtually edge-on to us. The first such *eclipsing binary* to be noticed, and still the most famous, is Algol in the constellation Perseus; its variability was discovered in the 1660s by Geminiano Montanari of Italy.

Eclipsing variables

Algol consists of a blue-white main-sequence star, from which most of the light comes, orbited by a fainter orange subgiant. Every 2.87 days Algol's brightness drops from magnitude 2.1 to 3.4 in five hours as the fainter star obscures the brighter one, returning to normal after another five hours as the star reemerges. Compare it at maximum with α (alpha) Persei, magnitude 1.8, and at minimum with δ (delta) Persei, magnitude 3.0 (see the chart on page 286). There is also a secondary minimum half an orbit later when the brighter star passes in front of the fainter one, but the drop in light is too small for the eye to notice. Most variable-star observers start their careers by following the changes in Algol's brightness; predictions of its eclipses are issued by astronomical societies and in astronomy magazines.

An extreme form of eclipsing binary is β (beta) Lyrae, in which the stars are so close together that they have been distorted into elongated shapes by each other's gravity. Beta Lyrae varies between magnitudes 3.3 and 4.4 every 12.9 days. Its brightness continually changes even outside of eclipses as the elongated stars move around their orbits. Compare it with nearby γ (gamma) Lyrae, constant at magnitude 3.2, and κ (kappa) Lyrae, magnitude 4.3.

Pulsating variables

Of the stars that vary intrinsically in brightness, most do so because of physical changes in their size. These are known as *pulsating variables* (not to be confused with pulsars). Of particular importance to astronomers are the so-called *Cepheid variables*, named after their prototype δ (delta) Cephei.

Cepheid variables are yellow supergiant stars that go through one cycle of pulsations in periods ranging from about 2 to 40 days, varying in brightness by up to one magnitude as they do so. δ Cephei itself ranges between magnitudes 3.5 and 4.4 every 5.4 days, and is thus an easy object for amateur observation. Good comparison stars are ε (epsilon) Cephei, magnitude 4.2, and ζ (zeta) Cephei, magnitude 3.4 (see the chart on page 287). Other bright Cepheids for easy amateur observation are η (eta) Aquilae (magnitude range 3.5 to 4.4, period 7.2 days), ζ (zeta) Geminorum (3.6 to 4.2, 10.2 days) and, in the southern hemisphere, l Carinae (3.3 to 4.2, 35.5 days).

The importance of Cepheid variables is that their period of pulsation is directly related to their absolute magnitude: the brighter the Cepheid, the longer it takes to

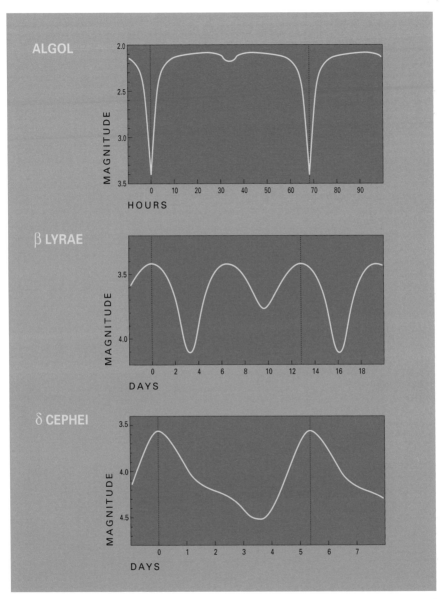

Light curves of three well-known variable stars: Algol, β (beta) Lyrae and δ (delta) Cephei. Their brightness rises and falls in a predictable fashion over the course of a few days. Note that the rise to maximum of δ Cephei is quicker than its subsequent decline. The small secondary minimum of Algol midway between the main fades is undetectable to the naked eye. (Wil Tirion)

complete one cycle of variation. Astronomers therefore have an easy way of finding these stars' absolute magnitude, simply by observing their period of fluctuation. When the absolute magnitude is compared with the apparent magnitude, the star's distance can easily be computed. Cepheid variables thus make important distance indicators within our own Galaxy and to other galaxies.

Another important class of pulsating stars are the *RR Lyrae variables*. These are old blue stars frequently found in globular clusters, and they vary by about 0.5 to 1.5 magnitudes in less than a day. Their prototype, RR Lyrae itself, varies from magnitude 7.1 to 8.1 and back in 0.57 days. Stars of this type all have similar absolute magnitudes, which makes them valuable distance indicators.

Minor types of pulsating variables include β (beta) Cephei stars and δ (delta) Scuti stars, both of which have short periods of only a few hours and ranges of variation too small to be noticeable to the naked eye.

Each type of variable falls in a specific area on the Hertzsprung–Russell diagram, and represents stars of different mass at various stages of their evolution. Variability of one kind or another seems to be an inevitable consequence of the ageing process of all stars – including, one day, our own Sun.

Long-period variables

Red giants and red supergiants are old stars that frequently prove to be variable. They pulsate, but not with anything like the same regularity as the types of stars mentioned above. The most abundant type of variables known are the Mira stars, also termed *long-period variables*, which have periods ranging from about three months to two years and amplitudes of several magnitudes. This is the type of variable our Sun is expected to evolve into, billions of years from now.

Their prototype is Mira, o (omicron) Ceti, in the constellation Cetus, the Whale, a red giant which ranges between about 3rd and 9th magnitude every 11 months or so, although the exact period and amplitude differ slightly from cycle to cycle. A finder chart is given on page 116. Another promi-

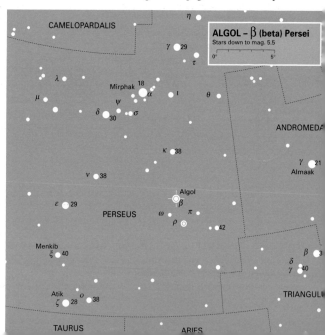

Comparison chart for Algol, β (beta) Persei, the prototype eclipsing variable which varies between magnitudes 2.1 and 3.4 every 2.87 days. The magnitudes of comparison stars are given without decimal points to prevent confusion between the points and faint stars. (Wil Tirion)

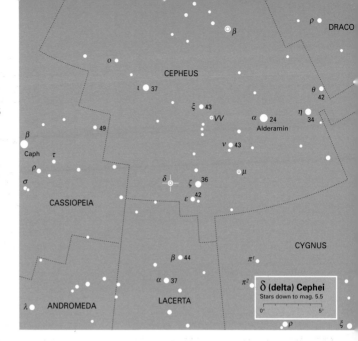

Comparison chart for δ (delta) Cephei, the prototype Cepheid variable which varies between magnitudes 3.5 and 4.4 every 5.4 days. The magnitudes of comparison stars are given without decimal points to prevent confusion between the points and faint stars. (Wil Tirion)

nent Mira-type variable is χ (chi) Cygni, in the neck of Cygnus, the Swan.

More erratic are the *semi-regular variables*, typically with periods of about 100 days and amplitudes of 1 or 2 magnitudes. Finally, the *irregular variables* have little or no discernible pattern at all to their fluctuations. All these stars are red giants and red supergiants that have reached a stage of instability, oscillating in size and brightness. Prominent examples of semi-regular and irregular variables are Antares, Betelgeuse, α (alpha) Herculis and μ (mu) Cephei.

Novae

Most spectacular of all variable stars are the *novae*, which suddenly and unexpectedly erupt by perhaps 10 magnitudes (10,000 times) or more, sometimes becoming visible to the naked eye where no star appeared before. Their name comes from the Latin meaning 'new', for they were once thought to be genuinely new stars. Now we know that they are merely old, faint stars under-

going a temporary outburst. Despite their name, they are not related to supernovae, which are massive stars that explode for different reasons, as explained on page 280.

According to current theory, novae are close binary stars, one member being a white dwarf. Gas spilling from the companion star onto the white dwarf is thrown off in an eruption. The star does not disrupt itself in a nova outburst. In fact, some novae have undergone more than one recorded outburst, notably RS Ophiuchi and T Pyxidis, and perhaps all novae recur, given time.

A nova rises to maximum brightness in a few days. Amateur astronomers are often the first to spot such eruptions and to notify professional observatories. After a few days or weeks at maximum comes a slow decline over several months, as the nova slowly sinks back to its previous obscurity, sometimes punctuated by minor additional outbursts. Following the progress of a nova is one of the most fascinating aspects of variable star observation. Prominent naked-eye novae occur only about once a decade, but many fainter ones are visible through binoculars.

M81, a near-perfect spiral galaxy in Ursa Major, is tipped at an angle to our line of sight and so appears foreshortened along one axis. (Nigel Sharp/NOAO/AURA/NSF)

The Milky Way, galaxies and the Universe

Our Sun and all the stars visible in the night sky are members of a vast aggregation of stars known as the Galaxy (given a capital G to distinguish it from any other galaxy). Our Galaxy is spiral in shape, with arms composed of stars and nebulae winding outwards from a central bulge of stars. It is about 100,000 light years in diameter; the Sun lies in a spiral arm some 30,000 light years from the Galaxy's centre. Astronomers estimate that the Galaxy contains at least 250,000 million stars.

Most of the stars in the Galaxy lie in a disk about 2000 light years thick. Seen from our position within the Galaxy, this disk of stars appears as a faint, hazy band crossing the sky on clear, dark nights. We call this band the Milky Way, and the name Milky Way is often used for our entire Galaxy. The star fields of the Milky Way are particularly dense in the region of Sagittarius, which is the direction of the Galaxy's centre. Note that the plane of the Milky Way is tilted at 63° with respect to the celestial equator. This results from a combination of the tilt of the Earth's axis, and the fact that the plane of the Earth's orbit around the Sun is tilted with respect to the plane of the Galaxy.

M3 in Canes Venatici, visible with binoculars and small telescopes, is one of the brightest and richest of the globular clusters that surround our Galaxy in a halo. Globular clusters contain some of the oldest stars in the Galaxy. (Maureen Van den Berg, Univ. Leiden/Nik Szymanek/ING)

M51, the beautiful Whirlpool Galaxy in Canes Venatici, and its small companion galaxy NGC 5195. This companion has apparently twisted the outstretched spiral arm of M51 during a recent encounter, and it now lies behind the arm. (Javier Méndez/ING and Nik Szymanek/SPA)

Globular clusters

Dotted around our Galaxy in a spherical halo are more than a hundred ball-shaped conglomerations of stars known as *globular clusters*. These each contain from a hundred thousand to several million stars, all bound together by gravity. Several globular clusters are visible with the naked eye or binoculars. The brightest are ω (omega) Centauri and 47 Tucanae, both in the southern hemisphere; in the northern hemisphere the best example is M13 in Hercules.

To the naked eye and in binoculars, these objects appear as softly glowing patches of light. Moderate-sized telescopes start to resolve some of the individual red giant stars, giving the clusters a speckled appearance. Globular clusters formed early in the history of the Galaxy and contain some of the most ancient stars known, 10,000 million years or more old, over twice the age of the Sun.

The Local Group

Our Galaxy has two small and irregularly shaped companion galaxies called the Magellanic Clouds. To the naked eye they appear like detached portions of the Milky Way, in the southern constellations Dorado and Tucana. The Large Magellanic Cloud contains about one-tenth the number of stars in our Galaxy and lies about 170,000 light years away. The Small Magellanic Cloud has only about a fifth as many stars as the Large Magellanic Cloud, and lies somewhat farther off, about 200,000 light years. Both Clouds contain numerous star clusters and bright

Galaxies are classified according to their shape: elliptical, type E; spiral, type S; barred spiral, type SB; and irregular. It was once thought that galaxies evolved from one type into another as they aged, but this is not so. (Wil Tirion)

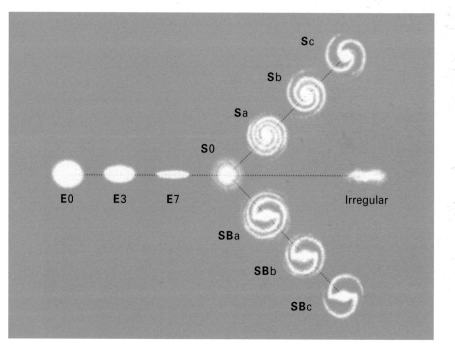

Sc

Sb

Sa

S0

E0   E3   E7   Irregular

SBa

SBb

SBc

nebulae, and are rich territories for sweeping with instruments of all sizes. One can only imagine the magnificent view of our Galaxy that any astronomers living in the Magellanic Clouds would have.

Countless galaxies are dotted like islands throughout the Universe as far as the largest telescopes can see. Most galaxies are members of clusters containing anything from a dozen to many thousands of galaxies. Our own Milky Way is the second-largest member of a small cluster of 40 or so galaxies known as the Local Group.

The largest galaxy in the Local Group is visible to the naked eye as a fuzzy, elongated patch in the constellation Andromeda. The Andromeda Galaxy, also commonly known as M31, is estimated to contain about twice

Right: NGC 5850 is an 11th-mag. barred spiral in the Virgo Cluster. Its arms are wound tightly around the central bar, forming the shape of a Greek letter theta (θ). (Jeff and Paul Neumann/Adam Block/ NOAO/AURA/NSF)

Below: NGC 4013 in Ursa Major is a spiral galaxy seen edge-on, crossed by a dark lane of dust in its central plane. (Blair Savage and Chris Howk, U. Wisconsin/ Nigel Sharp, NOAO/ WIYN/NSF)

Above: M87 is a
massive elliptical galaxy
in the Virgo Cluster,
surrounded by a swarm
of hundreds of globular
clusters, visible here as
hazy dots.
(NOAO/AURA/NSF)

Right: M82 in Ursa
Major is an edge-on
spiral galaxy undergoing
an immense burst of star
formation. (Nigel Sharp/
NOAO/AURA/NSF)

as many stars as our own Galaxy, and to be
about 25 per cent greater in diameter. It lies
about 2.5 million light years away.

Long-exposure photographs reveal that the
Andromeda Galaxy is a spiral, but tilted so
that we see it almost edge-on. Amateur telescopes show that the Andromeda Galaxy has
two small companion galaxies, M32 and
M110, its equivalent of the Magellanic

Clouds, although several additional fainter
ones are visible through large telescopes.

The only other member of the Local
Group within easy reach of amateur instruments is M33 in Triangulum, another spiral
galaxy, somewhat farther from us than the
Andromeda Galaxy and containing considerably fewer stars. M33 can be picked up in
binoculars under clear, dark skies.

The nearest rich cluster of galaxies to the Local Group lies in the constellation of Virgo, part of it spilling over into neighbouring Coma Berenices. Of the Virgo Cluster's 3000 or so known members, dozens are within the reach of amateur telescopes.

Types of galaxy

Astronomers classify galaxies into three main types: elliptical, spiral and barred spiral (see the diagram on page 291). *Elliptical galaxies* range in shape from virtually spherical, designated E0, to flattened lens shapes, designated E7. They include both the largest and the smallest galaxies in the Universe. Supergiant ellipticals composed of up to 10 million million stars are the most luminous galaxies known. An example is M87 in the Virgo Cluster. At the other end of the scale, dwarf ellipticals resemble large globular clusters. Dwarf ellipticals may be the most abundant

type of galaxy in the Universe, but their faintness makes them difficult to see.

Spiral galaxies (type S), such as the great Andromeda Galaxy, M31, have arms winding out from a central bulge. There are usually two arms, but sometimes more. In *barred spirals* (type SB), such as NGC 1300 in Eridanus, the arms emerge from the ends of a bar of stars and gas that runs across the galaxy's centre. Most of the galaxies we see in the Universe are large, bright spirals.

Spiral and barred spiral galaxies are subdivided according to how tightly their arms are wrapped around the galaxy's core: types Sa and SBa have the tightest-wound arms, while types Sc and SBc have the loosest-wound arms. The Andromeda Galaxy is type

M77 in Cetus is the most prominent example of a Seyfert galaxy, a type of spiral with an unusually bright and active nucleus. (ING Archive/Nik Szymanek)

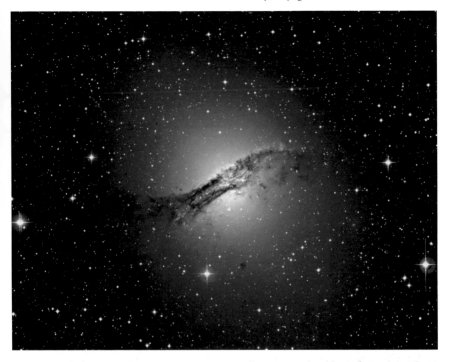

NGC 5128, also known as Centaurus A, is a supergiant elliptical galaxy with a prominent dust band. Elliptical galaxies do not normally contain dust, so NGC 5128 is thought to result from the merger of an elliptical galaxy with a spiral. (ESO)

Sb. Until recently, our own Galaxy was thought to fall midway between Sb and Sc, but it is now generally classified as a barred spiral, although the bar is only short.

In addition to these three main types there are certain galaxies classified as *irregular*; the two Magellanic Clouds are usually reckoned to fall into this class, although there is some semblance of spiral structure in the Large Magellanic Cloud.

Being faint, fuzzy objects, galaxies are best seen where skies are clear and dark, away from city haze and stray light. Low magnification is best, to increase the contrast of the galaxy against the sky background. Through a telescope you should see the nucleus of a galaxy as a starlike point surrounded by the misty halo of the rest of the galaxy. Do not expect to see prominent spiral arms, as on a long-exposure photograph, except through larger apertures at dark sites. Galaxies present themselves to us at all angles, so even spirals can seem to be elliptical in shape when viewed edge-on.

Peculiar galaxies

Strange things are going on within some galaxies. For example, certain galaxies give off vast amounts of energy as radio waves; these *radio galaxies* include the supergiant ellipticals M87 in Virgo and NGC 5128 (pictured above) in Centaurus. NGC 5128, in particular, looks like an elliptical galaxy that has been caught merging with a spiral. Some spiral galaxies have unusually bright nuclei. These are termed *Seyfert galaxies*, after the American astronomer Carl Seyfert, who was the first to draw attention to them, in 1943. M77 in Cetus (see the photograph on page 294) is the brightest Seyfert galaxy.

Most peculiar of all are the objects known as *quasars*, which emit as much energy as hundreds of normal galaxies from an area less than a light year in diameter. Despite their exceptional nature, quasars are not at all exciting visually, which is why they were overlooked until 1963. The brightest quasar, 3C 273, appears as an unimposing 13th-magnitude star in Virgo.

Observations with the Hubble Space Telescope have shown that quasars are actually the highly luminous centres of galaxies far off in the Universe. Radio galaxies, Seyferts and quasars are all now known to be related and are collectively termed *active galaxies.* Their central powerhouse is thought to be a massive black hole that gobbles up stars and gas from the surrounding galaxy, and its activity can be boosted by infalling gas when two galaxies collide. Most galaxies may have experienced such activity at their centres early in their evolution.

The expanding Universe

In 1929 the American astronomer Edwin Hubble made the most significant of all discoveries in cosmology: the galaxies are moving apart from one another, as though the Universe is expanding like a balloon being inflated. However, clusters of galaxies such as the Local Group do not expand – they are held together by their mutual gravitational attraction.

Hubble's discovery that the Universe is expanding came from a study of the spectrum of each galaxy's light. This revealed that the light from the galaxies was being lengthened

The redshift in the light from a distant galaxy is measured by the change in position of lines in its spectrum. Here, the change in position is shown in the visual window, the range of wavelengths to which the human eye is sensitive. All light from the galaxy is shifted by the same amount, so while some of the galaxy's light moves out of sight beyond the red end of the visual window, other light moves into the visible window from the ultraviolet. (Wil Tirion)

VISUAL WINDOW

Infrared   Red   Blue   Ultraviolet

Spectral lines

Infrared   Red   Blue   Ultraviolet in wavelength as a result of high-speed recession (this is called the Doppler effect). Such a lengthening of wavelength is called a *redshift*, because the light from the galaxy is moved towards the red (longer-wavelength) end of the spectrum.

Hubble found that the amount of redshift in a galaxy's light becomes greater with distance. Therefore, by measuring the redshift of a galaxy astronomers can tell how far away it is. Quasars, for instance, exhibit such enormous redshifts that they must be the most distant objects visible in the Universe, over 10,000 million light years away.

Since the Universe is expanding, it is logical to conclude that it was once smaller and more densely packed than it is now. According to the most widely accepted theory, the entire Universe was originally a compressed, superdense blob which, for some unknown reason, exploded in a cata-

Stephan's Quintet in Pegasus is a group of five galaxies: NGC 7317 (top left), NGC 7318A and B (centre), NGC 7319 (bottom right) and NGC 7320 (lower left). However, NGC 7320 has a lower redshift than the others and also appears larger, which suggests it is a foreground object and not a true member of the cluster at all. (Nigel Sharp/NOAO/AURA/NSF)

clysm known as the Big Bang. The galaxies are the fragments from that explosion, still flying outwards as the space between them expands.

As far as anyone can tell at present, the Universe will continue to expand for ever. According to the best current estimates the Big Bang took place 13,700 million years ago; that is the age of the Universe as we know it. It is impossible to tell what, if anything, happened before the Big Bang.

The Sun

Our Sun is a glowing sphere of hydrogen and helium gas 1.4 million km in diameter, 109 times the diameter of the Earth and 745 times as massive as all the planets put together. The Sun is vital to all life on Earth because it provides the heat and light that makes our planet habitable. To astronomers, the Sun is important because it is the only star that we can observe in close-up. Studying the Sun tells us a lot that we could not otherwise know about stars in general.

Whereas most celestial objects pose problems for observers because they are faint, the Sun presents entirely the opposite dilemma: it is so bright that it is dangerous to look at. **It cannot be emphasized too strongly that anyone glancing for an instant at the Sun through any form of optical instrument, be it binoculars or telescope, risks instant blindness.** Even staring with the naked eye at the Sun for a few seconds can permanently damage your sight. There is a

The Sun on 19 May 2000, during the rise to solar maximum showing several sunspot groups, each large enough to swallow the Earth many times over. Note the effect of limb darkening, against which brighter faculae may be seen. (Mees Solar Observatory, University of Hawaii)

A complex group of sunspots some 200,000 km long as it appeared on 4 September 1998. The black circle shows the Earth to the same scale. Filamentary structure in the outer penumbra of the spots gives them a sunflower-like appearance. Note also the granulated appearance of the surrounding solar photosphere. (Kiepenheuer-Institut für Sonnenphysik)

safe and easy way to observe the Sun, and that is by projecting its image onto a white surface such as a screen or sheet of card (see box on page 302).

Some telescopes come equipped with dark Sun filters to place behind the eyepiece, but these should never be used because the focused light and heat of the Sun can crack them, with inevitably disastrous results. Safe filters for telescopic solar observation do exist, made either of glass or plastic film with a metallic coating. These filters fit over the front of the telescope tube, reducing the amount of incoming light and heat to safe levels. Filters made from a thin plastic film called Mylar give a bluish image, whereas other types give yellow or orange images.

A telescopic view of the Sun, seen either by projection or through a safe filter, displays the brilliant solar surface, the *photosphere* ('sphere of light'), consisting of seething gases at a temperature of 5500°C. Although intensely hot by terrestrial standards, this is cool by comparison with the Sun's core, where the energy-generating nuclear reactions take place; there, the temperature is calculated to be about 15 million °C.

Seen at high magnification, the photosphere exhibits a mottled effect termed *granu-lation*, caused by cells of hot gas bubbling up in the photosphere like water boiling in a pan. Granules range from about 300 km to 1500 km in diameter.

Look carefully at a projected image of the Sun, and you will notice that the edges of the Sun appear fainter than the centre of the disk, an effect known as *limb darkening*. This is caused by the fact that the gases of the photosphere are somewhat transparent, so that at the centre of the disk we are looking more deeply into the Sun's hotter and brighter interior than at the limbs.

Brighter patches on the photosphere known as *faculae* may be visible against the limb darkening. Faculae are areas of higher-temperature gas. More prominent are dark markings known as *sunspots*, patches of cooler gas that come and go on the photosphere.

Sunspots and solar rotation

Sunspots arise where magnetic lines of force burst through the photosphere from within the Sun. The presence of a strong magnetic field blocks the outward flow of heat from inside the Sun, producing the cooler spot. Sunspots have a dark centre known as the *umbra*, of temperature about 4000°C, surrounded by a brighter *penumbra* at about 5000°C. The temperature of a sunspot's umbra is similar to that of the surface of a red giant star such as Aldebaran, so a spot would glow quite brightly if it could be seen in isolation; in practice, sunspots appear dark by contrast with the brilliance of the surrounding photosphere.

Sunspots range in size from small pores no bigger than a large granule to enormous, complex patches many times larger than the Earth. Spots of that size are visible to the naked eye when the Sun is dimmed by the atmosphere, shortly after sunrise or before sunset, or through special solar filters.

A major sunspot takes about a week to develop to full size, then slowly dies away again over the next fortnight or so. The largest sunspots tend to form in groups long enough to span the best part of the distance from the Earth to the Moon. Spot groups can persist for one or two solar rotations (a month or two).

As the Sun rotates, new spots are brought into view at one limb while others disappear around the opposite limb. By observing the progress of spots as they are carried across the Sun's face, we can measure the rotation period of the Sun.

Being gaseous and not solid, the Sun does not rotate at the same rate at all latitudes. It spins most quickly at its equator, in 25 days; this slows to about 28 days at latitude 45°; and near the poles it is slowest of all, 34 days. The average figure usually quoted, 25.38 days, refers to the rotation rate at latitude 17°. These rotation periods are relative to a fixed external point. Since the Earth is not fixed but is orbiting the Sun, a spot takes

SUN DATA

Diameter	1,392,000 km
Mass	1.99×10^{30} kg
Mean density (water = 1)	1.41
Volume (Earth = 1)	1,304,000
Apparent magnitude	−26.8
Absolute magnitude	+4.82
Axial rotation period (average)	25.38 d
Axial inclination	7°.25
Mean distance from Earth	149,600,000 km

about two days longer than this to rotate once with respect to the Earth.

Usually, a spot group contains two main components, aligned east–west. The spot that leads across the disk as the Sun rotates is termed the p (preceding) spot, and is usually larger than its follower, the f spot. The p and f spots have opposite magnetic polarities, like the ends of a horseshoe magnet. The horseshoe is completed by invisible lines of force looping between the spots.

Flares, CMEs and aurorae

Sometimes the intense magnetic field lines in a complex sunspot group become entangled, releasing a sudden flash of energy known as a *flare* that may last from a few minutes to an hour. Atomic particles are spewed out into space by the eruption of a flare. These particles reach Earth about a day later where they cause effects in the upper atmosphere such as radio interference and the ethereal displays known as *aurorae*.

In an aurora, the sky glows with patches of green and red light that can take the form of folded drapery or arches, shimmering and changing shape for hours on end. Aurorae occur near the Earth's magnetic poles and extend to lower latitudes only at times of extreme solar activity, so they are seldom visible outside far northerly or southerly latitudes.

Until the 1970s, flares were the only known cause of aurorae, but it is now realized that another form of activity called coronal mass ejections (CMEs) actually has a greater effect. CMEs are huge bubbles of hot gas thrown off from the Sun, sometimes at the same time as a flare but often without flares. The existence of CMEs was not recognized earlier because they can be detected only from space. They are now monitored routinely by spacecraft.

The solar cycle

The number of sunspots on view, along with other forms of solar activity, waxes and wanes in a cycle lasting 11 years on average, although the length of individual cycles has been as short as 8 years and as long as 16 years. At times of minimum activity the Sun may be spotless for days on end, whereas at solar maximum over a hundred spots may be visible at any one time.

However, the level of activity varies considerably from cycle to cycle; the average number of sunspots visible at maximum has ranged from 40 to 180, and even when the Sun is supposedly quiet some large spots and

Solar prominences extending 65,000 km into space look like enormous burning trees at the edge of the Sun. This photograph was taken at the wavelength of light emitted by hydrogen gas. (Big Bear Solar Observatory)

flares can break out. Solar activity is notoriously unpredictable, which adds to its fascination for observers.

A few general rules can be deduced, however. The first spots of each new cycle appear at latitudes of about 30–35° north and south of the Sun's equator. As the cycle progresses, spots tend to form closer to the equator. Sunspot numbers build up to a peak, then start to decay again.

As minimum approaches, the last spots of the cycle are found at latitudes between 5° and 10° north and south of the equator. At the same time – that is, around solar minimum – the first spots of the next cycle are starting to appear at higher latitudes. Sunspots are seldom found on the Sun's equator, and spots at solar latitudes greater than 40° are exceedingly rare.

The Sun's outer layers

Above the photosphere is a tenuous layer of gas known as the *chromosphere*, about 10,000 km deep. So faint is the chromosphere that it is normally invisible except with special instruments. It can, however, be seen for a few seconds at a total eclipse, when it appears as a pinkish crescent just before and after the Moon completely covers the face of the Sun. This pinkish colour is caused by light from hydrogen gas, and gives rise to the layer's name, which means 'colour sphere'.

Also visible around the edge of the Sun at total eclipses are huge tufts or loops of gas known as *prominences,* which extend from the chromosphere into space. They have the same characteristic rosy-pink colour as the chromosphere, caused by emission from hydrogen.

Like so many features of the Sun, prominences are controlled by magnetic fields. So-called *quiescent prominences* extend for 100,000 km or more across the Sun, often forming graceful arches tens of thousands of kilometres high. When seen silhouetted against the brighter background of the Sun's disk they are termed *filaments*. Quiescent prominences can last for months.

At the other end of the scale are *eruptive prominences*, with lifetimes of only a few hours. They are actually flares seen at the limb of the Sun, ejecting material into space at speeds of up to 1000 km per second. All forms of solar activity – sunspots, flares and prominences – follow the 11-year solar cycle.

Observing the Sun

Projecting the Sun's image is the recommended method of observation for all beginners. A small refractor is ideal for this purpose. If you have a reflector larger than 100 mm, stop down the aperture or heat may damage the optics. Take great care when aiming the telescope. Cover the main lens or mirror and the finder to ensure that no light accidentally enters your eyes. Squint along the tube or line up the telescope on its own shadow (do *not* look through the telescope).

Regular observers use a lightweight solar observing box attached to the eyepiece, into which the Sun's image is projected. Counting individual sunspots and sunspot groups to build up a record of changing solar activity is one of the main purposes of solar observation.

You may also wish to make a sketch of the Sun's appearance. A circle of 6 inches (150 mm) diameter is the standard size for solar disk drawings. Divide the circle into smaller squares 1 cm or so across so that the positions of sunspots can be accurately copied onto a chart.

In a projected image, north is at the top and west at the left for an observer in the northern hemisphere, but the other way round in the southern hemisphere. To find the exact orientation at any time, watch sunspots drift from east to west as the Earth turns.

To observe the Sun safely, project its image onto a white surface. Never look directly at the Sun through any form of optical instrument without a specialized solar filter. (ESO)

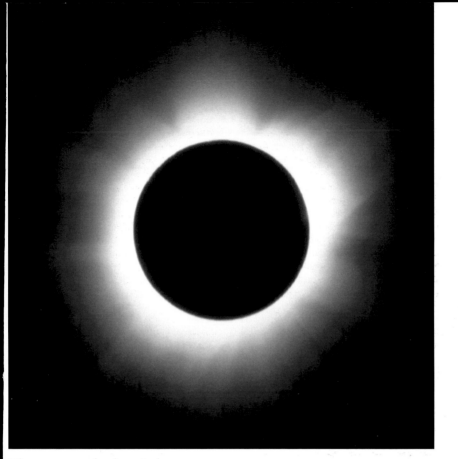

The corona and solar wind

The Sun's crowning glory is its *corona*, a faint
halo of gas that comes into view only when
the brilliant photosphere is blotted out at a
total solar eclipse. The corona is composed
of exceptionally rarefied gas at a temperature
of 1 to 2 million °C. Petal-like streamers of
coronal gas extend from the Sun's equatorial
zone, while shorter, more delicate plumes fan
out from the polar regions. The shape of the
corona changes during the solar cycle: at
solar maximum, when there are more active
areas on the Sun, the corona appears more
rounded than at solar minimum.

Gas from the corona is continually flowing
away from the Sun out into the Solar System,
forming what is known as the *solar wind*.

The spectacular corona of the Sun springs into
view at total solar eclipses. This was its
appearance at the eclipse of 21 June 2001, near
the time of solar maximum, photographed from
Zambia. (John Walker)

Atomic particles of the solar wind are
detected streaming past the Earth at a speed
of about 400 km per second. The most
obvious effect of the solar wind is to push
comet tails away from the Sun.

The solar wind extends outwards beyond
the orbit of the most distant planet, finally
merging with the thin gas between the stars.
In a sense, therefore, all the planets of the
Solar System can be said to lie within the
outer reaches of the Sun's corona.

The Solar System

In one sense our Sun is unusual, for it is not accompanied by another star, as are most stars, but has a family of eight major planets, various moons and countless smaller lumps of debris. This retinue of bodies, all held captive by the Sun's gravity, is called the Solar System.

All the objects in the Solar System shine by reflecting the light from the Sun. Several of the planets when at their brightest can equal or outshine the brightest stars. All the planets orbit the Sun in approximately the same flat plane, so they are always to be found near the ecliptic. A bright 'star' that disturbs a familiar constellation pattern along the ecliptic is therefore almost certainly a planet (although it could, just possibly, be a nova).

Planetary orbits

As seen from above the Sun's north pole, the planets orbit the Sun in an anticlockwise direction. Their orbits are slightly elliptical in shape, so that each planet's distance from the Sun varies somewhat during one orbit. The closest point of a body's orbit to the Sun is known as the *perihelion*; the farthest point is the *aphelion*.

In order of distance from the Sun, the eight major planets are: Mercury, Venus, Earth, Mars, Jupiter, Saturn, Uranus and Neptune. The four inner planets are rocky and relatively small. Then come four giants composed mostly of gas and liquid.

Between the orbits of Mars and Jupiter lies a belt of rocky bodies known as the minor planets or asteroids; some asteroids stray outside this belt across the paths of the planets, creating an impact hazard. At the outer limits of the Solar System is an assortment of small, frozen bodies including Pluto, once regarded as the ninth planet but downgraded to the status of a dwarf planet in 2006.

Venus has the most circular orbit of all the planets; its distance from the Sun varies by less than 1.4% (the Earth's distance varies by 3.4%). The planet with the most elliptical orbit is Mercury, which is over 50% farther from the Sun at aphelion than when at perihelion. Some of the asteroids have orbits that are more elliptical than this, while comets have highly distended orbits that can take them from the innermost part of the Solar System out to way beyond Neptune.

The basic unit of distance in the Solar System is the *astronomical unit*, the average distance of the Earth from the Sun; it is equivalent to 149,597,870 km. Light takes 499 seconds (8.3 minutes) to cross this gulf, so we on Earth see the Sun as it actually appeared 8.3 minutes ago. Although large compared with everyday distances on Earth, the astronomical unit is insignificant in comparison with a light year, the unit used to express distances between stars. There are 63,240 astronomical units in a light year.

How long a planet takes to orbit the Sun depends on its distance: the closest planets orbit the quickest, and the farthest orbit the slowest. The orbital periods of the planets are usually expressed in terms of Earth days and years; they range from 88 days for Mercury to 165 years for Neptune. These values are known technically as *sidereal periods*, and are measured with respect to a fixed point such as the distant stars.

Aspects of the planets

As the two inner planets, Mercury and Venus, move along their orbits they periodically pass between us and the Sun; when that happens, they are said to be at *inferior conjunction*. They are then closest to the Earth, although this fact is of little use to observers since they are lost in the Sun's glare and their illuminated hemispheres are turned away from us.

Above left: The sizes of the planets to scale, compared with a section of the Sun. Above right: The orbits of the planets to scale. (Wil Tirion)

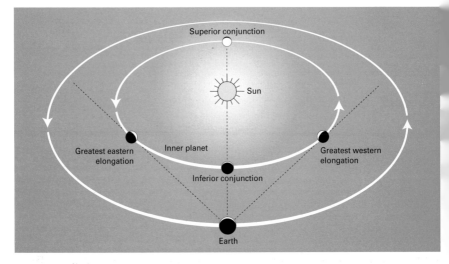

When Venus or Mercury lies directly between the Earth and the Sun it is said to be at inferior conjunction; on the far side of the Sun it is at superior conjunction. Maximum angular separation from the Sun is termed greatest elongation. (Wil Tirion)

The orbits of Mercury and Venus are tilted by 7° and 3°.4 respectively to the orbit of the Earth, sufficient to ensure that they usually pass above or below the disk of the Sun at inferior conjunction. But occasionally they do cross the face of the Sun as seen from Earth, an event termed a *transit*. When in transit, the planet appears as a tiny dark dot like a small sunspot against the glaring solar surface. Transits of Mercury are more common than those of Venus: the next are due in 2016, 2019, 2032 and 2039, whereas the transit of Venus in 2012 is the last one for over a century.

When Mercury or Venus lie on the far side of the Sun from us they are said to be at *superior conjunction*. However, when Mars or any of the planets beyond it lie on the far side of the Sun they are simply said to be at conjunction – being farther from the Sun than we are, outer planets can never come between Earth and the Sun, so there is no ambiguity about which sort of conjunction is

Transit: Multiple images of Mercury at the transit of 8 November 2006 create a dotted line across the face of the Sun in this composite view from the Solar and Heliospheric Observatory (SOHO) spacecraft. (ESA/NASA)

meant. Planets at conjunction are invisible in the Sun's glare.

The best time to view the two inner planets is when they are at their maximum angular separation from the Sun, known as *greatest elongation*. At greatest elongation Venus appears almost exactly half-phase; in the case of Mercury, though, the phase can vary noticeably from 50 per cent as a result of the planet's elliptical orbit. When at eastern elongation, planets set after the Sun in the evening sky; at western elongation, they rise before the Sun in the morning.

The time between one greatest eastern elongation of Mercury and the next, or one greatest western elongation and the next, is 116 days. With Venus, greatest western elongations or greatest eastern elongations recur every 584 days, although Venus is so prominent that it can be adequately observed well away from greatest elongation.

Incidentally, the interval between one given aspect of a planet and the next – be it conjunction, elongation or whatever – is known as its *synodic period*. This differs from the sidereal period because our observation platform, the Earth, is constantly moving in orbit around the Sun.

The best opportunities to see Mars and the other outer planets is when they are directly opposite the Sun in the sky; this is termed *opposition* (Mercury and Venus, being between the Earth and the Sun, cannot come to opposition). At opposition, a planet appears due south for northern hemisphere observers (due north for observers in the southern hemisphere) at midnight local time, or 1 a.m. if daylight-saving time is in operation. Opposition is the best time to observe the outer planets for this is when they lie closest to the Earth and hence appear at their biggest and brightest.

Other Solar Systems?

Planets are thought to be born from a disk of gas and dust left over after the formation of a star, a process that happened some 4.6 billion years ago around our Sun. Several stars are known to have such disks around them, including Vega, Fomalhaut and β (beta) Pictoris, so in these cases we may well be catching the process of planet formation in action. Astronomers have also detected signs of fully formed planetary systems, in most cases from the slight wobble in position of a star as a large planet orbits it. The first star with a planet detected in this way was 51 Pegasi, in 1995. Four years later, not just one but three planets were found around υ (upsilon) Andromedae, the first multi-planet system known.

Other techniques have since confirmed the existence of extrasolar planets, such as tiny dips in a star's light as a planet crosses in front of it and the detection of a star's light reflected from the surface of a planet. With large telescopes of the future, it may be possible to obtain direct images of planets around other stars.

A disk of dust around β (beta) Pictoris, photographed by the Hubble Space Telescope and depicted in false colour, presents persuasive evidence of planet formation. Beta Pictoris itself is obscured by the black circle at the centre. The dust disk is aligned almost edge-on to us, and is larger than the orbit of Neptune around the Sun. (NASA)

The Moon

The Moon, the Earth's natural satellite and nearest celestial neighbour, is an object of perennial fascination for observers. Despite its small size – 3475 km in diameter, roughly a quarter that of the Earth – it is so close to us, on average 384,400 km, that even the unaided eye can pick out the major markings that create the familiar 'man in the Moon' pattern, while ordinary binoculars reveal a wealth of detail on its cratered surface. Some of the most interesting objects to look out for with small instruments are described in the notes on pages 319–327.

The Moon undergoes the most obvious changes in appearance of any celestial body, termed its phases. In the course of a month we see it change shape from a thin crescent to fully illuminated and back again, a result of its orbital motion around us which affects how much of its sunlit side we can see.

When the Moon is in line with the Sun, all its sunlit side is turned away from us and the Moon is said to be *new*. A day or so later the Moon emerges as a thin crescent low in the evening sky. At this stage its night side can be seen faintly illuminated by light reflected from the Earth, a phenomenon known as *earthshine* or, more popularly, 'the old Moon in the young Moon's arms'.

Lunar phases: The Moon goes through a cycle of phases as it orbits the Earth and we see differing proportions of its illuminated side. (Wil Tirion)

The Moon at gibbous phase, 10 days after new. Dark lowland plains (the maria), relatively devoid of craters, contrast with bright, rugged highlands. The large mare towards the top is Mare Imbrium, with the horseshoe-shaped bay Sinus Iridum on its shore. South of Mare Imbrium is the prominent rayed crater Copernicus. Most of the features in this photograph can be picked out with small telescopes or even binoculars, if firmly mounted. (Lick Observatory)

After a week we see the Moon half-lit, a phase termed first quarter because the Moon is a quarter of the way around its orbit. The Moon then fills out, passing through the gibbous phase to become full, approximately 15 days after new Moon. At full the Moon lies on the opposite side of the sky from the Sun and rises around sunset. The phases then repeat in reverse order, reaching last quarter a week or so after full and passing through a crescent back to new Moon. As the Moon's phase increases towards full it is said to be *waxing*, and when the phases are decreasing it is said to be *waning*.

Motions of the Moon

One complete cycle of lunar phases takes 29.5 days, an interval called a *synodic month* or a *lunation*. However, there is a second type of lunar month, called a *sidereal month*, which lasts 27.3 days; this is the time the Moon takes to complete one orbit of the Earth relative to a fixed point such as a dis-

MOON DATA	
Diameter	3475 km
Mass (Earth = 1)	0.012
Mean density (water = 1)	3.35
Volume (Earth = 1)	0.02
Apparent magnitude (full)	−12.7
Axial rotation period (sidereal)	27.32 d
Orbital period (sidereal)	27.32 d
Orbital period (synodic)	29.53 d
Mean distance from Earth	384,400 km
Orbital eccentricity	0.055

tant star. The two differ because the Earth is on the move around the Sun, so the Moon must complete rather more than one orbit to return to the same phase as seen from Earth.

Since the Moon's orbit is somewhat elliptical, its distance from Earth changes during the month. When at its nearest to Earth, a point called *perigee*, it appears as much as 12% larger than when at its farthest, *apogee*.

The Moon spins on its axis in 27.3 days, the same as its orbital period. Such an arrangement is known as a *captured rotation* and comes about because of the effect of the Earth's gravity. With a captured rotation, the Moon keeps the same face turned towards us at all times – so whenever we look at the Moon we always see the same features.

In practice, though, we can see slightly more than half the Moon's surface because of an effect termed *libration*. The Moon's equator is tilted at about 6½° to the plane of its orbit, so that at times we can see as much as 6½° over the Moon's north or south pole; this is known as *libration in latitude*.

In addition, the Moon's speed of motion along its elliptical orbit changes rhythmically as it approaches and recedes from the Earth, while its axial rotation remains uniform. The Moon therefore seems to rock slightly to west and east as it orbits the Earth, so that we can peer up to 7¾° or so around each limb; this is known as *libration in longitude*. The net effect of these librations is that we can see 59 per cent of the Moon's surface at one time or another.

Perigee and apogee: The Moon appears noticeably larger when it is at its closest to Earth (left) than when it is at its farthest.

The lunar surface

The line dividing the lit and unlit portions of the Moon is known as the *terminator*. An astronaut standing on the terminator would see the Sun either rising or setting. Objects near the terminator are thrown into sharp relief by the low angle of illumination; with no atmosphere to soften the shadows, craters and mountains appear particularly rugged. As the Sun rises higher over the moonscape, details become more washed out, but at the same time the difference in contrast between the bright highlands and the dark lowland plains becomes more noticeable.

Near full Moon many individual formations are difficult to pick out. An exception is those craters that have bright ray systems, apparently made of pulverized rock thrown out from the crater during its formation; the rays become more prominent under high illumination. Notable examples of rayed craters are Copernicus, 93 km in diameter, and Aristarchus, 40 km across, the brightest area on the Moon. Most magnificent of all is Tycho, 85 km across, with rays that stretch for over 1000 km.

An oblique view of the rayed crater Copernicus, taken from orbit by the Apollo 12 astronauts in 1969, demonstrates the relative shallowness of such large formations. The keyhole-shaped feature in the foreground is Fauth. (NASA)

With the exception of a few small areas near the poles, each spot on the Moon is subjected to two weeks of unbroken daylight, during which surface temperatures reach more than 100°C, i.e. higher than the boiling point of water, followed by a two-week night when temperatures plummet to −170°C or less. At the poles, though, some crater floors remain permanently shaded from the Sun so temperatures never rise above freezing. Here, ice deposited by cometary impacts could exist beneath the surface.

After looking at the brilliant full Moon it comes as a surprise to realize that the lunar surface rocks are actually dark grey in colour; on average, the Moon's surface reflects only 12 per cent of the light that hits it. If the Moon were, for instance, covered in clouds like those of Venus, it would be over five times brighter.

Seas, mountains and craters

Formations on the Moon bear a variety of curious names. The dark lowland plains are termed *maria* (singular: mare), Latin for 'seas', because the first observers imagined them to be stretches of water; the name persists, even though it has been clear for centuries that the Moon possesses neither air nor liquid water.

Thus we have, for instance, Oceanus Procellarum (Ocean of Storms), Mare Imbrium (Sea of Rains) and Mare Tranquillitatis (Sea of Tranquillity). Less prominent lowlands are termed bays (Sinus), marshes (Palus) or lakes (Lacus). The mare areas were formed by giant impacts and were subsequently flooded, not by water but by lava – so perhaps the old names are not so inappropriate after all.

Mountains on the Moon are named after terrestrial ranges; hence we have the lunar

Mare Humorum, a lowland plain filled with solidified lava that has wrinkled in places. The large, partially flooded crater on its northern shore is Gassendi, noted for the complex clefts on its floor. (European Southern Observatory)

Apollo 12 made a precision landing in November 1969 alongside the Surveyor 3 probe that had arrived 2½ years earlier. Here, astronaut Pete Conrad prepares to remove part of the Surveyor's camera, which was brought back to Earth. In the background is the lunar module. (NASA)

Alps (Montes Alpes) and the lunar Apennines (Montes Apenninus). The craters have been named after philosophers and scientists of the past, although it is true to say that the selection has been somewhat arbitrary.

The term 'crater' is rather misleading as it conjures up the vision of a deep, circular bowl. A better term for the largest craters is walled plains. For example, the 150-km-diameter Ptolemaeus, near the Moon's centre is noticeably hexagonal in outline, as is the 115-km Purbach farther south. These distorted shapes presumably arose from stresses in the lunar crust.

Many large craters have terraced walls, caused by slumping, and central peaks, caused by rebound of the crater floor after the impact of the meteorite that formed them. Outstanding examples of craters that show both terracing and central peaks are Theophilus, Arzachel and Copernicus.

There is a near-continuous range of sizes from the largest craters to the smallest maria.

Most crater-like in appearance of the maria is Mare Crisium, a basin some 500 km across, twice the width of the largest craters. Some craters have been flooded with dark lava which has left them with mare-like floors; a notable example is Plato, 100 km wide, on the northern shore of Mare Imbrium.

Near Plato is a huge half-crater called Sinus Iridum (Bay of Rainbows), 250 km across, which bridges the gap between craters and maria. It opens into Mare Imbrium and only a few low ridges, visible under oblique sunlight, remain where its seaward wall was overrun by the advancing lava. The wall of Sinus Iridum that still stands forms an arcuate range called the Montes Jura (Jura Mountains), 4.5 km high. Sunrise on the Montes Jura, 10 days after new Moon, is one of the most impressive sights for lunar observers (see the photograph on page 309).

Controversy over the origin of the lunar features started almost immediately after Galileo turned his first telescope towards the

Moon in 1609. One school of thought held that they were caused by volcanic action; the opposition believed the craters and maria to have been blasted out by impacts of meteorites and asteroids. It was not until the late 1960s, when space probes and astronauts first reached the Moon, that the controversy was finally resolved in favour of the impact theory, although it is accepted that there has also been a certain amount of subsequent volcanic action.

Space-probe exploration

In 1964 and 1965, three American probes called Ranger 7, 8 and 9 zoomed towards the Moon, sending back a stream of photographs showing features down to a metre across, a hundred times smaller than Earth-based telescopes could see. They revealed that even the apparently smoothest parts of the Moon's surface were pitted with small craters caused by aeons of meteorite bombardment. Landing sites for the manned missions which were then being planned would have to be chosen

with particular care, to prevent the landing craft from toppling into a crater or hitting a boulder.

This initial cursory examination by the Rangers was followed by a two-pronged attack: the Surveyors, a series of craft which soft-landed automatically (the Rangers had simply crashed), and the Lunar Orbiters which, as their name implies, photographed the Moon from close orbit.

Between 1966 and 1968, these two series of probes revolutionized our knowledge of the Moon, paving the way for the manned Apollo landings. The Surveyors showed that the Moon's surface rocks had been eroded by micrometeorites into a compacted topsoil, known as the *regolith*, which fortunately was firm enough to bear the weight of astronauts and their spacecraft. Lunar Orbiter photographs allowed astronomers to compile their most detailed maps of the near and far sides of the Moon.

The Moon's far side, permanently turned away from the Earth, had first been glimpsed by Luna 3, a Soviet probe, in October 1959. Although poor by modern standards, its

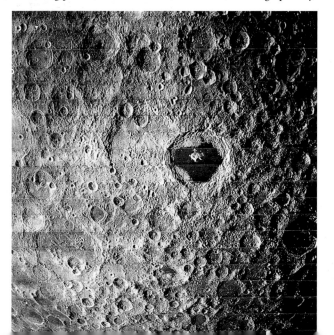

Tsiolkovsky, a 185-km-diameter crater on the far side of the Moon with a dark floor and a prominent central peak, photographed by Lunar Orbiter 3 in 1967. Five Lunar Orbiters were launched in all, photographing the entire Moon in detail and mapping potential landing sites for Apollo astronauts. Lunar Orbiter pictures were transmitted to Earth in strips and so appear banded. (NASA)

On the Apollo 17 mission in December 1972, the last of the Apollo Moon landings, astronauts Eugene Cernan and Harrison Schmitt spent three days exploring southeastern Mare Serenitatis.

Above: Schmitt is dwarfed by a rock that has rolled downhill. In the foreground is the electric Moon car, the lunar rover. (NASA)
Right: Schmitt scoops up pebbles from the lunar soil with a rake. (NASA)

photographs did at least reveal the main difference between the two hemispheres of the Moon: there are scarcely any mare areas on the Moon's far side. Instead, heavily cratered bright uplands dominate the scene. This asymmetry arises because the Moon's crust is some 15 km thicker on the far side.

Large lowland basins like Mare Imbrium do exist on the far side of the Moon, but they have not been flooded by dark lava. Volcanic lavas from the Moon's interior found it easier to leak out through the thinner crust of the Earth-facing hemisphere. The most prominent dark area on the Moon's far side is not a true mare at all, but a deep crater called Tsiolkovsky, 185 km in diameter, three-quarters the size of Sinus Iridum on the visible hemisphere.

The Apollo missions

Once the Lunar Orbiter probes had spied out potential landing sites, the stage was set for astronauts to follow. On 20 July 1969, the Apollo 11 lunar module called Eagle carried Neil Armstrong and Edwin Aldrin to the first manned lunar landing, in the south-western Sea of Tranquillity. The two astronauts spent two hours exploring the Moon's surface, setting up experiments and collecting samples for geologists to study. For the first time, humans had touched another world in space.

By the time Apollo 17 concluded the series of manned landings in December 1972, astronauts had brought back to Earth over 380 kg of Moon samples, most of which remains stored at NASA's Johnson Space Center in Houston, Texas. When divided into the overall cost of Apollo, one kilogram of Moon rocks is worth $100 million. In addition to the Apollo samples, three Soviet automatic lunar probes have returned with a few hundred grams of Moon soil.

What have we learned from these precious specimens? The most astounding fact about the Moon rocks is their immense age. The Apollo 11 samples, for instance, proved to have an age of 3700 million years, older than virtually any rocks on Earth – yet the site from which they came, Mare Tranquillitatis, is one of the youngest areas on the Moon. The youngest of all the rocks brought back came from the Apollo 12 site in Oceanus Procellarum; they are 3200 million years old.

As expected, the lunar maria turned out to be covered with lava flows similar in composition to volcanic basalt on Earth. They do not resemble a rugged lava field on Earth because, over the thousands of millions of years since the lavas were laid down, the sandblasting effect of micrometeorites has eroded the surface rocks to form a layer of soil, the regolith, several metres deep.

By contrast, the highlands, sampled by later Apollo missions, consist of a paler rock called anorthosite, rare on Earth. Rocks from the highlands proved older than those from the maria, mostly dating back 4000 million years or more. Their jumbled, fragmented nature bears witness to the violent bombardment from meteorites which the Moon suffered early in its history.

Tycho, 85 km wide, in the Moon's southern highlands, is the youngest large crater on the Moon, with an estimated age of around 300 million years. Bright rays, thrown out in the impact that created it, extend across the visible face of the Moon. Terraces on its inner slopes, also seen on other large impact craters, are caused by slumping. This photograph was taken by Lunar Orbiter 5 in 1967. (NASA)

Letronne, a crater flooded by lava on the southern shore of Oceanus Procellarum. One wall has been destroyed, although parts of its central peak remain. The low angle of illumination throws the wrinkles of solidified lava into sharp relief. This photograph was taken from orbit during the Apollo 16 mission in 1972. (NASA)

Origin and evolution of the Moon

At the time of Apollo there were three main theories of the Moon's origin – that it split from our planet shortly after its formation, that it was once a separate body that was captured by the Earth's gravity, or that the Earth and Moon formed side by side, much as they are now – but all had drawbacks.

Following the Apollo landings, a fourth proposal emerged, usually termed the Giant Impact theory, which combined aspects of the others. According to this now widely accepted view, a stray body the size of Mars struck the young Earth a glancing blow, spraying debris into orbit around the Earth where it coalesced into the Moon.

Precise dating of lunar samples suggests that this collision happened about 4.5 billion years ago, some 50 million years after the Earth formed. By then, iron had sunk to the Earth's centre, producing a core, so that the material thrown off the outer layers in the collision was predominantly rocky.

While many details of the Moon's origin remain obscure, we now have a much clearer picture of its subsequent history. Heat released by its rapid accumulation in orbit around the Earth (a process termed *accretion*) melted its outer layers. A scum of less dense rock formed a primitive crust which was then battered for several hundred million years by the infall of other debris left over from the Solar System's formation. Similar scars from this mopping-up operation can also be seen on other rocky bodies in the Solar System, notably lunar-like Mercury.

This heavy bombardment created the Moon's jumbled highlands and carved out the mare basins, in the process removing much of the crust on one hemisphere – the side that ended up facing the Earth when the Moon's rotation became tidally locked.

About 4000 million years ago, the storm of meteoric debris abated. Then, slowly, molten lava began to seep out from inside the Moon, solidifying to form the dark low-land maria. So-called *wrinkle ridges* of solidified lava, hundreds of kilometres long but only a few hundred metres high, are visible through binoculars and small telescopes when the maria are under low illumination. In particular, look out for the Serpentine Ridge in eastern Mare Serenitatis when the Moon is five to six days old, or six days past full. Mare Tranquillitatis and Mare Imbrium also have prominent wrinkle ridges. In fact, one 'crater' on Mare Tranquillitatis, called Lamont, consists of nothing more than low ridges of solidified lava.

In places, particularly in western Oceanus Procellarum, a number of blister-like domes can be found near the crater Marius, produced by the upwelling of molten lava. The largest lunar dome, or rather a complex of domes, is Rümker, 70 km wide, farther north on Oceanus Procellarum. Volcanic cones like Vesuvius are absent on the Moon, evidently because the lunar lavas were too runny to build up into mountains.

Further evidence of geological activity is provided by valleys of various kinds. One obvious example, Vallis Alpes (the Alpine Valley), slices through the lunar Alps near Plato. The Alpine Valley seems to be a result of crustal faults, as do some narrower grooves known as *rilles*. The crater Hyginus near the Moon's centre lies in the middle of a long, rimless cleft along which smaller craters have been formed by subsidence. Some lunar faults have produced shallow cliffs, as in the case of the Rupes Recta (popularly termed the Straight Wall), which stretches for 120 km down the eastern edge of Mare Nubium.

Another type of rille, called a *sinuous rille*, snakes over mare surfaces like a meandering river. Apollo 15 landed at the edge of one of these sinuous rilles, called Hadley Rille. Such winding rilles were not formed by running water but are probably collapsed tunnels through which lava streams once flowed.

By 1000 million years ago the volcanic outpourings had ceased, leaving the Moon cold and dead. Since then it has remained virtually unchanged, save for the arrival of an occasional meteorite to punch a new crater in its surface. For instance, the rayed crater Copernicus was formed roughly 1000 million years ago; Tycho was blasted out about 300 million years ago.

Meandering near the foothills of the Apennine Mountains at the eastern rim of Mare Imbrium, Rima Hadley (or Hadley Rille) is a dried-up lava channel that was visited by the Apollo 15 astronauts in 1971. (NASA)

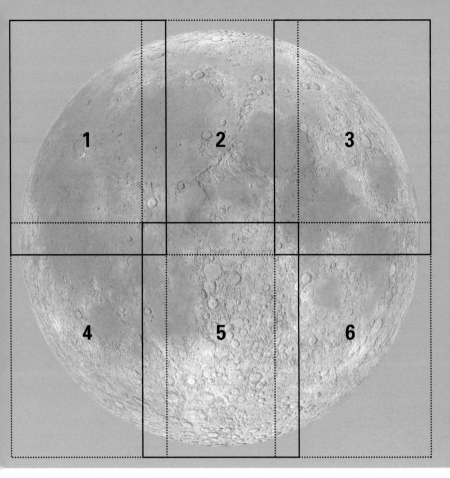

Key to the Moon maps on pages 328–339. These maps show the Moon with north at the top, as it appears to the naked eye and in binoculars. Through an astronomical telescope, south will be at the top, so telescope users must invert the maps. West is at the left, as on Earth, although in the rest of the sky west is in the opposite direction, towards the western horizon.

Moon map 1

Aristarchus Brilliant young crater 40 km in diameter with multiple terracing on its inner walls. The brightest area on the Moon, and the centre of a major ray system. Dark bands are visible on its inner walls under high illumination. Aristarchus has been the location of numerous red glows known as transient lunar phenomena (TLPs), possibly caused by outgassing from the surface. Mountains to the north exhibit fantastic sculpturing.

Copernicus Diameter 93 km. One of the most magnificent craters on the Moon. Centre of a major ray system. Terraced walls and numerous hummocky central peaks. Rays from Copernicus are splashed for more than 600 km across the Mare Imbrium and Oceanus Procellarum.

Encke Low-walled crater 28 km in diameter covered by rays from nearby Kepler. Brilliant under high illumination.

Euler   Small, sharp crater, 28 km across, in southwestern Mare Imbrium. Centre of a minor ray system.

Fauth   Keyhole-shaped crater south of Copernicus, nearly 2 km deep. Striking under low illumination.

Harpalus   Crater 39 km in diameter on Mare Frigoris, bright under high illumination. The smaller crater Foucault, 23 km in diameter, lies between it and Sinus Iridum.

Herodotus   Companion to Aristarchus, similar in size (diameter 35 km) but different in structure – Herodotus has a dark, lava-flooded floor and it is not a ray centre. The W-shaped Vallis Schröteri (Schröter's Valley), 150 km long, starts at a craterlet outside the northern wall of Herodotus.

Hevelius   Large, bright crater 115 km in diameter on western shore of Oceanus Procellarum, with clefts crossing the floor. The smaller, sharper crater Cavalerius (diameter 60 km) adjoins to the north.

Kepler   Major ray centre in Oceanus Procellarum. Brilliant crater, 31 km in diameter, with central peak and heavily terraced walls.

Mairan   Prominent crater in highlands west of Sinus Iridum. Diameter 40 km.

Marius   Dark, flat-floored ring 41 km in diameter on Oceanus Procellarum, notable because it lies on a wrinkle ridge. There are many dome-like structures in this area where lava has bubbled up through the surface.

Oceanus Procellarum   Vast dark plain with no clear-cut borders extending from Mare Imbrium southwards to Mare Humorum. Oceanus Procellarum has a maximum width of about 2000 km and occupies an area of more than two million square km. It is dotted with numerous craters and bright rays.

Prinz   U-shaped formation 52 km in diameter, remains of a half-destroyed crater flooded by lava

Copernicus, a large, bright ray crater on Oceanus Procellarum, is one of the most impressive sights on the lunar surface. (Consolidated Lunar Atlas)

from Oceanus Procellarum. Notable because of sinuous rilles to the north.

Pythagoras   Magnificent large crater 128 km in diameter with terraced walls rising nearly 5 km and prominent central peak, near the northwestern limb of the Moon.

Reiner   Sharp crater 30 km in diameter west of Kepler in Oceanus Procellarum. Particularly notable is a tadpole-shaped splash of brighter material on the dark plain to the west and north.

Reinhold   Prominent crater 42 km in diameter southwest of Copernicus, with terraced walls. A smaller, lower ring to the northeast is Reinhold B.

Rümker   Remarkable formation on northwestern Oceanus Procellarum, visible only under low illumination. Rümker is an irregular, lumpy dome 70 km wide.

Sinus Iridum   Beautiful large (diameter 250 km) bay on the Mare Imbrium. Its seaward wall has been broken down by invading lava and is reduced to a few low wrinkle ridges. Its remaining walls, known as Montes Jura (the Jura Mountains), are brilliantly illuminated by the morning Sun. The deep, 38-km-diameter crater on the northern rim is Bianchini.

Moon map 2

Agrippa Oval-shaped crater 46 km in diameter with central peak, forming a neat pair with Godin.

Anaxagoras Crater 51 km in diameter near the lunar north pole, centre of an extensive ray system.

Archimedes Distinctive, flooded ring 83 km in diameter in eastern Mare Imbrium, notable for its almost perfectly flat floor. Apollo 15 landed southeast of here, at the foot of the Montes Apenninus, in 1971.

Aristillus Prominent crater 55 km in diameter in eastern Mare Imbrium with terraced walls, numerous surrounding ridges and a complex central peak 900 m high. Under high illumination it is seen to be surrounded by a faint ray system. Forms a pair with Autolycus to the south.

Aristoteles Magnificent, partially lava-flooded crater 87 km in diameter, touching the smaller crater Mitchell to the east. Numerous ridges radiate from its outer walls. Makes a pair with Eudoxus to the south.

Aristillus (top right), Autolycus and the flooded Archimedes form a fascinating group on the eastern Mare Imbrium. (Consolidated Lunar Atlas)

Autolycus Prominent crater 39 km in diameter, south of Aristillus. It is the centre of a faint ray system, seen under high illumination.

Cassini Flooded ring of unusual appearance with partially destroyed walls, 56 km in diameter. Contains a bowl-shaped crater, Cassini A, 15 km wide.

Eratosthenes Prominent, deep crater on the edge of Sinus Aestuum at the southern end of Montes Apenninus (the Apennine Mountains). Terraced walls and craterlet on central peak. Its diameter is 58 km.

Eudoxus Rugged crater 67 km in diameter with small central peak and terraced walls. South of Aristoteles.

Godin Smaller but deeper companion to Agrippa; 35 km in diameter with central peak, bright walls and faint ray system.

Hyginus Rimless crater 9 km in diameter, centre of a cleft or rille 220 km long, visible in small telescopes, evidently formed by collapse of the surface. To the east is another cleft, Rima Ariadaeus.

Lambert Crater 30 km in diameter in Mare Imbrium, with central craterlet. Situated on a wrinkle ridge. Under low illumination a larger 'ghost' ring, Lambert R, is seen to the south.

Linné Bright spot on Mare Serenitatis, best seen under high illumination. At its centre is a small, young crater 2.4 km in diameter.

Manilius Bright crater 39 km in diameter in Mare Vaporum, with terraced walls and central peak. Develops a ray system as illumination increases, as does its neighbour Menelaus (diameter 27 km) on the rim of Mare Serenitatis.

Mare Imbrium Enormous circular plain 1150 km in diameter, dominating this section. Bounded by the Montes Alpes, Caucasus, Apenninus and Carpatus, but open on the southwest to Oceanus Procellarum. Mare Imbrium has a double structure: traces of a smaller inner ring are visible, marked by a few isolated mountains and wrinkle ridges. Note the different shades of lava on its surface. Isolated mountains protruding from its dark floor include Pico and Piton, plus the ranges known as Montes Recti, Spitzbergen and Teneriffe.

Mare Serenitatis   A major rounded lunar sea 660 × 600 km, bounded on the northwest by Montes Caucasus and on the southwest by Montes Haemus. A bright ray from Tycho crosses the dark lava plain, passing through the crater Bessel, 16 km in diameter. Under high illumination, Mare Serenitatis is seen to be rimmed with darker lava. Note also a major wrinkle ridge (the Serpentine Ridge) on the east side. Apollo 17 landed at the southeastern edge of Mare Serenitatis in 1972.

Plato   Unmistakable large, dark-floored crater in the highlands north of Mare Imbrium, 101 km in diameter; prominent under all conditions of illumination. Tiny craterlets pockmark its flat floor. Temporary obscurations of the surface, presumed to be caused by outgassing, have been observed in this area. Landslips appear to have detached part of the inner western wall.

Pytheas   Small (diameter 20 km) but deep and prominent crater on Mare Imbrium, rhomboidal in outline. Becomes brilliant under high illumination.

Stadius   'Ghost' ring to the east of Copernicus, outlined only by a few ridges and craterlets and visible only under low illumination. It is 69 km in diameter.

Timocharis   Bright crater 34 km in diameter on Mare Imbrium, with terraced walls and distinctive central crater. Faint ray centre.

Triesnecker   Crater 26 km in diameter with a surrounding system of clefts.

Vallis Alpes (Alpine Valley)   Flat-floored valley 180 km long through the lunar Alps, connecting Mare Imbrium with Mare Frigoris.

Moon map 3

Atlas   Large crater, 87 km in diameter, with terraced walls and complex floor. A ruined ring abuts to the northwest. Atlas and Hercules form one of the many crater pairs in this region.

Burckhardt   Complex crater 57 km across, overlapping older formations (Burckhardt E and F) on either side.

Bürg   Prominent crater despite its moderate size (diameter 40 km). Central peak. Lies in the centre of Lacus Mortis. Note nearby rilles.

Cleomedes   Large, irregular-shaped crater 126 km in diameter with partially flooded floor, to north of Mare Crisium. West wall is interrupted by 43-km-diameter crater Tralles.

Endymion   Large, dark-floored enclosure 125 km in diameter, with walls up to 4900 m high.

Franklin   Crater 56 km in diameter. Forms a pair with the 40-km-diameter Cepheus to the northwest.

Geminus   Prominent crater 86 km in diameter with central peak. Bright rays emanate from two smaller craters nearby, Messala B and Geminus C.

Hercules   Flat-floored enclosure 67 km in diameter containing sharp, bright bowl crater Hercules G.

Le Monnier   Old and flooded crater with its western wall washed away by lava from Mare Serenitatis. Diameter 61 km.

Mare Crisium   Unmistakable dark lowland plain ringed by high mountains like an oversized crater, 420 × 550 km. Main feature on its flat, lava-flooded floor is the crater Picard, 23 km across, with smaller crater Peirce to its north.

Posidonius, complex and partly ruined, lies east of Mare Serenitatis near a prominent wrinkle ridge, the Serpentine Ridge. (Consolidated Lunar Atlas)

Mare Tranquillitatis An irregular-shaped low-land, 540 × 780 km. Wrinkle ridges attest to numerous lava flows over the plain. The main crater on the mare is the distorted Arago, on the western side, 26 km in diameter. Apollo 11 made the first manned lunar landing in southwest Mare Tranquillitatis in 1969.

Plinius Distinctive crater 42 km in diameter with complex central peak combined with a central craterlet. Stands between Mare Serenitatis and Mare Tranquillitatis, along with Dawes to the northeast, 18 km in diameter.

Posidonius Large (diameter 100 km) partially flooded and ruined crater on the northeast shore of Mare Serenitatis. Its floor contains a curving ridge, several rilles and a small bowl crater. The ruined structure Chacornac, 51 km in diameter, abuts to the southeast.

Proclus Small (28 km wide) but brilliant crater on the western edge of Mare Crisium. High illumination shows it to be the centre of a fan-shaped ray system.

Taruntius Low-walled crater in northwestern Mare Fecunditatis, 56 km in diameter, with concentric inner ring. Crater Cameron (formerly Taruntius C) interrupts the northwestern wall. Centre of a faint ray system.

Thales Bright ray crater 32 km in diameter, northeast of Mare Frigoris.

Moon map 4

Bullialdus Handsome crater in Mare Nubium with terraced walls and complex central peak. Diameter 59 km. Two smaller craters, Bullialdus A and B, form a chain extending south.

Flamsteed Modest (diameter 21 km) crater on Oceanus Procellarum with a much larger ring of eroded hills to its north, Flamsteed P, which appears bright against the mare background at full Moon.

Gassendi Large, 110-km-diameter, partially flooded ring on northern border of Mare Humorum. Complex internal pattern of clefts, ridges and hillocks. A rich area for transient lunar phenomena (TLPs). Deeper crater Gassendi A interrupts the north wall, with the smaller Gassendi B lying farther north.

Grimaldi Vast, 220-km-diameter, dark-floored enclosure at the western limb of the Moon with broad, crater-strewn walls. Nearer the limb is a smaller dark patch, marking the floor of Riccioli, 140 km in diameter.

Hainzel Curious keyhole-shaped formation, composed of three craters fused together; the two smaller ones are Hainzel A and C.

Hippalus Lava-flooded bay 58 km in diameter on the shores of Mare Humorum, in an area with many rilles.

Lansberg Prominent crater 40 km in diameter with massive walls and a central peak, on Oceanus Procellarum. Apollo 12 landed southeast of Lansberg in 1969.

Schickard, the large formation at upper left, lies north of the lava-filled plateau Wargentin. Note the oddly shaped Hainzel and Schiller at upper and lower right. (Consolidated Lunar Atlas)

Letronne   Large bay, 120 km wide, on the south side of Oceanus Procellarum. Its seaward-facing wall has evidently been washed away by invading dark lava.

Mare Humorum   Rounded lowland plain 370 km across with Gassendi on its northern rim. Ringed by clefts and wrinkle ridges. On the south it invades the rings Doppelmayer and Lee, although Vitello escapes destruction. On the east is the ringed bay Hippalus, associated with much surface faulting.

Schickard   Major dark-floored enclosure 227 km across. South of Schickard are the overlapping craters Nasmyth (diameter 77 km) and Phocylides (114 km). Adjoining it to the southwest is the extraordinary plateau Wargentin, 84 km wide, evidently a crater filled to the brim with solidified lava.

Schiller   Curious footprint-shaped enclosure, 165 × 65 km.

Sirsalis and Sirsalis A   Twin craters 42 and 49 km wide near a 280-km-long cleft, Rima Sirsalis, which stretches towards 130-km Darwin.

Moon map 5

Abulfeda   Prominent smooth-floored crater with sculptured inner walls, 65 km in diameter. Apollo 16 landed in the highlands north of here in 1972.

Albategnius   Large (diameter 136 km) walled enclosure with central peak. The prominent crater Klein, 44 km in diameter, breaks its southwest rim.

Aliacensis   Prominent crater with irregular outline, 80 km in diameter. Forms a pair with the slightly smaller crater Werner.

Alpetragius   Bowl-shaped crater 3900 m deep on outer slopes of Alphonsus, with large central dome. Diameter 40 km.

Alphonsus   Large enclosure 118 km in diameter with complex walls and a ridge running through its centre. Numerous craterlets and clefts cover the floor; several dark patches are visible under high illumination. Alphonsus has been the site of reported obscurations, possibly caused by the release of gas from the surface.

Arzachel   Magnificent crater 96 km in diameter with terraced walls and prominent central peak. To its east is a noticeable crater 31 km in diameter with central peak, like a smaller version of Alpetragius, called Parrot C.

Barocius   Large formation southeast of Maurolycus, diameter 82 km. Its northeast wall is broken by Barocius B. To the southwest is Clairaut, 75 km in diameter, between Barocius and Cuvier.

Birt   Sharp, bright crater 17 km in diameter on eastern Mare Nubium. Telescopes reveal a smaller crater (diameter 7 km), Birt A, on the east wall, and under low illumination a rille to the west.

Blancanus   Crater 110 km in diameter, south of Clavius.

Clavius   Magnificent walled plain 225 km in diameter. Note the distinctive arc of smaller craters across its convex floor. Its south wall is interrupted by the 50-km crater Rutherfurd, and its northeast wall by 52-km Porter.

Delambre   Prominent crater 53 km in diameter with irregular interior, southwest of Mare Tranquillitatis.

Deslandres   Huge, low and eroded formation 235 km in diameter, southeast of Mare Nubium. The partially ruined ring Lexell, diameter 63 km, opens onto its southern side, and the prominent crater Hell (diameter 33 km) lies on its western floor.

Fra Mauro   Largest member, 94 km wide, of an old, eroded crater group north of Mare Nubium, also including Bonpland (diameter 60 km), Parry (47 km) and Guericke (60 km). Apollo 14 landed just north of Fra Mauro in 1971.

Heraclitus   Curious elongated formation 90 km long with central ridge, south of Stöfler. Its southern end is rounded off by the crater Heraclitus D. Between Heraclitus and Stöfler is the crater Licetus, diameter 75 km. Touching Heraclitus to the east is Cuvier, also 75 km in diameter.

Clavius is a magnificent formation in the Moon's southern uplands with an arc of smaller craters crossing its floor. (Consolidated Lunar Atlas)

Herschel   Deep (3900 m) crater with elongated central peak, north of Ptolemaeus; diameter 41 km. Farther north is the slightly smaller Spörer, a partially filled ring.

Hipparchus   Large, eroded enclosure 150 km in diameter north of Albategnius. Its central 'peak' is actually a small ruined crater. On its northeastern floor is Horrocks, 31 km in diameter and 2800 m deep. Between Hipparchus and Albategnius lies the crater Halley (diameter 36 km), east of which is Hind, 29 km wide and 2800 m deep.

Longomontanus   Large walled plain in the rugged southern uplands of the Moon. Diameter 145 km. A ridge to the east forms a crescent-shaped enclosure known as Longomontanus Z.

Maginus   Major walled plain 185 km in diameter north of Clavius. Convex floor with small peaks in centre. Southwest wall is interrupted by smaller crater, Maginus C.

Mare Nubium   Irregular dark lowland plain, covered with numerous wrinkle ridges and ghost craters. On its southern shore is the flooded crater Pitatus, and in the southwest are the dark-floored pair Campanus and Mercator (48 km and 47 km in diameter). Its most celebrated feature is a 120-km-long fault on the eastern side called the Rupes Recta, or Straight Wall, between the craters Birt and Thebit. The Rupes Recta appears to run north–south through the remains of an old flooded crater of which only the eastern half remains. At the southern end of the Rupes Recta are the Stag's Horn Mountains, apparently the remains of a flooded crater.

Maurolycus   Distinctive crater, 114 km wide, with twin central peak. Its walls rise to 5000 m. It partially obliterates a smaller, unnamed formation to the north.

Moretus   Crater with a central peak in the jumbled uplands southeast of Clavius. Diameter 114 km. Closer to the south pole are the craters Short (diameter 71 km) and Newton (diameter 79 km), heavily foreshortened.

Pitatus   Large, dark-floored ring 105 km across on the southern shore of Mare Nubium. Invading lava from Mare Nubium has partly destroyed the crater's walls and left only the vestige of a central peak. Note the rilles around its inner walls. A smaller, similarly flooded ring adjoining it to the east is Hesiodus.

Ptolemaeus   Vast walled plain 153 km in diameter, hexagonal in shape. Its ancient floor is heavily pockmarked with smaller craters, the most prominent being Ptolemaeus A.

Purbach   Battered but still prominent large crater, 115 km in diameter. Its floor contains ridges, and its north wall is interrupted by the oval crater Purbach G, while its south wall intrudes into Regiomontanus.

Regiomontanus   Flooded crater, 124 km across. Its central peak has a summit craterlet. Forms a pair with Purbach, both noticeably hexagonal in outline.

Scheiner   Crater 110 km in diameter southwest of Clavius. The largest of three craterlets on its floor is Scheiner A.

Stöfler   Large, flat-floored formation 126 km wide, west of Maurolycus. Its eastern wall is destroyed by the intrusion of several craters, the largest being Faraday, diameter 69 km. The southern wall of Faraday is disturbed by Faraday C, which itself intrudes into Stöfler P.

Thebit   Fascinating triple crater on southeastern Mare Nubium. The main crater, 55 km in diameter, is broken by the 20-km Thebit A, which in turn is broken by the still-smaller Thebit L.

Tycho   Magnificent crater in the Moon's southern uplands, 85 km in diameter. Prominent at all angles of illumination, and brilliant under high lighting. Massively terraced walls rising up to 4500 m, imposing central peak and rough floor. Tycho is the major ray crater on the Moon. Rays from Tycho extend for 1500 km or more in all directions. Note the dark 'collar' around Tycho under high illumination. Tycho is the youngest of the Moon's major features, some 300 million years old.

Walter   Large crater 128 km in diameter, considerably modified by landslips on the inner walls and by several interior craterlets. Appears almost square in outline.

Werner   Prominent crater 70 km in diameter with walls 4200 m high, notably sharper and more rounded than neighbouring craters. The floor of Werner is dotted with several hills.

Moon map 6

Capella   Prominent crater north of Mare Nectaris, 45 km in diameter, deformed by a surface fault. Large central peak. Isidorus, 42 km in diameter, adjoins it to the west.

Catharina   One of a curving trio of craters around western Mare Nectaris. Diameter 104 km. A faint ring, Catharina P, covers much of its northern floor.

Cyrillus   Crater 95 km in diameter, with complex terraced walls, multiple central peak and rugged floor. Overlapped by Theophilus.

Fracastorius   Horseshoe-shaped bay 124 km in diameter on the southern shore of Mare Nectaris. Dark lava has breached its northern wall and flooded its interior. The crater Fracastorius D distorts its western wall.

Janssen   Vast, irregular enclosure, 180 × 240 km across, heavily bombarded. In the north it is interrupted by Fabricius, 78 km wide, which has a central peak. On the west wall is a smaller (diameter 34 km) sharp crater, Lockyer. Southeast of Janssen are the twin craters Steinheil and Watt (67 km and 66 km in diameter respectively). Farther north of Fabricius is Metius, diameter 88 km.

Langrenus   Magnificent bright-walled plain on eastern Mare Fecunditatis with terraced walls, outer ridges and a complex central peak. Its diameter is 133 km and it is a ray centre. Northwest of it, on Mare Fecunditatis, is a trio of smaller craters called Langrenus F, B and K in order of decreasing size. South of Langrenus is the large, flooded formation Vendelinus, 155 km across.

Mädler   Prominent crater (diameter 28 km) on northwest Mare Nectaris, with a central ridge.

Mare Fecunditatis   Irregularly shaped 820 × 660 km dark lowland area, connecting with Mare Tranquillitatis. On its western border it invades several craters, notably Gutenberg (diameter 71 km) and Goclenius (diameter 55 × 75 km). Note the numerous clefts in this area.

Mare Nectaris   Rounded lowland plain 350 km wide, bordered by several large craters, notably Theophilus, Cyrillus, Catharina and Fracastorius. An outer mountain ring, Rupes Altai (the Altai Scarp), surrounds Mare Nectaris.

Messier and Messier A   Elliptical pair of craters on Mare Fecunditatis, prominent despite their small sizes, 11 km and 13 km. Two bright rays extend from the western member, Messier A. Both craters appear brilliant under high illumination. The pair may be the result of a glancing impact.

Palitzsch   A crater and valley on the eastern side of Petavius. The crater Palitzsch itself, 41 km wide, is at the southern end of the valley, Vallis Palitzsch, which is 110 km long.

Petavius   Magnificent walled enclosure 177 km in diameter. A prominent rille runs across the floor from the massive, complex central peak to the terraced walls, which appear double in parts. Ridges radiate from its outer walls. West of Petavius is Wrottesley, 57 km in diameter, and with a central peak.

Piccolomini   Beautiful crater on Rupes Altai (the Altai Scarp). Diameter 89 km, with a broad central peak and terraced walls.

Snellius   Eroded crater 83 km in diameter which straddles the shallow Vallis Snellius, running northwest towards Borda and southeast past Adams.

Theophilus   Imposing crater 110 km in diameter on the northwestern rim of Mare Nectaris, with a massive 2200-m central mountain. Terraced walls rise over 5000 m above the floor, with many external ridges.

Vallis Rheita (Rheita Valley)   Crater chain northeast of Janssen and Fabricius. It can be traced for a total length of about 500 km. The crater Rheita itself, 70 km in diameter and with a small central peak, lies at the northern end of the valley.

The curving crater chain of Theophilus (top), Cyrillus and Catharina on the western shore of Mare Nectaris is a distinctive feature of this section of the Moon. (Consolidated Lunar Atlas)

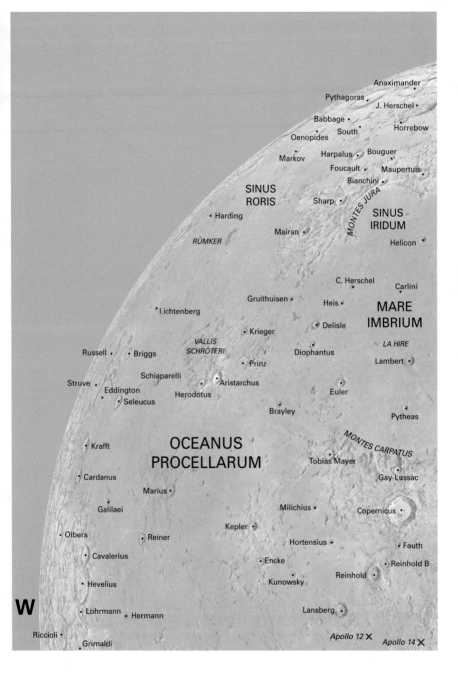

Anaximander

Pythagoras

J. Herschel

Babbage

Horrebow

Oenopides   South

Markov   Harpalus   Bouguer

Foucault   Maupertuis

Bianchini

SINUS RORIS

Sharp   **SINUS IRIDUM**

MONTES JURA

Harding

Mairan

RÜMKER

Helicon

C. Herschel   Carlini

Gruithuisen   Heis

Lichtenberg

Delisle   **MARE IMBRIUM**

Krieger

VALLIS SCHRÖTERI   *LA HIRE*

Russell   Briggs   Diophantus   Lambert

Prinz

Schiaparelli

Struve   Aristarchus

Eddington   Herodotus   Euler

Seleucus

Brayley

Pytheas

Krafft

OCEANUS PROCELLARUM   *MONTES CARPATUS*

Tobias Mayer

Cardanus   Gay-Lussac

Marius

Milichius   Copernicus

Galilaei

Kepler

Olbers   Reiner   Hortensius   Fauth

Cavalerius   Encke   Reinhold B

Reinhold

Hevelius   Kunowsky

Lohrmann   Hermann   Lansberg

W

Riccioli   *Apollo 12* ✕   *Apollo 14* ✕

Grimaldi

N

Scoresby

Anaximenes · Philolaus · Anaxagoras · Goldschmidt · Meton · Baillaud

Carpenter · Barrow Neison

· Anaximander Epigenes · Arnold

· Pythagoras Birmingham · W. Bond · C. Mayer · Democritus Strabo

J. Herschel · Fontenelle Timaeus Kane Thales ·

Horrebow MARE FRIGORIS Archytas Sheepshanks Gärtner

Protagoras · Galle

Bouguer · La Condamine

Maupertuis Plato · Aristoteles · Mitchell Baily

MONTES Egede

· Bianchini MONTES TENERIFFE PICO ALPES Eudoxus LACUS Bürg

SINUS RECTI VALLIS ALPES MORTIS Mason

IRIDUM Plana

Piazzi Smyth ·

Helicon · · Le Verrier Kirch · PITON Cassini · Alexander

Theaetetus · Calippus

MONTES MONTES

SPITZBERGEN Aristillus CAUCASUS

Carlini

MARE Archimedes · Autolycus

IMBRIUM PALUS

PUTREDINIS Linné · MARE

· LA HIRE Apollo 15 × SERENITATIS

Lambert · Timocharis

· Aratus Bessel ·

Wallace · Conon · Sulpicius Gallus

Pytheas MONTES HAEMUS Menelaus

MONTES CARPATUS Manilius · Auwers Plinius

Gay-Lussac Eratosthenes MARE Ross ·

VAPORUM Julius Maclear ·

SINUS Boscovich · Caesar

Stadius AESTUUM Ukert · Sosigenes

Copernicus Hyginus RIMA ARIADAEUS Arago

· Fauth Bode · Silberschlag · Manners

Pallas · Murchison Agrippa · Whewell · Cayley

· Reinhold B Chladni Triesnecker Tempel · Dionysius

· Reinhold Godin · d'Arrest Ritter · Sabine

Gambart Schröter · SINUS Rhaeticus Schmidt ·

Sömmering · MEDII Theon

Turner · Mösting · Oppolzer Senior

Apollo 14 × Lalande Flammarion · Réaumur Pickering Lade Theon · Delambre

Spörer Horrocks · Junior

· Saunder

Democritus • Strabo
Thales •
• Gärtner   de la Rue

MARE HUMBOLDTIANUM

MARE FRIGORIS
Endymion

Baily •

Hercules Atlas   Mercurius •
LACUS   Chevallier •   • Zeno
Bürg   • Carrington
MORTIS   Mason   Oersted • Shuckburgh • Schumacher
Plana Williams •   Hooke •
Grove •   Cepheus •

Franklin •
LACUS SOMNIORUM   • Messala
Maury •   Berzelius •   Bernoulli •   • Gauss
Daniell •   Geminus •
Hall •   • Berosus
• G. Bond   Burckhardt •   • Hahn
Posidonius •   Kirchhoff • Newcomb
Chacornac •   Debes •
MONTES   Tralles •
TAURUS   Cleomedes •   • Delmotte
MARE   Le Monnier •
SERENITATIS
Römer •   Eimmart • Plutarch
Macrobius
• Bessel   • Littrow   • Hill • Tisserand   MARE
X Apollo 17   ANGUIS
Maraldi •   Carmichael •
• Dawes   Vitruvius •   • Peirce
MARE   Alhazen •
Auwers • Plinius   Franz • PALUS   Proclus •   CRISIUM
SOMNI   Yerkes   Hansen •
Ross •   • Jansen   Lyell •   • Picard
Maclear •   Glaisher •   • Lick
• Sosigenes   MARE   • Cauchy • da Vinci   Condorcet •
Arago •   TRANQUILLITATIS   Shapley •   • Auzout
Manners • • Lamont   Lawrence •   Firmicus •   MARE
Dionysius •   Daly •   UNDARUM
Ritter • Maskelyne   Taruntius • Apollonius •   • Dubyago
• Sabine
Schmidt   X Apollo 11   • Secchi   MARE
Censorinus •   MARE   SPUMANS
Delambre •   Messier A • • Messier FECUNDITATIS   Webb • • Maclaurin
• Hypatia

MARE MARGINIS
MARE SMYTHII

E

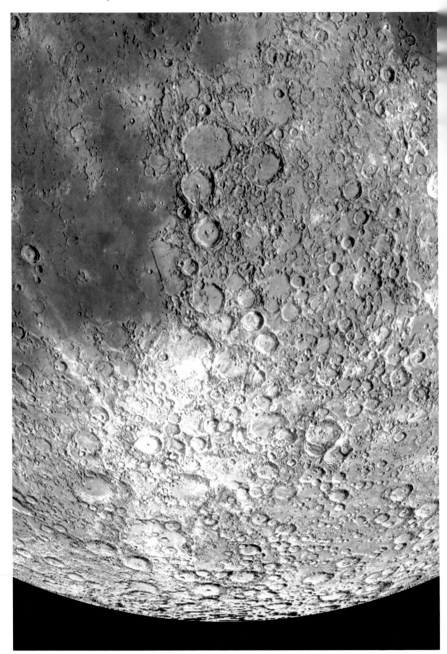

Reinhold
Schröter
Chladni Triesnecker
Agrippa
Tempel Whewell
Cayley
Dionysius
Gambart
Godin
d'Arrest
Ritter
SINUS
MEDII
Schmidt
Sabine
Sömmering
Rhaeticus
Theon
Senior
Turner
Mösting
Oppolzer
Réamur
Pickering
Lade
Theon
Junior
Flammarion
Spörer
Horrocks
Saunder
Delambre
Apollo 14 ✕
Lalande
Taylor
Fra Mauro
Herschel
Gyldén
Hipparchus
Alfraganus
Bonpland
Parry
Müller
Hind
Apollo 16
✕
Zöllner
Palisa
Ptolemaeus
Halley
Andĕl Dollond
Kant
MARE
COGNITUM
Guericke
Davy
Albategnius
Ritchey
Descartes
Darney
Alphonsus
Klein
Burnham
Abulfeda
Cyrillus
Lassell
Parrot
Vogel
Tacitus
Opelt
Alpetragius
Argelander
Almanon
Lubiniezky
MARE
NUBIUM
Arzachel
Airy
Catharina
Bullialdus
Geber
Bullialdus A
Donati
Faye
Abenezra
Fermat
Nicollet
Birt
Thebit
Delaunay
Azophi
Sacrobosco
Bullialdus B
La Caille
Playfair
König
Purbach
Blanchinus
Pons
Kies
Regiomontanus
Werner
Apianus
Pontanus
Wilkins
Mercator
Hesiodus
Pitatus
Aliacensis
Poisson
Zagut
Lindenau
Weiss
Hell
Walter
Goodacre
Cichus
Gauricus
Deslandres
Gemma
Frisius
Celsius
Rabbi Levi
Capuanus
Wurzelbauer
Lexell
Nonius
Kaiser
Riccius
Ball
Haidinger
Heinsius
Sasserides
Miller
Fernelius
Buch
Büsching
Epimenides
Orontius
Stöfler
Wilhelm
Huggins Nasireddin
Faraday Maurolycus
Nicolai
Lagalla
Tycho
Pictet
Saussure
Montanari
Brown
Barocius
Spallanzani
Street
Proctor
Licetus
Clairaut
Breislak
Dove
Longomontanus
Heraclitus
Ideler
Pitiscus
Bayer
Maginus
Cuvier
Baco
Hommel
Vlacq
Schiller
Porter
Deluc
Lilius
Asclepi
Rosenberger
Rost
Jacobi
Tannerus
Nearch
Segner
Clavius
Zach
Kinau
Hagecius
Zucchius
Bettinus Scheiner
Rutherfurd
Pentland
Mutus
Bailly Kircher
Blancanus Gruemberger
Cysatus
Curtius
Manzinus
Helmholtz
Wilson
Klaproth
Moretus
Simpelius
Boussingault
Casatus
Short
Boguslawsky
Newton
Schomberger

RUPES RECTA

RUPES ALTAI

S

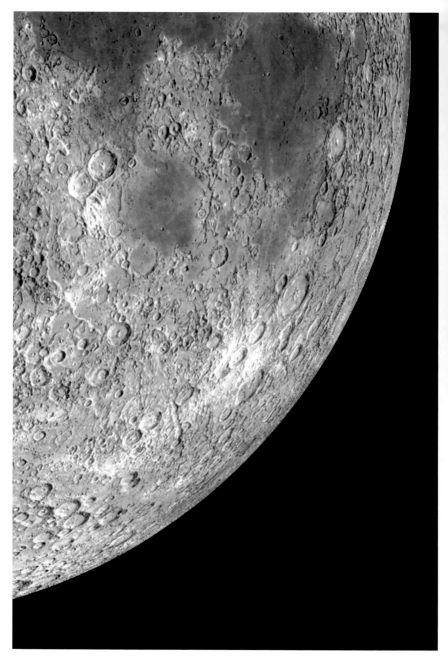

Dionysius
Ritter
Sabine
Schmidt
✕ Apollo 11
Delambre
Hypatia
Taylor
Alfraganus
Torricelli
Kant
Theophilus
Mädler
Gaudibert
Daguerre
Cyrillus
MARE
NECTARIS
Tacitus
Beaumont
Catharina
Rosse
Fracastorius
Fermat
Polybius
Pons
RUPES ALTAI
Wilkins
Zagut
Rothmann
Piccolomini
Weinek
Lindenau
Neander
Rabbi Levi
Stiborius
Riccius
Wöhler
Brenner
VALLIS RHEITA
Metius
Nicolai
Fabricius
Spallanzani
Lockyer
Janssen
Dove
Pitiscus
Steinheil
Watt
Vlacq
Hommel
Biela
Nearch
Rosenberger
Hanno
Hagecius
Pontécoulant
Helmholtz
Boussingault

Maskelyne
MARE
TRANQUILLITATIS
Censorinus
Capella
Isidorus
Zöllner
Gutenberg
Goclenius
Magelhaens
Magelhaens A
Colombo A
Colombo
Bohnenberger
Santbech
Borda
Snellius
Hase
Reichenbach
Stevinus
Adams
Rheita
Fraunhofer
Young
Vega
Brisbane
MARE AUSTRALE

Secchi
MARE
FECUNDITATIS
Messier A • Messier
Langrenus F
Langrenus
Bellot
Crozier
Lohse
McClure
Cook
Monge
Wrottesley
Petavius
Palitzsch
Furnerius
Marinus
Oken
Peirescius

Apollonius
Dubyago
MARE
SPUMANS
Webb
Maclaurin
Kästner
Kapteyn
La Pérouse
Ansgarius
Lamé
Vendelinus
Behaim
Holden
Balmer
Phillips
Humboldt
Legendre
MARE SMYTHII
E

Eclipses of the Sun and Moon

Occasionally the Sun, Moon and Earth line up exactly to cause an eclipse of either the Sun or the Moon. The Sun is eclipsed when the Moon passes directly between it and the Earth; the Moon's shadow falls on part of the Earth, and from within the shadow part or all of the Sun is obscured from view. Lunar eclipses occur when the Moon is on the opposite side of the Earth to the Sun; the Moon can then enter the Earth's shadow and be darkened.

If the Moon orbited the Earth in the same plane that the Earth orbits the Sun, a solar eclipse would occur at each new Moon and a lunar eclipse at each full Moon. But the Moon's orbit is inclined at 5° to that of the Earth, just enough to ensure that the alignment of the three bodies is seldom exact. Only on those occasions when the Moon crosses the Earth's orbit at the time of new or full Moon does an eclipse occur.

Each year at least two solar eclipses are visible somewhere on Earth, and there can be as many as five, while there can be up to three eclipses of the Moon. The maximum number of eclipses possible in one year, both solar and lunar, is seven. Whereas a lunar eclipse is visible from anywhere on Earth that the Moon is above the horizon, an eclipse of the Sun can be seen only from within the narrow band along which the Moon's shadow passes. Hence, from any one place, lunar eclipses are about twice as frequent as solar ones.

Scientifically, total eclipses of the Sun are by far the more important. At a total solar

Eclipses: When the Moon passes in front of the Sun as seen from the Earth, a solar eclipse occurs. When the Moon enters the Earth's shadow, a lunar eclipse results. (Wil Tirion)

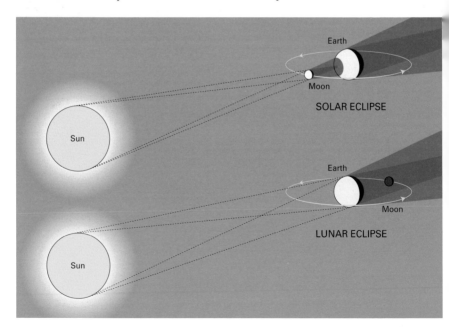

clipse the Moon completely blots out the brilliant disk of the Sun, allowing astronomers to observe the Sun's faint outer halo of gas, the corona. To see the Sun totally eclipsed we must be in the darkest central portion of the Moon's shadow, the umbra, as it sweeps over the Earth. The band of totality is usually no more than a few hundred kilometres wide, but outside this is a much wider area that experiences a partial eclipse.

Astronomers travel across the globe for the few precious moments that totality affords. The longest that a total eclipse of the Sun can last in theory is just over 7½ minutes, but in practice the usual duration is 2–4 minutes. Total solar eclipses demonstrate one of the most extraordinary coincidences in nature: the Sun and Moon appear virtually the same size in the sky. This results from the fact that the Moon, despite being 400 times smaller than the Sun, is also 400 times closer.

Sometimes, though, when the Moon is at its farthest from us in its elliptical orbit, it appears slightly too small to cover the Sun completely, and a ring of bright sunlight remains visible around the Moon's obscuring disk. Such an event is termed an *annular* eclipse (from the Latin word 'annulus' meaning a ring, not because they occur yearly). Partial and annular eclipses are of curiosity value only; they have none of the scientific importance of a total eclipse.

Safe ways to watch solar eclipses

Suitable filters for looking at the Sun with the naked eye are now produced commercially, so there is no excuse for risking your eyesight by using unsafe materials. One popular type of filter specifically intended for solar observation is made of aluminized plastic film, known by the trade name Mylar, while a second type consists of a thicker, black plastic. Mylar filters give a bluish image but black polymer filters give a more natural-looking orange image. These types of filter are often mounted in cardboard to make simple lightweight eclipse-watching glasses, as shown in the photograph.

A welder's glass of shade 13 or 14 makes a safe solar filter for naked-eye observing. Two or three layers of heavily fogged black-and-white photographic negative film are also safe, because the silver in the film absorbs the Sun's heat as well as its light. Unsuitable materials include sunglasses, neutral-density photographic filters, colour film and compact discs (CDs); although these and other materials dim the light from the Sun, the Sun's heat will still get through to damage your eyes.

A simple arrangement not involving filters is to make a pinhole in a piece of card and allow the Sun's light to shine through this hole onto a

Eclipses can be watched safely using goggles with specially designed filters. Ordinary sunglasses are completely inadequate. (ESO)

white surface. You are in fact using a pinhole camera to observe the Sun. Drawbacks with this arrangement are that the image so formed is small and faint and requires strong sunlight, undimmed by clouds, to work successfully.

For more detailed information see http://sunearth.gsfc.nasa.gov/eclipse/SEhelp/safety2.html

If you prefer to use a telescope, the safest method is to project the Sun's image onto a white surface (see the box on page 302).

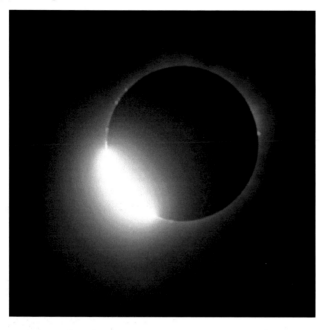

The 'diamond ring' effect, plus pink prominences visible against the pearly light of the Sun's inner corona, at the end of the total solar eclipse of 21 June 2001, viewed from Zambia. For a view of the full corona at the same eclipse, see page 303. (John Walker)

Progress of a solar eclipse

A solar eclipse gets underway at *first contact*, when the edge of the Moon begins its progress across the face of the Sun. Totality is still 1½ hours away. The partial phases of the eclipse can be observed by looking at the Sun through a special filter (see the box on page 341), or by projecting the Sun's image through binoculars or a telescope onto a white surface. When you do this, compare the jet-black outline of the Moon with the umbra of sunspots. This will confirm that sunspots are not totally black but appear somewhat brownish in colour.

Not until about 20 minutes before totality, when the Sun's disk is over two-thirds covered, does the sky start to darken noticeably. An eerie half-light covers the landscape; animals act as though night is falling. Totality itself rushes up as the last crescent sliver of sunlight is blotted out by the Moon.

In the final seconds, chinks of sunlight peep between the mountains at the Moon's rugged edge, producing *Baily's beads*, named after the English astronomer Francis Baily who described them at the eclipse of 1836. Often one bead shines brighter than the others, giving the appearance of a dazzling diamond in a ring.

Then, second contact: the Moon completely covers the Sun, and the pearl-coloured corona springs into view. Now it is safe to watch without eye protection.

Feather-like plumes and streamers of the corona extend from the polar and equatorial regions of the Sun for several solar diameters. The shape of the corona changes during the solar cycle, being more rounded near solar maximum. Pinkish-red prominences can be seen looping out from the Sun's chromosphere around the dark outline of the Moon. Bright stars are visible in the darkened sky.

All too soon, the beautiful spectacle is over. The diamond ring flashes out at third contact, signalling the end of totality. At fourth contact, the Moon moves completely clear of the Sun. The eclipse is over.

Lunar eclipses

Eclipses of the Moon are far less spectacular than solar ones, although they last longer. The Moon takes several hours to move completely through the dark inner part of the Earth's shadow, the umbra. The outer part of the shadow, the penumbra, is so light that it produces little noticeable darkening of the Moon's surface.

A total lunar eclipse can last up to 1¾ hours. Even when totally eclipsed the Moon rarely disappears because light is scattered into the Earth's shadow by the Earth's atmosphere, becoming reddened in the process.

In the early stages of the eclipse the shadow appears dark grey by contrast with the brighter lunar surface, but as the eclipse progresses and more of the Moon is covered the shadow seems to lighten and turn orange or coppery in colour. Binoculars are ideal for observing these colourful effects.

The appearance of the Moon during totality can vary markedly from one eclipse to another. The darkest lunar eclipses occur when the Earth's atmosphere is unusually cloudy or dusty (as after a volcanic eruption). Although an interesting natural spectacle, a lunar eclipse is of little scientific importance.

Eclipsed Moon: Unusual colour effects were seen at the very dark total lunar eclipse of 9 December 1992, in which most of the Moon became almost invisible to the naked eye, apart from a brighter crescent with a bluish tinge. (Eric Hutton)

Mercury

Mercury is a disappointing object for observers. Small telescopes show its changing phase as it orbits the Sun every 88 days, but little else is visible because the planet is small, it is usually viewed close to the horizon where seeing is bad and its surface markings are of low contrast. Most observers must therefore be content with simply catching a glimpse of this elusive object during one of its periodic excursions into the evening or morning sky.

Being the innermost planet, Mercury keeps closest to the Sun; hence would-be observers are restricted to the handful of times each year when it is at or near its maximum angular separation from the Sun, known as *greatest elongation*. Since Mercury's orbit is markedly elliptical, taking it between 46 and 70 million km from the Sun, the separation between the two at greatest elongation varies from nearly 28° to less than 18°. At greatest elongation, a magnification of 250 times is needed to show Mercury the same size as the full Moon appears to the naked eye.

Circumstances dictate that there are two favourable times to look for Mercury each year. From mid-northern latitudes, the planet is most easily seen when setting after the Sun (eastern elongation) in March and April, and when rising before the Sun (western elongation) in September and October. For those at mid-southern latitudes the best times are mornings in March and April and evenings in September and October.

Even when it is best placed, a low, clear horizon will be needed to see it; binoculars help to pick the planet out of the twilight glow, for Mercury can never be seen against a truly dark sky. With all these complications, it is little wonder that many town dwellers have never seen the planet. Nevertheless it is worth looking for, because at its best it can become almost as bright as Sirius.

Rarely, Mercury can be seen as a tiny black dot passing in front of the Sun, an event termed a *transit*. Transits of Mercury in the first half of this century take place on 9 May 2016, 11 November 2019, 13 November 2032, 7 November 2039 and 7 May 2049.

Mercury's rotation

The difficulty in observing Mercury led to a long-standing misconception about the time it takes to rotate on its axis. Towards the end of the 19th century, the Italian astronomer Giovanni Schiaparelli proposed that the planet spins on its axis in 88 days, the same time as it takes to orbit the Sun. It would therefore keep one face turned permanently towards the Sun, as the Moon does to the Earth. In the 1920s a respected planetary observer, Eugène Antoniadi, compiled a map based on an assumed 88-day rotation which seemed to settle the matter once and for all.

Then, in 1965, came a surprise. Astronomers at Arecibo Radio Observatory in

MERCURY DATA	
Diameter	4879 km
Mass (Earth = 1)	0.06
Mean density (water = 1)	5,43
Volume (Earth = 1)	0.06
Axial rotation period (sidereal)	58.65 d
Orbital period (sidereal)	87.97 d
Orbital period (synodic)	115.88 d
Axial inclination	0°.0
Mean distance from Sun	57.91×10^6 km
Orbital eccentricity	0.206
Orbital inclination	7°.0
Number of moons	0

Mercury's lunar-like landscape, photographed by the American space probe Mariner 10 in March 1974. The bright ray crater just above centre is called Kuiper, 62 km across, shown in more detail in the inset. It intrudes onto a crater twice its size, called Murasaki. (USGS)

Puerto Rico bounced radio waves off the planet. From the change in frequency of the reflected radio waves, they deduced that Mercury spins once every 59 days, two-thirds of the time it takes to orbit the Sun.

A rotation period that is such a large fraction of the planet's orbital period has some odd consequences. The Sun does rise and set on Mercury, but it does so very slowly. For the Sun to go once around the sky as seen from the planet's surface – say, from one noon to the next – takes 176 Earth days, during which time Mercury orbits the Sun twice, spinning three times on its axis.

As seen from Mercury the Sun would appear two and a half times larger than it does from Earth. Lethal doses of high-energy solar radiation blast Mercury's daytime side. The Sun's intense heat roasts the surface rocks to over 400°C at noon on the equator, hot enough to melt tin and lead. Yet, without an atmosphere to hold in the heat, the planet's surface drops to a frigid −180°C during the long night.

With a diameter of 4879 km, Mercury is only 40 per cent larger than our Moon and smaller than every other major planet in the Solar System. Given that Mercury is rocky, airless and waterless like the Moon, it is no surprise that it strongly resembles our Moon in appearance.

The surface of Mercury

Our first close-up views of Mercury came in 1974 when the space probe Mariner 10 flew past the planet, photographing a surface heavily pockmarked with craters of all sizes that look almost identical to their lunar counterparts. There are deep, young craters, eroded ancient craters, craters with terraced walls, central peaks and bright rays.

Many of the features on Mercury have been named after artists, composers and writers. For instance, we now find memorials to Bach, Mozart, Van Gogh, Shakespeare and Chekhov on Mercury. Its largest crater, 640 km across, is named Beethoven.

At first glance it is difficult to distinguish between a picture of Mercury and one of the Moon. Craters on both bodies have been formed in the same way, by the impact of asteroids, meteorites and comets early in the history of the Solar System.

One noticeable difference on Mercury is that material ejected from the impacts has not travelled as far as around lunar craters, because Mercury has a stronger surface gravity – over twice that of the Moon, although still only 38 per cent of the Earth's. Another result of the higher gravity on Mercury is that craters tend to be shallower for a given diameter than on the Moon.

Verdi, a crater 160 km in diameter on Mercury, has similar characteristics to many lunar craters: terraced walls, a complex central peak and a surrounding blanket of secondary craters. The shadow-filled crater at upper left is Martial, 50 km wide. (NASA/JPL/Mark Robinson)

Santa Maria Rupes, a surface fold on Mercury of the type known as a lobate scarp, runs nearly vertically for 200 km through this picture from Mariner 10. It cuts across old craters and intercrater plains, evidence that Mercury has shrunk slightly in size since these features were formed. (NASA/JPL/ Mark Robinson)

Scarps, plains and basins

Meandering cliffs called *lobate scarps*, several hundred kilometres long and a kilometre or so high, are a feature of Mercury's surface unlike anything on the Moon. These features are believed to have been caused by a shrinking of the planet as its core cooled early in its history, leading to compression and faulting in the crustal rocks.

The surface rocks of Mercury are slightly darker in tone than those of the Moon, reflecting a mere 11 per cent of the sunlight hitting them, compared with an average 12 per cent for the Moon; overall, Mercury in fact has the darkest surface of any planet in the Solar System, although the lunar maria are darker than Mercury.

Between many of the large craters in the highlands of Mercury are areas peppered only by small craters. These areas, termed the *intercrater plains*, have no real counterpart on the Moon. They clearly pre-date the large craters, but whether they were formed by volcanic action or are deposits of ejecta from large impacts is uncertain. Future Mercury orbiters should resolve the question.

Other areas of particular interest for closer study are deep craters near the poles whose interiors are permanently shaded from the Sun, thereby preserving in deep freeze any gas that has seeped out from within the planet or ice deposited by cometary impacts.

Of all the features seen by Mariner 10, the most prominent is an enormous bull's-eye structure, partly hidden by shadow, named the Caloris Basin. It is 1300 km across, larger than Mare Imbrium on the Moon and fully one-quarter the diameter of Mercury.

The Caloris Basin contains several concentric rings of mountains and is surrounded by radial ridges and grooves. Its interior, and much of the low-lying land around it, has been flooded by molten rock. Whether this is lava from inside the planet that oozed out after the impact, or rock melted by the heat of the impact, remains uncertain.

Geological activity died out on Mercury over 3000 million years ago, as it did on the Moon. Since then, little has changed except for the random arrival of a stray meteorite.

Despite its outward resemblance to the Moon, inwardly Mercury is believed to be more like the Earth. Mercury has a relatively large mass for its small diameter, implying that it has a large iron core three-quarters of its own diameter. A core that size would be as big as the Moon.

One explanation for such a disproportionately large core is that Mercury was originally much bigger, but had most of its rocky outer layers blasted off in a collision or collisions with one or more asteroid-sized bodies, possibly being knocked into its current elliptical orbit in the process. In its own way, Mercury turns out to be a fascinating world containing many clues to the origin and development of the Solar System.

Venus

Many people have seen Venus without realizing it. The planet appears as the brilliant evening or morning 'star', the most prominent object in the twilight, outshining every genuine star with its cold white light. So striking is Venus when at its best that it is frequently reported as a hovering UFO.

Venus orbits the Sun at a distance of 108 million km and can approach closer to Earth than any other planet, within 40 million km.

With a diameter of 12,100 km, only 5 per cent smaller than the Earth, it is almost a twin of our own planet in size. The main

The clouds of Venus circulate around the planet every 4 days, spiralling from equator to pole and creating V- or Y-shaped patterns that are most prominent in ultraviolet light, as in this image from the Pioneer Venus orbiter. (NASA/Ames)

VENUS DATA	
Diameter	12,104 km
Mass (Earth = 1)	0.82
Mean density (water = 1)	5.24
Volume (Earth = 1)	0.86
Axial rotation period (sidereal)	243.02 d
Orbital period (sidereal)	224.70 d
Orbital period (synodic)	583.92 d
Axial inclination	177°.4
Mean distance from Sun	108.2×10^6 km
Orbital eccentricity	0.007
Orbital inclination	3°.4
Number of moons	0

occurring in pairs over a century apart. In the 21st century the dates are 2004 and 2012, while the next pair will not happen until 2117 and 2125.

Observing Venus

Through a telescope, Venus appears like a white billiard ball that exhibits a cycle of phases as it orbits the Sun. At half phase, around the time of greatest elongation, it appears half the size of Jupiter; a magnification of 75 times will then show it the same size as the Moon appears to the naked eye. As Venus approaches the Earth its apparent size grows to equal or exceed that of Jupiter.

Seen from Earth, Venus goes through one cycle of phases (say, from one superior conjunction to the next) in 584 days This, its synodic period, is two and a half times longer than it takes to orbit the Sun (the sidereal period). The two intervals are very different because the Earth and Venus move rapidly relative to each other in their respective orbits around the Sun.

When it is a crescent, Venus is close enough to the Earth for its phase to be picked out in modest binoculars; some people even claim to have seen Venus as a crescent with the naked eye. The planet appears most brilliant when about 27 per cent of its disk is illuminated as seen from Earth, this being the most favourable combination of distance and phase.

Venus can reach a maximum magnitude of −4.7, nearly seven times brighter than the next most prominent planet, Jupiter. At such times, Venus is best observed telescopically against a twilight sky to reduce dazzle.

Only the vaguest markings can be made out in the clouds through telescopes, usually dusky shadings shaped like a sideways V or Y. The clouds often appear brighter at the poles, with a surrounding darker collar. These polar brightenings are termed *cusp caps.*

Observers draw the disk of Venus to a standard diameter of 50 mm, and estimate the brightness of features on a scale ranging

reason for the brilliance of Venus in the sky is not its size or proximity but its cloak of unbroken clouds that reflect two-thirds of the light hitting them. While making Venus so prominent, these clouds also conceal its surface from observers.

As with Mercury, visibility of Venus is restricted to the evening or morning sky but it can be seen over a much wider arc of its orbit than Mercury. Its elongations from the Sun are greater than Mercury's, up to 47°, and at best it can be observed for several hours after sunset or before sunrise. Transits of Venus across the face of the Sun are even rarer than those of Mercury,

Phases of Venus: At full phase Venus is farthest from the Earth and only about 10 arc seconds across; when at half phase (dichotomy) the apparent diameter has grown to 24 arc seconds and by the time of greatest brilliancy it is nearly 40 arc seconds across. (Wil Tirion)

from extremely bright to unusually dark. Observing forms are issued by the relevant sections of the main astronomical societies. Different societies use different scales for making intensity estimates.

The terminator of Venus (the edge of the illuminated portion) can appear irregular, including extensions to or blunting of the cusps, due not so much to differences in the height of the clouds as to differences in brightness. These effects are caused by the corkscrew circulation of clouds around Venus from its equator to its pole.

One curiosity is that the observed time of half phase, known as *dichotomy*, does not coincide with its predicted date. This is termed the *phase anomaly* or *Schröter effect* after a German astronomer, Johann Schröter, who discovered it in the late 18th century. Dichotomy at eastern (evening) elongations, when the planet's phase is decreasing, or waning, is usually a few days early, while dichotomy at western (morning) elongations,

when the phase is increasing, is correspondingly late. It is usually attributed to the fact that the clouds near the terminator are darker than the rest of the disk.

An even odder phenomenon is the *ashen light*, an apparent brightening of the night side of the crescent Venus when seen against a dark sky. To observe the ashen light reliably the illuminated portion of the planet must be blocked out, otherwise any reports will probably be spurious. The reality of the ashen light remains controversial, but if it exists it may be due to scattering of sunlight or electrical effects in the atmosphere.

Large meteorites can penetrate the thick atmosphere of Venus to produce impact craters. In the foreground is Howe, 37 km wide, surrounded by a blanket of bright ejecta. This image, created from Magellan radar data, has been coloured to match the hues recorded by Soviet Venus landers. (NASA)

Beneath the clouds

Unable to see the planet's surface because of the enveloping clouds, astronomers could only guess at the rotation period of Venus until the 1960s – and they guessed incorrectly. As with Mercury, radar observations provided the surprising truth. It turns out that Venus rotates on its axis from east to west, the opposite direction from the Earth and other planets, and it does so very slowly: once every 243 days, longer than the 225 days the planet takes to orbit the Sun. Its clouds, though, rotate every four days, also retrograde (east to west), a result of high-speed winds in the upper atmosphere.

Before space probes arrived there, no one came close to anticipating the uniquely hostile conditions on Venus. First to experience them was a Soviet probe, Venera 4, which parachuted down through the clouds in October 1967. It found that Venus' atmosphere is made almost entirely of unbreathable carbon dioxide gas, but the intense heat and crushing pressure destroyed it long before it could reach the surface.

Later probes, better reinforced, reached the surface intact, registering temperatures of 460°C on both the day and night sides and

Maat Mons, a volcano on Venus 8 km high, in a view reconstructed from Magellan radar imagery. Lava flows extend across the plains in the foreground, where an impact crater can also be seen. The vertical scale has been exaggerated tenfold, making the terrain seem much rougher than it is in reality. (NASA)

an atmospheric pressure over 90 times that on Earth. Why should Venus be so hot – hotter even than the day side of Mercury, despite the fact that its clouds reflect over three-quarters of the incoming sunlight?

The answer lies in the *greenhouse effect*, which works far more effectively on Venus than on Earth. About 1 per cent of the incoming sunlight penetrates to the planet's surface, so that it is as gloomy there as on a heavily overcast day on Earth. That incoming sunlight is absorbed by the surface and is re-radiated at longer wavelengths, in the infrared. Although the carbon dioxide of the atmosphere is transparent to visible light, it traps infrared; since infrared is heat energy, the temperature of the atmosphere rises.

An additional contribution to the greenhouse effect comes from the clouds themselves. These are made not of water vapour, as are the clouds of Earth, but of sulphuric acid of 80 per cent concentration, stronger than in a car battery. Sulphuric acid is another absorber of infrared. Taken together, the carbon dioxide atmosphere, the sulphuric acid clouds and a trace of water vapour turn Venus into a perfect trap for solar heat.

The uppermost layer of sulphuric acid clouds, the one which we view through telescopes, starts about 65 km above the planet's surface and is a few kilometres thick. Around 55 km altitude is a thin haze layer apparently consisting of sulphuric acid particles, which give the clouds their yellow tinge.

The densest cloud layer of all is at about 50 km altitude, and from this layer falls a corrosive rain of sulphuric acid. Below the clouds, the gloom is broken by flashes of lightning. Despite its heavenly name, Venus turns out to be an incarnation of hell.

The surface of Venus

Although the clouds of Venus mask its surface from view, astronomers have nevertheless been able to map the planet's features by radar, which penetrates the clouds. Detailed maps of the entire planet have been made by spacecraft, notably NASA's Magellan probe that went into orbit around Venus in 1990.

Venus is mostly rolling plains, but there are three main continental areas. One, called Ishtar Terra, the size of the United States, has a mountain range, Maxwell Montes, which towers 12 km above the mean surface level, higher than Mount Everest on Earth. The largest continental area of all, Aphrodite Terra, the size of South America, is cut by a system of rift valleys that extends for thousands of kilometres.

Magellan's radar spotted impact craters ranging in size from over 100 km across down to 3 km, demonstrating that large meteorites can get through the dense atmosphere without burning up. Most exciting of all were volcanic mountains with fresh-looking lava flows on their flanks, notably Maat Mons, at 8½ km high the second-highest peak on the planet, which lies near the equator in Aphrodite Terra. Evidently Venus is still an active planet, with highlands formed by volcanic action.

Other types of volcanic feature on Venus include flattened, pancake-like domes, thought to be formed by the outpouring of sticky lava, and ringlike formations of cracks and ridges termed coronae, hundreds of kilometres across, apparently caused by subsidence after upwelling of magma from below.

Photographs from Soviet lander probes resting on the planet's surface show a rocky wasteland bathed in a sulphurous glow. Chemical analyses made by these probes confirm that the surface rocks of Venus are similar in composition to volcanic basalts on Earth. Venus is a tantalizing vision of what the Earth might have been had it been born closer to the Sun – and a terrifying demonstration of what it might yet become if the greenhouse effect were to take hold here.

Under its clouds, Venus is bathed in a sulphurous yellow glow, as photographed by the Russian Venera 13 probe which landed there in 1982. Part of the spacecraft is visible at the bottom. (Brown University/Russian Academy of Sciences)

Mars

Mars is distinguishable by its intense reddish or orange hue, stronger than the colour of any star and the cause of its association with the god of war. Potentially it is the most exciting planet of all to observe, although not the easiest. It is a rocky world about half the size of the Earth with a day about 40 minutes longer than our own. Its year lasts nearly twice as long as ours, 687 Earth days, during which it undergoes seasonal changes, including melting of its polar caps and variations in the size and shape of its surface markings. Its atmosphere is so thin that we can usually see straight through to the surface, but occasionally white clouds or dust storms interrupt our view. Studies of the red planet are all the more important because one day humans will walk on its surface.

Mars in 1997, photographed by the Hubble Space Telescope. The dark feature at the centre is Syrtis Major. South of it is the lowland basin Hellas, filled with frost and clouds. At right, white clouds also lie around Elysium Mons, a volcano. (Steve Lee/Jim Bell/Mike Wolff/NASA)

Observing Mars

As with all the outer planets, the best time to observe Mars is near opposition. Because the orbit of Mars lies beyond that of the Earth it does not go through a cycle of phases like the inner planets, although when it is at an elongation of around 90° from the Sun it displays a distinctly gibbous phase, like the Moon a couple of days from full.

Mars comes to opposition every 26 months but some oppositions are much more favourable than others because its orbit is markedly elliptical, taking it between 207 and 249 million km from the Sun. When oppositions occur at perihelion, as in 2003 and 2018, Mars can come as close as 56 million km to us and we get our best views of the planet.

At such times Mars shines like a brilliant red star at magnitude −2.9, rivalling Jupiter, and a magnification of ×75 will show the planet the same size as the full Moon to the naked eye. (Incidentally, the date of closest approach may differ by a week or more from

MARS DATA	
Diameter (equatorial)	6792 km
Mass (Earth = 1)	0.11
Mean density (water = 1)	3.94
Volume (Earth = 1)	0.15
Axial rotation period (sidereal)	24.62 h
Orbital period (sidereal)	686.98 d
Orbital period (synodic)	779.94 d
Axial inclination	25°.2
Mean distance from Sun	227.9 × 10⁶ km
Orbital eccentricity	0.093
Orbital inclination	1°.9
Number of moons	2

the date of opposition because of the eccentricity of the planet's orbit.)

But at aphelic oppositions, as in 2012 and 2027, Mars is 100 million km from Earth, almost twice the distance of the closest approaches, and it appears unimpressive even in powerful telescopes. Hence astronomers do not waste their chances of observing Mars when it is at its closest.

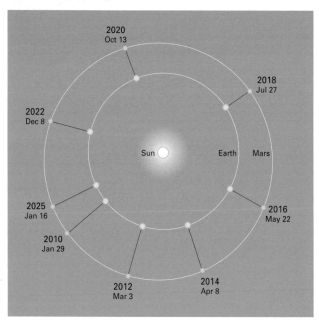

Oppositions of Mars over a 15-year period. The actual distances on each occasion shown are: 2010, 99.4 million km; 2012, 100.9 million km; 2014, 92.9 million km; 2016, 76.2 million km; 2018, 57.7 million km; 2020, 62.6 million km; 2022, 82.3 million km; 2025, 96.2 million km. (Wil Tirion)

Albedo markings on Mars, as mapped from Earth. North is at the top. (Wil Tirion)

Left: Mars from longitude 300° to 180°. Note the dark wedge-shaped Syrtis Major just north of the equator and the bright Hellas basin in the southern hemisphere.

Right: Mars from longitude 180° to 60°. The white spot of Nix Olympica in the northern hemisphere is actually a large volcanic mountain better known as Olympus Mons.
NASA's Spirit rover landed in 2004 at roughly longitude 175° and latitude 15°south.

Left: Mars from longitude 60° to 300°. This face of the planet is crossed by the elongated feature called Sinus Sabaeus, which stretches westwards from Syrtis Major and ends in the clawlike Sinus Meridiani. NASA's Opportunity rover landed in 2004 at the western end of Sinus Meridiani at approximately longitude 354° and latitude 2°S.

Through a telescope, the most obvious features on Mars are the white polar caps, which stand out in stark contrast to its ochre-coloured deserts. The caps shrink noticeably during summer in each hemisphere, leaving detached 'islands'. The southern cap shrinks more than the northern one because the southern hemisphere of Mars is tilted towards the Sun at perihelic oppositions, and so gets warmer than the northern hemisphere, which leans sunwards at aphelic oppositions.

The dark markings, once thought to be vegetation, are now known to be areas of rock and dust. These markings change in appearance from year to year as winds blow the dust around, so any map of Mars can be only an approximation of the planet's true appearance. Such markings are termed *albedo features* as they are simply differences in surface brightness and do not necessarily correspond to actual physical features.

The most prominent dark albedo feature is a large triangular area named Syrtis Major. This should be visible in a modest amateur telescope when it is on the Earth-facing side, as should a bright circular lowland to the south of it called Hellas.

Sometimes, yellow clouds appear on the planet, particularly in the areas of Hellas and Solis Lacus. These clouds are in fact dust storms which, with the increased temperatures around the time of perihelion, can be whipped up by high winds to envelop the entire planet and obscure its features.

Different parts of the surface will be seen as the planet rotates, turning through nearly 15° every hour. Since Mars rotates slightly more slowly than the Earth, the same features will be on view just over half an hour later each night.

Drawings of Mars are usually made with a disk size of 50 mm, the same as Venus. As well as making drawings, observers estimate the brightness of features on a scale of 0 to 10 from the white polar caps to the blackness of the sky. Report forms are issued by amateur astronomy organizations.

Dust storm on Mars in 2001, observed by the Hubble Space Telescope. In June the atmosphere was mostly clear, apart from some water-ice clouds around the north polar region and at lower left. But a storm was brewing in the Hellas basin (lower right) and by August the planet's surface was almost entirely obscured (second image). The dark spot at upper right is the peak of the volcano Olympus Mons. Mars is gibbous in the image at right because our viewing angle has changed. (James Bell/Michael Wolff/STScI/AURA)

The illusory canals

The temptation to assume too many similarities between Mars and the Earth led early astronomers astray. The dark areas, which range in colour from brown to grey-green, were termed seas and lakes in the belief that they really were filled with water, while the orange areas were named after places on Earth – there is an Arabia, Libya, Syria and Sinai on Mars.

By the end of the 19th century it was realized that there were no oceans on Mars after all, but in their place arose a much more intriguing explanation for the dark areas: that they were covered with primitive vegetation, such as moss or lichen. In support of this view, observers noted that when the polar caps melted in the Martian summer, the surface markings became larger and darker; this was interpreted as vegetation growing in the milder, wetter conditions.

The most extreme proponent of the life-on-Mars idea was an American astronomer,

Clouds of ice crystals over Olympus Mons (top left) and the chain of three volcanoes of the Tharsis Bulge: Ascraeus Mons (top), Pavonis Mons (middle) and Arsia Mons, in a mosaic of images from the Mars Global Surveyor orbiter. These clouds are visible from Earth. The cracked terrain at lower right is the Valles Marineris fault complex. (NASA/JPL/MSSS)

Percival Lowell. His ideas were inspired by an Italian observer, Giovanni Schiaparelli, who in 1877 reported seeing long, straight lines that appeared to criss-cross the planet's surface. Schiaparelli called these lines *canali*, a word which in Italian means 'channels'; but inevitably it was translated as 'canals', implying that they were artificial.

For Lowell, the canals were evidence of an advanced civilization clinging to existence on an arid planet, reliant on the canals to bring meltwater from the polar caps to irrigate their crops at the equator. His beliefs gave rise to a whole generation of science-fiction stories, including the famous *War of the Worlds* by H. G. Wells.

Most other astronomers failed to see the canals, or in their place could detect only broad, irregular smudges. It is now clear that Lowell's canals were illusory, testimony to the fallibility of human observers straining at, or past, the limits of vision.

Space-probe exploration of Mars

Mars is currently the subject of a vigorous programme of exploration by space probes, leading to a sample-return mission and an eventual human landing. The first close-up pictures of the planet were sent back in July 1965 by NASA's Mariner 4 probe as it swept past at a distance of 10,000 km. It photographed craters on Mars, looking like more eroded versions of those on the Moon. Since then, a succession of probes has examined the planet in increasing detail, both from orbit and by landing on the surface.

Mars turns out to be a planet of two contrasting halves. The northern hemisphere is the lower and smoother of the two, flooded by lava from volcanoes. The youngest eruptions occurred within the past few hundred million years. The southern hemisphere is higher and has been heavily cratered by meteorite impacts, giving it a broad similarity to the lunar highlands.

Among the major formations on Mars is a chain of three volcanic mountains atop a highland area known as the Tharsis Bulge which straddles the equator; these volcanoes are named Arsia Mons, Pavonis Mons and Ascraeus Mons ('Mons' means mountain). Other types of Martian landforms and the names given to them include lowland plains (Planitia), highland plateaus (Planum), valleys (Vallis), canyons (Chasma), and eroded craters (Patera).

Northwest of the chain of Tharsis volcanoes is an even bigger individual volcanic mountain, Olympus Mons, visible from Earth as a white ring and originally named Nix Olympica ('the snows of Olympus'). Olympus Mons, 600 km wide and 25 km high, is the largest volcano in the Solar System, larger even than the volcanic islands of Hawaii on Earth. The whole Tharsis area is noted for the frequent appearance of white clouds, often described as being shaped like a letter W. The presence of mountains explains the preference of clouds for this region.

An immense system of canyons, up to 500 km wide and 4 km deep, extends eastwards from the Tharsis Bulge. This massive rift in the surface of Mars, the Valles Marineris (Mariner Valleys, named in the plural because of their multiplicity), not merely dwarfs the Earth's Grand Canyon, but at 4000 km long could span the entire United States. Valles Marineris is visible from Earth as a feature called Coprates; it is prominent because dark dust collects on its floor.

In the southern hemisphere are two large impact basins, Hellas and Argyre. Argyre, 900 km in diameter, is similar in size to the Moon's Mare Imbrium, while Hellas is

Above: The rocky red surface of Mars inside the crater Gusev as seen by the Spirit rover in 2006. Spirit's wheel tracks, visible on the left, have churned up bright underlying material, thought to consist of salty minerals created by water. Many dark, porous volcanic rocks can be seen around the rover. Among them is a smoother, lighter one (inset, foreground) which may be an iron meteorite. (NASA/JPL/Cornell)

nearly three times larger, equivalent in size to Oceanus Procellarum. Unlike the lunar maria, however, Argyre and Hellas are filled with light-coloured dust. Many of the dust storms that periodically obscure the surface of Mars after perihelion start in Hellas.

Much of the surface of Mars consists of dusty red deserts liberally scattered with rocks of all sizes, thrown out by meteorite impacts. The rusty redness of the soil is due to iron oxide. Mars may be the richest source of iron ore in the Solar System. Even the sky is pink, a result of fine particles of dust suspended in the wispy atmosphere.

Astronauts who land on Mars will need to wear spacesuits to explore the surface, as they did on the Moon, because the atmosphere is unbreathable and temperatures are sub-zero. About 95 per cent of the atmosphere consists of carbon dioxide, the remainder being mostly nitrogen and argon with a trace of water vapour. The average atmospheric pressure is a mere 6 millibars (600 pascals), equivalent to the pressure 35 km above the Earth's surface. Surface temperatures can reach 20°C on a summer's afternoon, but the temperature of the atmosphere drops rapidly with height to well below freezing. At the poles, surface temperatures are about −130°C.

Each polar cap has a permanent core, consisting of water ice a few metres thick, augmented each winter by carbon dioxide. Carbon dioxide freezes out of the atmosphere to produce a thin polar 'hood' of frost that can extend more than halfway to the equator. During each Martian year the atmospheric pressure changes by 25 per cent or more as carbon dioxide evaporates from one polar cap with the coming of spring, migrates to the opposite hemisphere and then freezes out again as winter arrives at the other pole. Such atmospheric motions may help set off the dust storms that sweep across Mars.

Water on Mars

Although the Lowellian canals are not in evidence, space probes in orbit and on the surface have found convincing signs that water once flowed on Mars. Sinuous channels, looking like dried-up river beds, snake across parts of the planet's surface, and some of the lowlands appear to have been inundated by flash floods.

Direct evidence for past water on Mars has been found by NASA's two Mars Exploration Rovers, Spirit and Opportunity, which landed on the planet in January 2004. Spirit touched down in an impact crater called Gusev (see photograph below), thought to have been the site of a former lake, while Opportunity descended in Meridiani Planum, on the western edge of the albedo feature Sinus Meridiani. Both craft found minerals in the surface rocks which indicated that liquid water had once been present.

Mars is not short of water today, although it is in frozen form because because the atmospheric pressure is too low for it to exist as a liquid. Apart from traces of water vapour in the atmosphere, most of the water on Mars is in the polar caps and in a subsurface permafrost layer polewards of about 30° latitude north and south.

If watercourses and lakes really did exist on Mars in the past, the atmosphere must then have been denser – and a denser atmosphere would keep the planet warmer. Perhaps enough gas gushed out of the volcanoes to change the climate temporarily. Alternatively, subsurface ice could have been melted by volcanic heating or meteorite impacts. Either way, life might have had a chance to arise on Mars at some time in the past, and microscopic organisms such as bacteria might still cling to existence underneath the red sands of the planet.

Looking for life on Mars

The first search for Martian life was carried out in 1976 by two American space probes called Viking. Each probe came in two halves: a lander and an orbiter. The Viking 1 lander touched down on a lowland plain called Chryse in the northern hemisphere, over which water appears to have flowed during the wet times on Mars. The second lander descended on the opposite side of the planet in an area known as Utopia, which is encroached upon by the outer fringes of the north polar cap during winter.

The landers carried colour cameras as well as instruments to analyse the soil and the atmosphere. Each Viking lander, like those

Ma'adim Vallis, over 800 km long, leads northwards into a crater called Gusev, 165 km wide. at the top of this Viking image. NASA's Spirit rover landed in Gusev in January 2004. Water may once have flowed through Ma'adim Vallis, creating a lake within Gusev. Dark areas on the crater floor are where wind has swept away light dust to reveal underlying rock. (USGS)

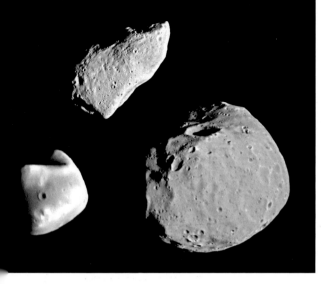

Phobos (lower right) and Deimos, the moons of Mars, are thought to be captured asteroids. Here they are compared with the asteroid Gaspra, top, all at the same scale and under similar lighting conditions. Gaspra was photographed by the Galileo probe, Phobos and Deimos by the Viking orbiter. (NASA)

that have landed subsequently, found itself on a rock-strewn desert without any visible signs of life – no plants, insects or animal tracks. A mechanical arm on each Viking picked up samples of the soil and tipped them into an on-board biological laboratory which set to work in search of Martian microorganisms. Neither Viking found life in the soil of Mars, despite exhaustive tests.

That is not to say that Mars is completely sterile; maybe life exists in places that the Vikings were unable to sample, such as deeper underground or inside rocks. Alternatively, perhaps life once got started but then died out when the climate changed; there may be fossil life, rather than current life, to be found on Mars. Future probes will examine such possibilities.

One final puzzle from the pre-space-probe era has now been explained: what causes the seasonal changes in the dark areas if there is no vegetation on Mars? Wind-blown dust provides the answer. Orbiting probes have recorded many examples of changes in surface markings caused by light and dark dust being blown by seasonal winds. Syrtis Major, for example, is a gently sloping area of dark volcanic rock which periodically becomes partly covered by paler dust and is later swept clear again. Dust blown by the winds on Mars, which can reach 200 km per hour, is also a powerful erosive agent, sandblasting the surface rocks.

The moons of Mars

No discussion of Mars would be complete without reference to its two tiny moons, Phobos and Deimos, discovered in 1877 and both beyond the range of normal amateur telescopes. Space probe photographs show that they are are cratered chunks of rock shaped like lumpy potatoes. Phobos, larger and closer to Mars, measures about 27×18 km; Deimos is about 15×10 km.

Phobos lies closer than any other moon to its parent – a mere 6000 km above the surface of Mars, at which distance it orbits Mars three times a day. Probably Phobos and Deimos are former asteroids which strayed too close to Mars and were captured by the planet's gravity. They will make a fascinating sight in the sky for the first astronauts to set foot on Mars, the planet of red, frozen deserts which was so nearly suitable for life of its own.

• Charts showing the positions of Mars over a five-year period can be found at www.collins.co.uk/starsandplanets

Jupiter

Jupiter is the king of the planets, and the most fascinating of all to study with small instruments. Humble binoculars reveal the planet's cream-coloured disk and four main moons, called the Galilean satellites after the Italian scientist Galileo Galilei who discovered them in 1610. Some people with exceptionally acute vision can see the Galilean satellites with the naked eye as faint starlike points either side of the planet.

A small telescope brings into view the main features of Jupiter's disk: dark belts parallel to the equator, along with an eye-shaped mark-ing in the southern hemisphere known as the Great Red Spot. This seems to have been first observed in 1831, and a similar feature was seen as early as 1664. Careful study of these features reveals that Jupiter's rotation period varies with latitude, averaging 9 hours

Jupiter as seen by the Cassini space probe in 2000. Light-coloured zones of cloud alternate with darker belts. In the southern hemisphere is the swirling Great Red Spot, with white ovals to the south of it. (NASA/University of Arizona)

50 minutes in the bright equatorial zone (System I, the fastest rotation of any planet in the Solar System) and about 5 minutes longer at higher latitudes, called System II. Hence the equatorial zone gains one rotation on the regions to the north and south every 48 days. In addition, the Great Red Spot drifts slightly in relation to its surroundings.

These effects demonstrate that the visible surface of Jupiter is not solid. We are looking at clouds, constantly seething and swirling, changing in colour and shape. Jupiter never appears the same twice. That is its attraction.

Jupiter is easy to spot with the naked eye, looking like a bright cream-coloured star. It comes to opposition every 13 months and can appear as bright as magnitude −2.9 at the closest oppositions, when it is within 600 million km of us. At such times, a magnification of only 40 times will show it as large as the full Moon appears to the naked eye.

Even well away from opposition Jupiter still outshines every star in the sky. Its bright-

Jupiter's four main satellites can be easily seen through small telescopes and even binoculars. Here, from left, are Europa, Io, Callisto and Ganymede. All four are not always visible at one time. (Chris Walker/Calvin College Observatory)

ness results from its highly reflective clouds and its imposing size – it is the largest planet in the Solar System.

Examination of Jupiter's outline gives added confirmation that it is not a solid planet: Jupiter has a bulging midriff. At the equator it is nearly 143,000 km across, but from pole to pole it measures some 9000 km less. A line of 11 Earths would be needed to equal Jupiter's equatorial width. Even though Jupiter is composed largely of the lightest elements in the Universe, hydrogen and helium, its bulk is such that its mass is two and a half times as much as all the other planets put together.

Observing Jupiter

Observers draw Jupiter on pre-printed blank outlines with an equatorial diameter of 64 mm and a polar diameter of 60 mm; forms are issued by amateur astronomy societies or can be drawn on a computer. After sketching in the main features, details can be added starting at the leading limb, where the planet's rotation of 6° every 10 minutes rapidly carries the features out of sight. Intensities can be estimated on a scale from 0 to 10 (European observers rate 0 as the brightest and 10 the blackest, but in the US the scale runs the

JUPITER DATA	
Diameter (equatorial)	142,984 km
Mass (Earth = 1)	317.83
Mean density (water = 1)	1.33
Volume (Earth = 1)	1321
Axial rotation period (sidereal)	9.84 h
Orbital period (sidereal)	11.86 y
Orbital period (synodic)	398.88 d
Axial inclination	3°.1
Mean distance from Sun	778.4×10^6 km
Orbital eccentricity	0.048
Orbital inclination	1°.3
Number of moons	60+

other way). By continuing to watch Jupiter as it rotates, the observer can build up a strip chart of the entire planet.

Most valuable of all are timings of the passage of various spots across the central meridian, the imaginary line running north–south down the centre of the planet's visible disk. Repeated timings will establish the rotation rate of spots, and reveal any drift in their longitude caused by winds in Jupiter's atmosphere. Even in this age of space probes and space telescopes, long-term series of such observations are useful.

Because Jupiter's cloud features are so impermanent and mobile, it is impossible to give more than a generalized description of the planet's appearance. Its disk is crossed by alternating bright and dark bands, termed

zones and belts respectively, named according to their latitude: equatorial, tropical, temperate, or polar. The main belts and zones, along with their names and abbreviations, are shown on the diagram below.

Frozen ammonia crystals form high, cold clouds in the bright zones, where gas ascends; the darker belts, where the gases descend, are lower and warmer (although 'warm' is only a relative term, for the temperature of the cloud tops is around −150°C). The colours of the belts can vary from yellow and brown

The main belts and zones in the clouds of Jupiter. This outline has a diameter of 64 mm and a depth of 60 mm, the size used for making disk drawings of the planet. (Wil Tirion)

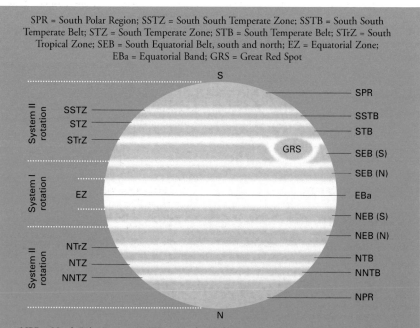

SPR = South Polar Region; SSTZ = South South Temperate Zone; SSTB = South South Temperate Belt; STZ = South Temperate Zone; STB = South Temperate Belt; STrZ = South Tropical Zone; SEB = South Equatorial Belt, south and north; EZ = Equatorial Zone; EBa = Equatorial Band; GRS = Great Red Spot

NPR = North Polar Region; NNTZ = North North Temperate Zone; NNTB = North North Temperate Belt; NTZ = North Temperate Zone; NTB = North Temperate Belt; NTrZ = North Tropical Zone; NEB = North Equatorial Belt, north and south; EZ = Equatorial Zone. System I and System II are the rotation periods of the planet's equatorial regions and higher latitudes, respectively.

Dark stains in Jupiter's clouds south of the Great Red Spot mark the sites where fragments of Comet Shoemaker–Levy 9 impacted the planet in July 1994. This photograph was taken by the Hubble Space Telescope. (STScI/NASA)

to orange, red or even purple as a result of complex chemicals in the atmosphere of Jupiter, sulphur being one of the prominent colourants.

High-speed winds of up to 500 km per hour whip the edges of the zones and belts into turbulent eddies, giving them a scalloped appearance. The weather on Jupiter is unpredictable. Dark and light spots can suddenly erupt in the clouds, lasting for weeks or even decades before fading away. One of the main roles of amateur observers is to track these storms as they erupt and circulate around the planet.

The Great Red Spot

Of all the markings on Jupiter, the most famous, and by far the longest-lived, is the Great Red Spot, centred at about latitude 22° south. It certainly is great: 14,000 km wide and as much as 40,000 km long, enough to swallow three Earths. But it is not always red: most often it is pinkish, and sometimes it can fade to a colourless grey. Its colour is believed to be due either to red phosphorus or sulphur.

Even now its nature is not fully understood, but it appears to be an upwardly spiralling column of gas similar to a hurricane on Earth, its top spreading out about 8 km above the surrounding cloud deck. Jupiter's other, smaller spots, including a series of long-lived white ovals, are believed to be similar swirling storm systems. As if to emphasize the storminess of the planet, the Voyagers and the more recent Galileo probe photographed massive flashes of lightning on Jupiter's night side, far larger than in any thunderstorm on Earth.

In 2005 a mini red spot, about half the width of its older relative, formed at latitude 34° south from the merger of three large white oval clouds in the South Temperate Belt. The increased size of the merged ovals is thought to have created a sufficient disturbance to dredge up reddish material from deeper in the atmosphere. Here we may be seeing a re-enactment of the process by which the Great Red Spot itself originally formed. No one knows how long this mini red spot may last.

The key to Jupiter's meteorology is the fact that it gives off twice as much heat as it receives from the Sun. The planet was hot when it formed, and still retains some of that heat today. This internal store of heat drives the complex cloud systems of Jupiter, keeping the Great Red Spot and its smaller relatives alive far longer than any storms can persist on Earth.

Interestingly, Jupiter has virtually the same chemical composition as the Sun: mostly hydrogen and helium. There is thought to be a rocky core about twice the size of the Earth at Jupiter's centre, but no space probe could ever land on it. Beneath the wispy high-altitude clouds of frozen ammonia are complex chemicals that give the dark belts their colour.

Deeper still, temperatures are similar to those on Earth and clouds of water vapour condense. About 1000 km below the visible cloud tops, temperatures and pressures have increased to the point where hydrogen is compressed into a liquid. Jupiter's liquid hydrogen seas are about 20,000 km deep. Below, under the crushing pressure of 3 million Earth atmospheres, hydrogen is compressed into a superdense state with the properties of a metal; such a substance is known as metallic hydrogen.

Convection within the hot metallic hydrogen interior of Jupiter is thought to be responsible for the planet's intense magnetic field, 10 times stronger than the Earth's and extending out for 100 times Jupiter's radius into space. If the magnetosphere around Jupiter were visible to the naked eye to us on Earth, it would appear over twice the size of the full Moon.

Family portrait of Jupiter's four largest satellites as seen by the Galileo space probe, to scale. From left: the sulphurous surface of volcanic Io; icy, cracked Europa; dark markings and bright craters on Ganymede; and battered Callisto, with the Valhalla impact basin at centre. (NASA)

Cometary impact

One of the most remarkable events in the history of planetary observation took place in 1994 when Comet Shoemaker–Levy 9 collided with Jupiter. The comet had been captured into orbit around the planet some time in the 1920s but was not discovered until 1993 by when it had broken into over 20 pieces. These crashed into Jupiter during the week of 16–22 July 1994, leaving dark markings in its clouds that were visible from Earth through small telescopes.

Over the ensuing months, the dark patches spread out into a band around Jupiter that took over a year to fade away. Crater chains on Jupiter's moons Callisto and Ganymede suggest that they, too, have been struck by fragments of disintegrated comets in the past.

Jupiter's moons and rings

Jupiter possesses a fascinating collection of over 60 satellites, like a mini Solar System. With the simplest optical aid the four largest, known as the Galilean satellites, can be seen performing a merry dance around Jupiter, changing position from night to night – sometimes out of sight behind the planet, sometimes transiting across its face and sometimes being eclipsed in its shadow.

The closest to Jupiter of the Galilean satellites is Io, 3643 km in diameter (slightly larger than our own Moon), orbiting every 42½ hours. Io is the most volcanically active body in the Solar System. The Voyager 1 probe in 1979 photographed eight volcanoes erupting simultaneously on it. Hundreds of other volcanic vents were visible, although not actually erupting. Those volcanoes erupt not just molten rock (lava) as on Earth but also sulphur which solidifies to form the garish red, orange and yellow of Io's surface.

Io is molten because it is caught in a gravitational tug of war between Jupiter and the other Galilean satellites; their opposing pulls release tidal energy that melts Io's interior. Io recycles its interior onto its surface, endlessly turning itself inside out. Some of the sulphur escapes and showers onto the innermost moon of Jupiter, Amalthea, giving it an orange coating. Amalthea is an irregularly shaped lump of rock only about 200 km in diameter, too faint to be seen in amateur telescopes.

Within the orbit of Amalthea is a faint ring of dust, extending inwards to a mere 30,000 km above Jupiter's cloud tops. This tenuous ring of Jupiter is believed to result from the break-up of one or more tiny moons.

Moving outwards past Io we come to Europa, smallest of the Galilean satellites with a diameter of 3122 km. Europa is encased in a white shell of ice, veined with fine cracks probably produced by tidal forces. Underneath its fractured icy crust, Europa may have an ocean of liquid water.

Next in line from Jupiter is Ganymede, the largest and brightest of the satellites; what's more, at 5262 km in diameter it is the largest moon in the Solar System, larger even than the planet Mercury. Ganymede and the fourth of the Galilean satellites, Callisto, 4821 km in diameter, are both balls of rock and ice, rather like giant muddy snowballs.

Callisto is saturated with impact craters, the largest 300 km in diameter and called Valhalla; it is similar to the large basins on the Moon and Mercury, surrounded by wave-like ridges. Ganymede is also cratered by impacts, but less heavily than Callisto, and exhibits a strange grooving on much of its surface, apparently formed by faulting and stressing of the moon's icy surface layers. Bright spots on Ganymede and Callisto mark places where recent impacts have exposed fresh ice.

The rest of Jupiter's moons are small and insignificant. Most of them – particularly the outermost ones, which move in retrograde orbits – were probably passing bodies that were captured by Jupiter's gravity.

• Charts showing the positions of Jupiter over a five-year period can be found at www.collins.co.uk/starsandplanets

Blowing its top – a volcano called Loki erupts at the limb of Jupiter's moon Io, sending clouds of sulphur 150 km into space, as photographed by the Voyager 1 probe in 1979. The volcanoes of Io are named after fire gods. (USGS)

Saturn

Bright rings girdle Saturn's equator, making it the most beautiful of the planets. Those distinctive rings can be spotted clearly in a small telescope; good binoculars, mounted steadily, will show the small outline of the planet elongated into an ellipse by the rings. Mounted binoculars should also pick out Saturn's largest moon, Titan, which orbits the planet every 16 days.

Saturn's rings can reflect more light than the body of the planet itself, so that at its best Saturn appears as a bright yellowish star of magnitude −0.3, outshone only by Sirius and Canopus. Without the rings, Saturn would be less than half as bright, no more than magnitude 0.7.

Strange to say, from time to time Saturn really can appear to be without rings. The reason is the tilt of the planet's axis. As Saturn orbits the Sun, the rings are presented to us at an angle that varies from 27° to 0°. So thin are the rings that when they are edge-on (which happens about every 15 years, as in 2009 and 2025) they disappear from view in even the largest telescopes on Earth. The brightness of Saturn therefore depends not only on its distance from us, but also on the aspect of the rings.

The ringed planet orbits the Sun every 29½ years at an average distance of 1430 million km, 9½ times farther than the Earth. Being so slow-moving, Saturn returns to opposition about two weeks later each year. At a typical

Saturn photographed by Voyager 2 in 1981; note the Cassini Division and subtle detail in the rings. Three moons, Tethys, Dione and Rhea, are visible as tiny dots beneath the planet, with the shadow of Tethys on the clouds. (USGS)

opposition, a magnification of 90 times will show the globe of Saturn as large as the full Moon viewed with the naked eye.

In many ways, Saturn is a smaller version of Jupiter. Its equatorial diameter is second only to that of Jupiter; its rotation period is second-fastest to that of Jupiter; and, like Jupiter, it is composed mostly of hydrogen and helium. In one way, though, Saturn is unique among the planets: its average density is a mere 70 per cent that of water. This remarkable fact comes about because the planet's mass is less than a third that of Jupiter, so its gravity is less and hence its central regions are not compressed as densely.

Saturn's low density is apparent from its outline, which is even more squashed than that of Jupiter. Saturn's pole-to-pole diameter is fully 10 per cent smaller than its equatorial diameter, against 6 per cent for Jupiter.

Observing Saturn

Through a telescope Saturn appears as a tranquil, ochre-coloured disk, darker at the poles and with some dusky horizontal bands. The pattern of belts and zones is much the same as on Jupiter (see diagram on page 364). In addition, Saturn throws its shadow onto the rings, and the rings throw their shadow onto the globe of the planet, creating the impression of a dark equatorial belt.

Pleasing views of Saturn and its rings can be obtained with small telescopes, but serious study requires at least 200 mm aperture. When drawing Saturn, observers use a range of pre-printed blank forms to match the changing tilt of the rings. On such blanks the polar diameter of Saturn is 50 mm, the equatorial diameter 55 mm and the diameter of the rings 126 mm. As with Jupiter, the main features are sketched in quickly and detail added before it rotates out of view.

Observers can estimate the intensities of various features; in Europe, the scale runs from 1 (the brightest part of the rings) to 10 (black sky), whereas the US scale runs the

SATURN DATA	
Diameter (equatorial)	120,536 km
Mass (Earth = 1)	95.16
Mean density (water = 1)	0.69
Volume (Earth = 1)	764
Axial rotation period (sidereal)	10.23 h
Orbital period (sidereal)	29.45 y
Orbital period (synodic)	378.09 d
Axial inclination	26°.73
Mean distance from Sun	1426.7×10^6 km
Orbital eccentricity	0.054
Orbital inclination	2°.5
Number of moons	55+

opposite way, from 0 (black sky) to 8 (brightest part of the rings). Markings in the clouds can be timed as they cross the planet's central meridian, in the same way as with Jupiter.

There are none of the swirling, multi-coloured storm clouds that make Jupiter's globe so interesting, and no equivalent of the Great Red Spot. However, every 30 years or so a large white spot erupts in Saturn's northern hemisphere. Evidently, these white spots are storm clouds produced by solar heating, as they occur when the planet's north pole is tilted at its maximum towards the Sun. Such an outbreak occurred in 1990 and lasted a few months, with a succession of smaller spots appearing over the next few years.

White spots apart, this is not to say that Saturn lacks activity in its atmosphere; it is just that the cloud patterns are usually concealed by high-altitude haze. Low-contrast cloud swirls resembling those on Jupiter have been recorded by space probes.

The meteorology of Saturn and Jupiter is similar, for they both have internal sources of heat. Like Jupiter, Saturn radiates twice as much heat as it receives from the Sun, a legacy of its birth. However, since Saturn is farther from the Sun, its clouds are about 30°C colder than on Jupiter and form lower in the atmosphere. Tracking of cloud systems has revealed gales blowing at up to 1800 km per hour on Saturn, over three times faster than the winds of Jupiter.

8°

16°

24°

Rings of Saturn

Inevitably, much of the interest in Saturn lies in its magnificent rings. As seen from Earth, they look like a continuous disk encircling the planet, but appearances are deceptive. The Dutch scientist Christiaan Huygens in 1655 was the first to realize that the rings are not solid, but consist of a swarm of tiny particles in orbit around Saturn.

The central part of the rings, known as Ring B, is the widest and brightest. It is separated from the outer, fainter Ring A by a 5000-km-wide gap called Cassini's Division, visible with a 75-mm telescope. Extending inwards from Ring B towards the planet is the faintest ring of all, the transparent Ring C, also known as the crêpe ring.

A narrow gap called Encke's Division can be seen in the outer part of Ring A, while observers using large telescopes under good conditions have reported apparent ripples within the rings, a sign that the density of ring material varies from place to place.

But even the best telescopic views did not prepare astronomers for the astounding wealth of detail that has been revealed by space probes – firstly the two Voyager craft that flew past Saturn in 1980 and 1981, and then Cassini which began orbiting the planet in 2004. Under the close-up scrutiny of these probes, the rings break up into thousands of narrow ringlets and gaps, like the ridges and grooves of a gramophone record.

Some ringlets are not perfectly circular, but are elliptical in shape. Even Cassini's Division is not empty, but contains thread-like ringlets. An outer ring, called the F Ring, first shown by the Voyager probes, is distorted by the gravitational pull of satellites orbiting close to it, giving it a twisted appearance like strands of rope.

Saturn's rings are presented to us at a range of angles from 0° to a maximum of 27° as it orbits the Sun, so a range of printed blanks is used for drawing the planet. (Wil Tirion)

Saturn's rings are extraordinarily thin in relation to their 275,000 km diameter, no more than 100 m thick. On the same ratio of thickness to diameter, a CD or DVD disk would be about 5 km in diameter.

The particles that make up the rings come in a wide range of sizes, from dust specks to lumps the size of a house or larger. Their composition is mostly frozen water, possibly mixed with dust, resembling loosely compacted snowballs.

Saturn's rings could have originated from material that was prevented from forming into a moon by the overpowering force of Saturn's gravity. Alternatively, they may be the remains of a former moon that strayed too close to the planet and broke up, or a moon that was broken up by the impact of a comet. In fact, the rings may be replenished from time to time by icy material from impacting comets.

In some places, fine dust overlays the rings, producing transient darker features known as *spokes*. Spoke-like features were first reported by ground-based observers, but it took space probes to establish their reality.

Saturn's moons

In recent years, a host of small moons of Saturn have been discovered by space probes and Earth-based observations, bringing the known total to over 50 and climbing. The small innermost moons are intimately associated with the rings. One, Pan, orbits in Encke's Division in Ring A. Another, Atlas, patrols the outer edge of Ring A. Two of the moons, Prometheus and Pandora, orbit either side of the F Ring, shepherding its particles.

Slightly farther out from Saturn, Janus and Epimetheus move along the same orbit, which initially confused astronomers who first spotted them from Earth in 1966. Orbit-sharing is common among Saturn's moons. Tethys, a satellite visible from Earth, has two small siblings, Telesto and Calypso, moving on the same path. Dione, visible from Earth, shares its orbit with tiny Helene, another Voyager discovery.

The gravitational effect of some of these satellites helps maintain the gaps in Saturn's rings. For instance, the gravitational tug of Mimas pulls particles out of the Cassini Division. Mimas itself, like most of Saturn's moons, is a dirty snowball of frozen water and rock. It sports a remarkable crater 135 km wide, larger than the crater Copernicus on the Moon and fully a third of its own 400-km diameter. The impact that caused this grotesque feature, called Herschel, must have nearly shattered Mimas.

Iapetus, Saturn's third-largest moon, is another strange body: one side is five times darker than the other. This harlequin effect is probably caused by dust knocked off a more distant moon, Phoebe, which is the darkest of Saturn's moons. The dark dust from Phoebe falls inwards to be swept up by Iapetus, coating its leading face while bright ice on its trailing side remains exposed.

The largest of Saturn's moons, Titan, 5150 km in diameter, deserves to be considered a planet in its own right. Bigger than Mercury (but slightly smaller than Jupiter's main moon Ganymede), it is the only moon to have a substantial atmosphere – in fact, the atmospheric pressure on the surface of Titan is 50 per cent greater than at sea level on Earth. But being so far from the Sun, Titan's surface temperature is very low, about −180°C.

Titan's atmosphere is 90 per cent nitrogen, with methane making up most of the rest.

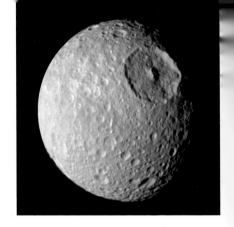

Mimas, one of the smaller moons of Saturn, sports a huge crater with a central peak, formed by a major impact. The crater is called Herschel, after William Herschel, the astronomer who discovered Mimas in 1789. The rest of the moon's surface is saturated with smaller craters. This photograph was taken by the Cassini spacecraft in 2005. (NASA/JPL/Space Science Institute)

Clouds of orange smog blanket its surface from view at visual wavelengths, but it has been possible to map most of Titan at infrared wavelengths and by radar. In January 2005 the Huygens probe parachuted down to the surface of Titan, landing on an area covered with frozen water and dotted with icy pebbles. It is thought that methane rain falls on Titan from time to time.

• Charts showing the positions of Saturn over a five-year period can be found at www.collins.co.uk/starsandplanets

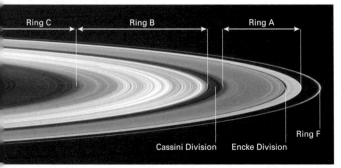

Saturn's rings as seen by the Cassini probe. From left are the faint Ring C; Ring B (the brightest section); the Cassini Division, with faint ringlets within it; Ring A with the narrow Encke Division towards its rim; and the thin F Ring outermost. (NASA/JPL/Space Science Institute)

Uranus, Neptune and beyond

Uranus should in theory be visible to the naked eye: at its brightest it is of magnitude 5.5. But it is so insignificant that it was never noticed by ancient astronomers, for whom the Solar System stopped at Saturn. Uranus was not discovered until March 1781, when William Herschel spotted it through his telescope during a systematic survey of the skies.

Once its position is known, Uranus can easily be followed in binoculars as it moves against the background stars. There is a certain fascination in seeing the blue-green disk of this distant planet through a telescope, but even with large apertures Uranus displays no detail.

Uranus is one of the four 'gas giants' of the outer Solar System, the others being Jupiter, Saturn and Neptune. Uranus has an equatorial diameter less than half that of Saturn but four times larger than the Earth. It lies 2900 million km from the Sun, 19 times farther than the Earth. Seasons pass slowly on Uranus, for the planet takes 84 years to

The green, featureless disk of Uranus, with its encircling rings, photographed by Voyager 2 in 1986. The brightest ring is the outermost one, the Epsilon ring. Background stars and moons can also be seen. (Erich Karkoschka/NASA)

complete one orbit, and they are extreme, for Uranus appears to have been knocked over onto its side: the planet's axial tilt is 98°, meaning that its axis of rotation lies almost in the plane of its orbit.

As a result, every 42 years one of the poles of Uranus is pointing towards the Sun, while its opposite pole is in darkness for decades. In between times the planet's equatorial region faces sunwards. During each 84-year orbit the Sun can appear overhead at every latitude on Uranus, something that never happens on any other planet. No one knows why Uranus should be lying on its side in this unique way; perhaps it suffered a collision with another large body long ago.

Uranus is celebrated for another reason: it was the second planet discovered to have rings. In 1977, astronomers watched as Uranus passed in front of a star. Unexpectedly, they noticed that the star winked on and off a number of times before and after it was obscured by the disk of Uranus, from which they deduced that Uranus was encircled by nine faint rings. The existence of these rings was confirmed by the Voyager 2 space probe which flew past the planet in January 1986. Since then the rings have been photographed from Earth at infrared wavelengths and the known number has increased to 13.

The rings are narrow, mostly only a few kilometres wide, separated by broad gaps. They lie between 12,000 and 72,000 km

URANUS DATA	
Diameter (equatorial)	51,118 km
Mass (Earth = 1)	14.54
Mean density (water = 1)	1.27
Volume (Earth = 1)	63
Axial rotation period (sidereal)	17.24 h
Orbital period (sidereal)	84.02 y
Orbital period (synodic)	369.66 d
Axial inclination	97°.77
Mean distance from Sun	2871.0×10^6 km
Orbital eccentricity	0.047
Orbital inclination	0°.8
Number of moons	27

NEPTUNE DATA	
Diameter (equatorial)	49,528 km
Mass (Earth = 1)	17.15
Mean density (water = 1)	1.64
Volume (Earth = 1)	58
Axial rotation period (sidereal)	16.11 h
Orbital period (sidereal)	164.8 y
Orbital period (synodic)	367.49 d
Axial inclination	28°.33
Mean distance from Sun	4498.3×10^6 km
Orbital eccentricity	0.009
Orbital inclination	1°.8
Number of moons	13

above the cloud tops of Uranus, and are composed of dust and other debris from tiny moonlets that orbit among them.

In addition to the rings, Uranus has five large moons visible from Earth: Miranda, Ariel, Umbriel, Titania and Oberon, ranging in diameter from 500 to 1500 km. Eleven smaller moons were discovered by Voyager 2 and other, more distant ones have since been discovered from Earth. The inner moons, like the rings, move in nearly circular orbits around Uranus's crazily tilted equator, while the outer ones have elliptical and inclined orbits, suggesting they have been captured since the planet's formation.

Even to the close-up gaze of Voyager, the gaseous surface of the planet itself was disappointingly bland, with scarcely any cloud features. In structure, Uranus is believed to have a rocky core surrounded by a mantle of ice, topped by an atmosphere of hydrogen and helium with about 2 per cent methane, which gives the planet its greenish colour.

Neptune, the outermost giant

Neptune, the outermost major planet of the Solar System, is in some ways a twin of Uranus. It is slightly smaller and shows the same featureless disk through amateur telescopes, with a stronger blue-green tinge because there is more methane in the atmosphere above its clouds. Magnitude 7.8 at its brightest, Neptune is far too faint to be seen with the naked eye, but can be followed in binoculars when one knows where to look.

Unlike the accidental discovery of Uranus, Neptune's existence was predicted in advance. Astronomers found that Uranus was not keeping to its expected course, and some suspected that it was being pulled by the gravity of an as-yet-unseen planet. In France, mathematician Urbain Le Verrier calculated the new planet's position in 1846, beating the Englishman John Couch Adams who had been working on the same problem. In September that year, astronomers at the

The icy blue of Neptune's clouds is disturbed by the Great Dark Spot, an anticyclone the size of the Earth, fringed with white methane cirrus. This photograph was taken by the Voyager 2 probe in August 1989. (NASA)

Berlin Observatory found Neptune close to Le Verrier's predicted position.

Neptune crawls around the Sun in 165 years at an average distance of 4500 million km, 30 times farther away than the Earth. Being so distant from the Sun, it is very cold and dark. Unlike bland Uranus, Neptune has markings in its clouds more reminiscent of

Jupiter. One such feature called the Great Dark Spot, similar in nature to Jupiter's Great Red Spot, was revealed by the cameras of Voyager 2 when it reached Neptune in August 1989. This spot gradually drifted towards the planet's equator and had disappeared by the time the Hubble Space Telescope looked at Neptune in 1994, although similar spots have appeared since.

As with the other giant planets, Neptune's moons are as interesting as the planet itself. The largest of its 13 known satellites, Triton, is 2700 km in diameter, over three-quarters the size of our own Moon. It has one of the coldest known surfaces in the Solar System, −235°C, and is covered with frozen nitrogen and methane. Geysers of nitrogen or methane gas spout from beneath the surface, leaving dark streaks on the bright ice.

Triton is in a retrograde (east to west) orbit around Neptune which suggests that it was captured. Tidal forces from Neptune mean that Triton's orbit is gradually shrinking, so that the moon will spiral closer to the planet until it is broken up in the distant future. A shattered Triton will form a far more substantial set of rings around Neptune than the faint, narrow ones that currently encircle it. The most prominent of Neptune's rings are named Galle, Le Verrier, Lassell, Arago and Adams, after those involved in the discovery and early study of the planet.

Trans-Neptunian objects

Beyond Neptune lies a swarm of icy bodies known collectively as trans-Neptunian objects (TNOs). The first of these was discovered in 1930 by Clyde Tombaugh at the Lowell Observatory in Arizona as part of a search for a new planet, and was named Pluto. Pluto was regarded as the ninth planet of the Solar System until 2006, when the International Astronomical Union reclassified it as a *dwarf planet*.

Pluto is far smaller than any of the eight major planets, being under 2400 km in diameter, smaller than Neptune's largest moon, Triton. Even less planet-like is Pluto's eccentric orbit which crosses the path of Neptune, as happened between February 1979 and February 1999; no other major planet crosses another's orbit, although many asteroids do.

During the 1990s astronomers began to discover many other trans-Neptunian objects; over a thousand are now known. One of these, Eris, discovered in 2005, is larger than Pluto. Eris and Pluto are both now classified as dwarf planets, a category which also includes the largest member of the asteroid belt, Ceres. Possibly more dwarf planets will be added to the list as further large TNOs are discovered.

Pluto orbits the Sun every 248 years at an average distance nearly 40 times that of the Earth. At 15th magnitude, it is difficult to spot, let alone study. Eris, even farther and fainter, takes over 500 years to complete one orbit.

Pluto has three known moons and Eris has one. By far the largest of them is Pluto's main moon, Charon, discovered in 1978, which is half Pluto's size. Charon orbits Pluto every 6.4 days, the same time that the planet takes to spin on its axis. Charon therefore hangs over one spot on the surface of Pluto, visible permanently from one hemisphere of the planet but invisible from the other. In close-up, Pluto's surface probably looks much like that of Triton.

Pluto's largest moon, Charon, is shown above and below the planet in these two images taken three days apart. (Gemini Observatory)

Comets and meteors

Comets are insubstantial bodies, loosely knit assemblages of frozen gas and dust that loop around the Sun on highly elongated orbits. They return to the inner Solar System at intervals ranging from a few years to many thousands of years, becoming visible to us on Earth as ghostly, glowing apparitions for a few weeks or months before receding back into distant obscurity.

When far from the Sun, a comet shines only by reflecting sunlight. At that stage it is small – usually no more than a few kilometres across – and faint. Approaching the Sun, the comet warms up, turning its icy surface into gas. Under the influence of the Sun's radiance the gases of the comet begin to fluoresce, in similar fashion to the gas in a neon tube, thereby considerably increasing the comet's brightness.

Halley's Comet pursues an elliptical path around the Sun, steeply inclined to the orbits of the planets. It travels from between the orbits of Mercury and Venus out to beyond Neptune and back again every 76 years. (Wil Tirion)

Gas and dust released from the warming comet produce a halo or *coma* 100,000 km or so in diameter. At the centre of the coma is the *nucleus*, the only solid part of the comet, consisting of a 'dirty snowball' of ice, dust and perhaps some rock. In a large comet the nucleus may be a few tens of kilometres across, but most are only a kilometre or so wide. Well over a thousand million comet nuclei would be needed to equal the mass of the Earth.

Not all comets develop tails, but many do. A typical tail stretches for 1°–2° (two to four Moon diameters) but in exceptional cases can be much longer. Comets are brightest – and their tails longest – near perihelion, so they are usually most prominent shortly after sunset or before sunrise when they are inevitably low down in the twilight.

One part of a comet's tail consists of gas blown away from the comet's head by the solar wind of atomic particles streaming from the Sun. The other part of the tail is made up of dust particles liberated from the head. The gas tail appears bluish in colour because of the fluorescing of the gas molecules,

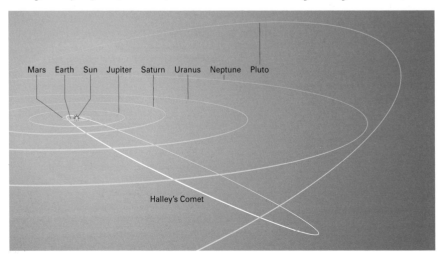

Mars Earth Sun Jupiter Saturn Uranus Neptune Pluto

Halley's Comet

whereas the dust tail is yellowish because it shines by reflecting sunlight.

Comet tails always point away from the Sun. Gas tails are mostly straight but dust tails tend to fan out more as the dust particles lag behind the comet. Some comets seem to have an *anti-tail* pointing towards the Sun, but this is a perspective effect resulting from the curving dust tail being seen from an unusual angle.

A comet's tail can extend for 100 million km or so, farther than the distance from the Earth to the Sun, as did the gas and dust tails of of Comet Hale–Bopp in 1997. Yet, for all its glorious appearance, the tail is less dense than a laboratory vacuum – stars shine through it undimmed. The tail of a comet gives it the appearance of speeding across the sky, but actually its movement against the stars is barely noticeable during the course of a night.

Being so large and diffuse, comets are best observed with binoculars and wide-field telescopes. Draw the comet in negative with a soft pencil on white paper. Record the shape of the coma, its central condensation and any tail. Plot the brightest background stars

Comet Hale–Bopp seen over Stonehenge in April 1997. Cassiopeia is at upper right and the Pleiades at upper left. The comet previously appeared some 4200 years ago, about the time that Stonehenge was built. (Paul Sutherland)

so that the dimensions of the coma and tail can later be measured from a star atlas.

Comet magnitudes are difficult to estimate. One way is to defocus your binoculars or telescope and compare the image of the comet with that of nearby stars. This technique works well only with brighter, condensed comets. For more diffuse comets, fix the in-focus image of the coma in your mind before defocusing your instrument, and compare this image with the defocused images of nearby stars.

Each year's batch of comets is a mixture of known specimens returning to the Sun and completely new discoveries. Dozens may be visible through an amateur telescope each year, and several can come within the range of binoculars, but only occasionally does one become bright enough to be prominent to the naked eye. Over 2500 comets have well-known orbits, and more are being discovered all the time. Dedicated amateur astronomers sweep the skies to discover new ones; each new comet is given its discoverer's name.

Comets with the longest orbital periods – more than a few centuries – are thought to come from an unseen swarm of thousands of millions of comets, the Oort Cloud, that envelops the Solar System at its dim outer edges, about a light year from the Sun. The gravitational influence of passing stars nudges comets from this cloud into new orbits that bring them in towards the Sun.

An inner comet cloud, termed the Kuiper Belt, lies just beyond the orbit of Neptune. Most so-called *periodic comets*, which have orbital periods of a few hundred years or less and have been observed on more than one return, are thought to come from the Kuiper Belt rather than the Oort Cloud.

The comet of shortest known period is Encke's Comet, which orbits the Sun every 3.3 years. It is so old that it has lost most of its gas and dust, and is too faint to see with the naked eye. Most famous of all is Halley's Comet, named after the English astronomer Edmond Halley, who calculated its orbit in 1705. Halley's Comet returns every 76 years or so, and last appeared in 1985–6. Its orbit takes it from 88 million km from the Sun (between the orbits of Mercury and Venus) out to 5300 million km (beyond Neptune).

Meteors and meteor showers

Dust lost from a comet disperses into space. The Earth and other planets are continually sweeping up cometary dust. When a particle of cometary dust comes whizzing into the atmosphere, it burns up by friction at a height of about 100 km, producing a streak of light known as a shooting star or meteor. The whole event is over quite literally in a flash, usually lasting less than a second. On

Members of a meteor shower appear to diverge from a small area of sky termed the radiant. This diagram shows the radiant of the Lyrid meteors, which lies in the constellation Lyra, near the bright star Vega. (Wil Tirion)

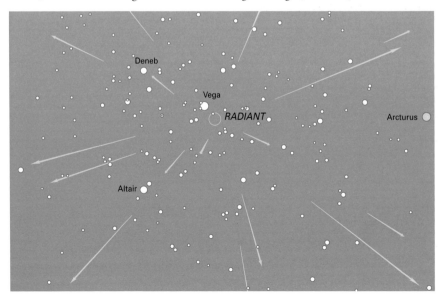

any clear night, a few meteors are visible each hour as particles of dust dash at random to their deaths in the atmosphere.

Such random meteors are termed *sporadic*. Occasionally, though, the Earth crosses the orbit of a comet and encounters a dense swarm of dust. This gives rise to a so-called meteor shower, in which meteors may be visible coming from one direction in the sky at the rate of dozens per hour.

The area of sky from which the meteors seem to come is known as the *radiant*. However, this is not the best point to look at, since the meteors are approaching head-on from this direction and so their trails are shortest; the longest trails will be seen up to 90° from the radiant.

A meteor shower is named for the constellation in which the radiant lies. For instance, the Perseids, an abundant shower of bright meteors which the Earth encounters each August, seem to radiate from Perseus; the Geminids from Gemini; and so on. One historical oddity concerns the Quadrantids, which come from an area in Boötes that was once part of the now-defunct constellation of Quadrans Muralis, the wall quadrant.

The strength of a meteor shower is measured by its *zenithal hourly rate* (ZHR), which is the number of meteors that an individual observer would see if the radiant were directly overhead in a dark sky. Since the radiant is seldom, if ever, at the zenith, the number of meteors actually seen per hour will be less than the theoretical ZHR. In addition, bright moonlight and sky glow from artificial lights will wash out the fainter meteors, again reducing the observed ZHR.

Amateur astronomers make valuable observations of meteor showers, counting the number of meteors visible with the naked eye and estimating their brightness. Comfort is essential for a meteor watch, which may last several hours: wrap up warmly and recline in a deckchair or on a sunbed.

Typical meteors are of magnitude +2 or +3, but some are brighter than the brightest stars and the occasional spectacular example, termed a *fireball*, can cast shadows. Some meteors seem to split up as they fall, and some leave a train of glowing gas which takes several seconds to fade. The table shows the main meteor showers visible each year. The ZHR is only a guide, and can vary considerably from year to year.

One extreme case is the Leonids, normally a modest shower, which bursts into life at 33-year intervals when its parent comet, Tempel–Tuttle, returns to perihelion. An intense storm of Leonids was seen over the United States in 1966 with as many as 100,000 in an hour, like celestial snowflakes, while rates of up to 4000 an hour were seen in 1999, 2001 and 2002.

MAJOR METEOR SHOWERS

Shower	Limits of activity	Date of maximum	Maximum rate (ZHR)
Quadrantids	January 1–6	January 3–4	100
Lyrids	April 19–25	April 21–22	10
Eta Aquarids	May 1–10	May 5	35
Delta Aquarids	July 15–August 15	July 28–29	20
Perseids	July 23–August 20	August 12–13	80
Orionids	October 16–27	October 20–22	25
Taurids	October 20–November 30	November 4	10
Leonids	November 15–20	November 17–18	10
Geminids	December 7–15	December 13–14	100

Asteroids and meteorites

Between Mars and Jupiter orbits a belt of rubble known as the asteroids or minor planets. Their existence was unknown until 1801, when the Italian astronomer Giuseppe Piazzi discovered Ceres, the largest of them, although astronomers had previously speculated that an unknown planet might exist in the suspiciously wide gap that separates Mars and Jupiter.

Well over 100,000 asteroids are now known, and there are estimated to be at least a million of them of 1 km diameter or larger in the asteroid belt. However, even if all the asteroids were rolled together, they would make a body less than half the size of the Moon. The asteroids are not, as was once suggested, the remnants of a former planet that was disrupted; they are merely the leftovers from the formation of the other planets, although many asteroids have been broken into smaller pieces by collisions.

Ceres itself is 940 km in diameter and takes 4.6 years to orbit the Sun. In 2006 the International Astronomical Union included Ceres in a new category of objects called *dwarf planets* on account of its size and the fact that it is rounded in shape. Ceres is composed of dark rock and, despite its size, is not the brightest asteroid. That honour goes to Vesta, 530 km in diameter, which is composed of paler rock; at times it is just

visible to the naked eye. Ceres, Vesta, Pallas and several other asteroids are bright enough to be followed in binoculars as they move from night to night against the background stars. If you are uncertain which is the asteroid, draw the star field and look again the following night.

Our first close-up view of an asteroid came in 1991 when the space probe Galileo, on its way to Jupiter, flew past the asteroid Gaspra, showing it to be an irregularly shaped, rocky body about 17 km long, pitted with impact craters. Gaspra resembles the moons of Mars, Phobos and Deimos, thus strengthening the suspicion that those moons are captured asteroids (see photograph on page 361). Asteroids can have their own moons, as was confirmed in 1993 when Galileo flew past asteroid Ida, 55 km long, photographing a tiny companion, now named Dactyl.

More recently, extraordinary views of the highly eroded surface of the irregularly shaped asteroid Itokawa, 500 m long, were obtained by the Japanese probe Hayabusa (see photograph below). The probe also took samples of the asteroid for return to Earth.

Ninety-five per cent of asteroids orbit in the main belt between Mars and Jupiter, but there are some notable exceptions. One group of asteroids, known as the Trojans, moves along the same orbit as Jupiter. Most

Asteroid Itokawa, seen from the Hayabusa spacecraft in 2005, resembles a pile of rubble, as though it was formed from the fragments of a former asteroid that was broken apart by a collision. (JAXA)

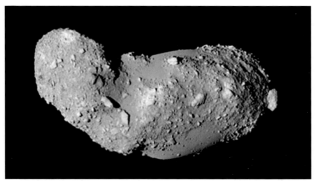

significant from our point of view are the near-Earth asteroids, which have orbits that bring them close to or even across the path of the Earth. Some of these objects may be the nuclei of extinct comets.

Asteroids of this type must have hit the Earth in the past, and others may do so in future, with devastating effect. Searching for such potentially hazardous asteroids is now a major international project.

Meteorites and meteorite craters

Objects that do reach the surface of the Earth are termed meteorites. Most meteorites are believed to be chips off asteroids, but a few rare specimens have compositions that show they come from the Moon and Mars.

Over 10,000 meteorites are estimated to land on Earth each year, but only about a dozen are actually picked up; the rest fall in uninhabited areas or into the sea. Most meteorites observed to fall are of the stony kind, and if not recovered immediately they soon weather away – or at least they do under normal conditions. But in Antarctica, scientists have found large numbers of ancient meteorites, including many rare types, pre-served in pristine condition by the natural deep-freeze of the ice cap. Another rich source is the deserts of Earth, where the arid conditions preserve the meteorites.

Stony meteorites are usually termed *chon-drites*, because they contain mineral-rich blobs known as chondrules. The small per-centage of stones without chondrules are called *achondrites*. The most interesting class of stony meteorite is the *carbonaceous chon-drites*; these have a high carbon content, and are believed to be among the most primitive rocks in existence, virtually unchanged since the formation of the Solar System. Some carbonaceous chondrites may be fragments of comets.

A small intermediate group of meteorites is the stony-irons, roughly half-and-half iron and rock. But the other main group of

meteorites apart from the chondrites is the irons. These contain 90 per cent iron and about 10 per cent nickel.

The largest known meteorite, the Hoba West meteorite, is an iron, and weighs about 60 tonnes. It must have hit the Earth at a relatively slow speed, for it did not dig a crater and did not break up. It lies where it fell in ancient times, near Grootfontein in Namibia. By contrast, the largest stony meteorite weighs 1.7 tonnes, and is one of a group of meteorites that fell near the city of Jilin, China, in 1976.

One particularly large meteorite, estimated to have weighed a quarter of a million tonnes, crashed into the Arizona desert about 25,000 years ago, digging out the now-famous Meteor Crater, 1.2 km in diameter. (Strictly, it should be called Meteorite Crater, but the old name is too firmly entrenched to change now.) Most of the meteorite was destroyed on impact, but enough fragments remain scattered around the crater to show that the meteorite was made of iron.

Over 150 meteorite impact sites are now known on Earth, including one 170 km in diameter at Chicxulub in Mexico which was formed 65 million years ago, at the time when the dinosaurs died out on Earth.

Meteor Crater, 1.2 km across and nearly 200 m deep, was formed by the impact of an iron meteorite in the Arizona desert about 50,000 years ago. Most of the meteorite was vaporized in the heat of the impact. (Ian Ridpath)

Astronomical instruments and observing

Binoculars and telescopes serve two main purposes: they collect more light than the human eye, and they magnify objects. In astronomy, the first of these two aspects is paramount. Astronomers seek large telescopes not because they magnify more, but because their increased light-gathering power brings fainter objects into view and allows finer details to be distinguished.

Telescopes come in two main types: refractors, which collect light through a main lens (known as the *object glass* or *objective*); and reflectors, which use a mirror. All the smallest telescopes are of the refracting type. Astronomical telescopes have interchangeable eyepieces to give different magnifications. Binoculars are a modified form of refractor in which the light path is folded by prisms to make them shorter and more portable.

Telescopes that combine lenses and mirrors are now very popular; these are known as *catadioptric* systems, and are best thought of as a modified form of reflector. Each type of instrument is dealt with in turn below.

Binoculars

The first optical equipment that almost every astronomer acquires is a pair of binoculars. Indeed, they are virtually indispensable for any observer, even those who own a substantial telescope. The advantages of binoculars are that they are portable, have a wide field of view and are relatively inexpensive.

Binoculars usually have a central focusing wheel that allows you to home in rapidly on objects of interest. One eyepiece should be individually adjustable to allow for differences in vision between your eyes. Avoid binoculars that can be focused only by adjusting each eyepiece separately as these are cumbersome to use. Zoom binoculars with a variable range of magnifications are made, but for best optical quality it is advisable to choose binoculars of fixed magnification.

Most binoculars used for astronomy incorporate what are known as Porro prisms to fold the light as shown in the diagram below. An alternative design uses so-called roof

Binoculars can be adjusted to suit the observer's eyes, using the following steps:

1. Alter the separation between the eyepieces by pivoting the two halves of the binocular around the central bar.

2. Turn the central focusing wheel until the image you see through the non-adjustable eyepiece is in focus.

3. Then turn the individually adjustable eyepiece to bring that eye into focus. Note the setting for future reference.

4. Turning the central wheel will then bring both eyepieces into focus at once.

Light emerges

Focus wheel

Eyepiece

Prisms

Objective lens

Light enters

prisms to provide a straight-through light path. While these are more compact they have smaller apertures and hence are less suitable for astronomy.

Binoculars bear markings consisting of two figures, such as these: 8 × 30, 8 × 40, 7 × 50, 10 × 50. In each case, the first figure is the magnification, and the second figure is the aperture (in millimetres) of the front lenses which gather the light. A magnification of 8 times, for example, means that the object appears 8 times larger than to the naked eye and hence 8 times closer.

There is a double price to be paid for magnification, though, and it applies to optical instruments of all kinds. Firstly, when an image is enlarged it becomes dimmer because the light is spread out over a larger area. So the greater the magnification, the more light is required to maintain an acceptably bright image. That can be obtained only with a larger collecting area – in other words, a wider aperture.

Secondly, as the magnification applied to a particular instrument goes up, its field of view shrinks. To get a satisfactory field of view and an acceptably bright image in bin-

oculars you should look for a ratio of at least 1:5 between the magnification and aperture.

Another factor to consider is that binoculars with powers greater than about ×10 are difficult to hold steady, because the image vibrates with each tremble of the hand; powerful binoculars need to be mounted on a stand, but this negates their portability.

Image-stabilizing binoculars are now made which use either microprocessors, gyros or a mechanical suspension system to compensate for hand tremors, but they are heavier and cost more than normal binoculars; an additional disadvantage of the first two varieties is that they require batteries.

Good binoculars have fields of view of 3°–6° (six to ten times the apparent diameter of the Moon) and give breathtaking wide-angle views of the heavens that telescopes cannot match. For comparison, the typical field of view of a telescope with a low-power eyepiece is about 1°.

Binoculars are ideal for scanning constellations and the Milky Way; observing variable stars, star clusters, nebulae and large galaxies; and comets, asteroids, the Moon and lunar eclipses – but *never* turn them on the Sun.

Choosing a telescope

Which telescope is best for you depends on the type of observing you intend to do, as well as your budget. Many amateurs start with small refractors of 50–60 mm (2–2.4 inches) aperture. These are sufficient to show details on the Moon and planets, to find the brighter galaxies and planetary nebulae, and to separate many well-known double stars. They are particularly suitable for projecting the image of the Sun, which is so bright that large apertures are unnecessary (but heed the safety precautions on page 302 and never look directly at the Sun through any form of optical equipment without an approved filter). Refractors are preferable for solar observing since the Sun's concentrated heat and light can damage a reflector's secondary mirror.

For larger sizes, the cost advantage of a simple Newtonian reflector usually outweighs all other considerations. Newtonians in the 150 mm (6 inch) to 250 mm (10 inch) range are commonly used by amateurs. Their only real drawback is that periodically the mirror needs its reflective coating renewed, although each coating should last many years.

If you intend to specialize in areas such as planetary observation or close double stars, the slightly better definition of a refractor may be worth the additional expense. If compactness and portability are the governing factors, then catadioptric telescopes have the advantage.

Whichever type of telescope you eventually decide on, remember that the bigger the aperture the more you will see, so get the largest aperture you can afford.

Refracting telescopes, also known simply as refractors, collect light through a lens and focus it to an eyepiece at the end of a long, narrow tube. Almost all telescopes with apertures smaller than 100 mm are refractors. In each of these diagrams, D is the telescope's aperture and F is the focal length of the main lens or mirror.

Reflecting telescopes, or reflectors, bounce light off a concave main mirror onto a secondary mirror near the entrance to the tube, which diverts the light into the eyepiece. In the Newtonian design shown here, the most common type used by amateurs, the eyepiece is in the side of the tube.

Catadioptric telescopes such as the popular Schmidt–Cassegrain design shown here have a thin lens at the front of the tube to improve the telescope's field of view. After passing through the lens the incoming light bounces off the main mirror onto a secondary and then back through a hole in the main mirror where the eyepiece is placed. (Wil Tirion)

Telescopes for beginners

A small telescope – one with an aperture of 50–60 mm (2–2.4 inches) – can cost several times more than a pair of binoculars, with little gain in light grasp. What telescopes do offer over binoculars is higher magnification and a mounting for support. Unfortunately, in some mass-produced telescopes the mounting is not as steady as might be wished, and the image can vibrate for several seconds each time the telescope is moved.

As a selling point, some small telescopes offer eyepieces with impressively high powers of over ×200. But such high magnifications

applied to a small telescope produce images so faint that little can be seen, and are therefore a complete waste.

As a good general rule, the maximum usable magnification on a telescope is ×20 for each 10 mm of aperture (×50 per inch). These pitfalls aside, there are many serviceable small telescopes available which will bring into view a wide selection of the celestial sights mentioned in this book.

The smallest telescopes are all refractors (top diagram above), which collect and focus light with a lens. For more ambitious observers, a telescope with an aperture of 75 mm (3 inches) aperture is the minimum required to begin serious work. Telescopes greater than 75 mm in aperture are usually of the reflecting type, because size for size they are cheaper to make than refractors. Popular sizes of reflectors are 150 mm (6 inch) and 220 mm (8½ inch) aperture, which should bring into view virtually all the targets for observation listed in this book.

Reflectors used by amateurs are usually built to the design invented in 1668 by Isaac Newton. In the Newtonian reflector, light collected by the concave main mirror is bounced back up the tube to a smaller secondary mirror, which diverts the light into an eyepiece at the side of the tube (centre diagram on previous page). Inevitably, the secondary mirror blocks some of the incoming light from the main mirror, but this shadowing effect of the secondary is not significant and does not adversely affect the image. An extreme form of Newtonian is the *rich-field telescope* or RFT, which uses a mirror of very short focal length and a low-power eyepiece to give a field of view of around 3°, not as good as binoculars but suitable for general stargazing.

Another design of reflecting telescope is the Cassegrain, in which the light is reflected back from the secondary through a hole bored in the main mirror where the eyepiece or other detectors can be mounted. Large professional telescopes commonly use the Cassegrain design or variants of it.

In *catadioptric* systems, which combine lenses and mirrors, the incoming light passes through a thin glass plate at the front of the telescope before falling onto the main mirror and being reflected as in a Cassegrain. The most commonly encountered variant in amateur astronomy is known as a Schmidt–Cassegrain telescope or SCT (bottom diagram on previous page, and at right).

The advantage of catadioptric designs is that, because the light path is folded back within the tube, the overall length of the telescope is much shorter than in a conventional reflector. The consequent saving in weight and space, allied with increased port-

A Newtonian reflector on an altazimuth mount, the simplest form of support for a telescope. The telescope is held in a cradle which allows it to rock up and down (in altitude) and swivel from side to side (in azimuth). (Wil Tirion)

A Schmidt–Cassegrain telescope on an altazimuth mount. In equatorial versions the base is tilted so that the arms of the fork holding the telescope tube point at the celestial pole. (Wil Tirion)

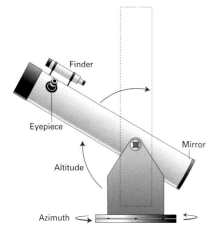

The Dobsonian is a popular type of altazimuth mounting for reflectors of all sizes where low-power viewing is all that is required. (Wil Tirion)

pivoted up and down (in altitude) and to pan from side to side (in azimuth). In an altazimuth mount, the telescope is usually held in a fork on top of a tripod, although in the case of small, lightweight telescopes the 'fork' can be reduced to a single arm.

Unless a refractor is mounted on a very tall tripod, the eyepiece will be too low down to look through comfortably when the telescope is pointed upwards. A common solution is to insert a *star diagonal* at the eyepiece end. Star diagonals use a mirror or prism to turn the light through 90° so that the observer can look downwards into the eyepiece, which is a more convenient observing position. For Newtonian reflectors, short tripods suffice since the eyepiece is near the top of the tube, and so is normally quite accessible from a standing position.

Convenient additions to an altazimuth mount are small knobs, known as slow-motion controls, which can be turned to move the telescope slightly in each axis. This

ability, compensates for the higher cost of a catadioptric telescope. Catadioptrics are particularly popular for astronomical imaging because cameras can be mounted behind the main mirror without unbalancing them.

As for cost, remember that a telescope is a precision optical instrument, and so you must expect to pay at least as much as you would for any similar instrument such as a good-quality camera.

Telescope mountings

A telescope requires a mounting on which it can be supported and swivelled to point at various places in the sky. The quality of the mounting is just as important as the quality of the optics, for even the best telescope cannot be expected to show much if it is shaking all the time, or if you have difficulty steering it to track objects as the Earth rotates. Sturdiness and smoothness of movement are all-important in a telescope mounting.

The simplest form of telescope mounting is the *altazimuth* (see diagrams on the facing page). This has two axes, one horizontal and one vertical, which allow the telescope to be

In a German-type equatorial mounting, one axis (termed the polar axis) is set up so that it is parallel to the Earth's axis and thus points towards the celestial pole; the other axis (the declination axis) is at right angles to it. (Wil Tirion)

is useful both for centring an object in the field of view, and for tracking it as the Earth spins. When high powers are in use, the Earth's rotation can carry an object out of the telescope's field of view in a remarkably short time.

A popular version of the altazimuth for large Newtonian reflectors is the *Dobsonian* mount, named after the American amateur John Dobson who devised it. In a Dobsonian, the telescope tube is made of lightweight material so that its centre of gravity lies near the mirror end. The tube is supported at its lower (mirror) end by a wooden box with a Formica base that swings around in azimuth on pads made of Teflon, the slippery plastic material used in non-stick frying pans. Formica glides smoothly on Teflon, making a simple yet steady bearing. The tube pivots up and down in altitude on a shaft supported by more Teflon pads (see the diagram at the top of the previous page).

Despite its deceptive simplicity, the Dobsonian has proved a very effective mounting. Cheapness and portability have ensured its widespread adoption by observers requiring wide-angle, low-power views of the heavens that do not involve precise tracking.

Best of all for tracking objects accurately as the Earth turns is an *equatorial* mount. In this, one axis, called the polar axis, is aligned so that it points directly at the celestial pole (either north or south, depending which hemisphere the observer is in). The sky appears to rotate around the celestial pole, so simply turning the polar axis will keep an object in view. A second axis at right angles to the first, called the declination axis, allows the telescope to pivot up and down (i.e. in declination).

There are several varieties of equatorial mounting, the most usual being the German type in which the telescope is counterbalanced by weights on the declination axis. In fork mountings as used by catadioptric telescopes the arms of the fork point towards the celestial pole, and the telescope tube pivots between them on the declination axis.

Which way round?

An astronomical telescope produces an image that is inverted top to bottom and reversed left to right by comparison with the naked-eye or binocular view. A 90° star diagonal, used on some telescopes to make observing more convenient, flips the image upright again but leaves it reversed left to right as in a mirror. These different orientations must be taken into account when trying to match up the view through a telescope with a star chart or a map of the Moon.

The Moon as seen with the naked eye or through binoculars

The Moon as seen through an astronomical telescope: upside down and reversed left to right

The Moon as seen through a 90° star diagonal: upright but still reversed left to right

Equatorial mounts are usually fitted with a small motor that slowly turns the equatorial axis at exactly the same speed as the Earth rotates. Once the telescope is pointed at a celestial object and the drive motor is running, the object will remain fixed firmly in the field of view for as long as the observer wishes, essential for high-power observing or photography. Experience soon demonstrates the desirability of a stationary image if you are trying to split a faint double star or make a drawing of a planet.

To help with locating objects, a smaller sighting telescope called a *finder* is usually mounted on the side of the main tube. Simple finders for small telescopes consist of a red dot from a light-emitting diode (LED) which reflects off a small glass window; the telescope is lined up by superimposing the dot on the required star. For larger telescopes, small low-power refractors are used as finders. Cross-wires in its eyepiece like the telescopic sights of a rifle, or sometimes a concentric series of faintly illuminated rings, allow the user to centre the target.

Computer-controlled telescopes are now widely available that will automatically align themselves on the sky and then point to any required object, selected from a list of many thousands whose positions are programmed into a handset. Such GOTO telescopes, as they are commonly termed, can also be controlled by a home computer with suitable software. Unlike traditional telescopes the altazimuth mounts of GOTO telescopes are motorized for tracking objects, although the accuracy is not as good as with an equatorial mounting.

Through the telescope

A few words need to be said about what you can expect to see with telescopes of different sizes, and why. Anyone looking through an astronomical telescope for the first time is usually surprised to find that the image is upside down. There is a simple practical reason for this: to turn the image the right way up, an extra lens would have to be inserted in the eyepiece.

Every time light passes through a lens or is reflected off a mirror, some light is lost. Celestial objects are usually so faint that any loss of light is undesirable. For astronomical purposes it does not really matter which way up the image appears, so the extra lens is left out and the image remains inverted. So-called terrestrial eyepieces or erecting prisms will turn the image the right way round for everyday viewing if required.

When you make an observation you should record the date, the time, the size of the instrument, the magnification and the sky conditions. Astronomers record dates in the form year/month/day. Times of astronomical events are usually given in Universal Time (UT), which is the same as GMT (Greenwich Mean Time). If you are in a different time zone from Greenwich and prefer to use your

The eye and observing

The human eye is an optical instrument in its own right which forms an image upon its rear surface, the retina. At night, the retina becomes much more sensitive to light, a process known as *dark adaptation*, but the increased sensitivity builds up only slowly. Hence, when going outdoors at night from a brightly lit room you should allow time for your eyes to adjust before beginning to observe.

The main change in sensitivity comes in the first ten minutes, but dark adaptation continues to improve for half an hour or more. Green and blue light rapidly destroys dark adaptation but red light does not, so when observers need to see to read or write at night they use a red light such as a torch with a red filter.

Images on the retina are sensed by two types of nerve endings, termed rods and cones on account of their shape. The central part of the retina consists only of cones, whereas rods are predominant in the outer regions. Rods are far more light-sensitive than cones. As a result, it is easier to see faint objects if you look slightly to one side of them so that the image falls on the rods in the outer regions of the retina. This technique is known as *averted vision*. Tapping the telescope so that the image gently vibrates is another way to make elusive objects easier to see.

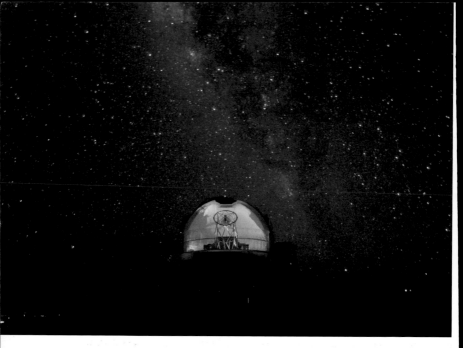

local time (e.g. Eastern Standard Time for the eastern seaboard of the USA) then be sure to note which time system you are using. Also remember to allow for the hour's difference when summer time (daylight saving time) is in operation.

Incidentally, your eyes need time to become accustomed to the dark before you can hope to see the faintest objects; when you go out at night allow at least 10 minutes for your eyes to become dark adapted. A useful trick when trying to glimpse faint objects is to use *averted vision* – that is, to look to the side of the object under study so that its light falls on the outer, more sensitive part of the retina.

Eyepieces and magnification

Unlike binoculars and spyglass telescopes, which have fixed eyepieces, the eyepieces in astronomical telescopes are interchangeable, offering a range of magnifications to suit the object under study. For instance, a star cluster or galaxy will require low magnification; planets are best viewed with medium magni-

The 8-m Gemini North reflector in Hawaii is one of the most powerful telescopes in the world. A series of exposures were taken as the dome was turned and combined to give the appearance of transparency. (Peter Michaud & Kirk Pu`uohau-Pummill/Gemini Observatory)

fication; and to split a close double star you will need your highest powers.

The magnification of an eyepiece depends on two factors: its own focal length and the focal length of the telescope. To work out the magnification of an eyepiece in a particular telescope requires no more than a little straightforward arithmetic: you simply divide the focal length of the telescope by the focal length of the eyepiece. The result tells you the power of that eyepiece.

You will soon realize that the shorter the focal length of the eyepiece, the higher its magnification. Eyepieces have their focal length marked on them in millimetres; a typical range of focal lengths is 6 mm to 40 mm. An additional lens, called a *Barlow lens*, can be inserted in the drawtube to increase the magnification of any eyepiece (usually doubling it), thereby offering a wider range of powers from a given set of eyepieces.

Manufacturers often describe their telescopes in terms of *focal ratio*, such as *f/*6 or *f/*8. The focal ratio is the focal length of the lens or mirror divided by its aperture. If you don't know the focal length of your telescope you can easily find out by multiplying its aperture by the focal ratio stated by the manufacturers. For instance, a 100-mm telescope with a focal ratio of *f/*6 has a focal length of 600 mm; if its focal ratio is *f/*8, its focal length is longer, at 800 mm. For a 150-mm telescope of *f/*6 or *f/*8, the focal length would be 900 mm or 1200 mm respectively.

Now, assume that you have an eyepiece of focal length 20 mm. In a telescope of 600 mm focal length it will give a magnification of 600 divided by 20, which is 30 times, usually expressed as ×30. That is quite a low power for astronomical purposes. In a telescope of 1200 mm focal length, the same eyepiece will give a magnification of ×60. An eyepiece of half the focal length, 10 mm, will give twice the magnification. Note that the aperture of the telescope is irrelevant in these calculations; the focal length is the sole figure governing magnification.

The most popular type of eyepiece for general use is the Plössl, named after its inventor, the Austrian optician Simon Plössl. Various other designs are used for wide-field views of the deep sky. Different designs of eyepiece have different fields of view, even though they may be of the same focal length.

A field of view is sometimes stated on the eyepiece itself, and is usually between about 30° and 80°, but this is the *apparent field of view* – it is not the field of view you will actually see, for the final result also depends on the focal length of the telescope. To find the true field of view for an eyepiece, divide the apparent field by the magnification it produces. For example, an eyepiece with 50° apparent field of view that magnifies 40 times will give a true field of 1¼°, just over twice the apparent diameter of the full Moon.

An easy way to measure your telescope's field of view is to time a star near the celestial equator as it crosses the field. The Earth rotates through 0.25 minutes of arc in 1 second of time so if the star takes 60 seconds to cross, then the field is $60 \times 0.25 = 15$ arc minutes wide.

Another feature of eyepieces to consider is their *eye relief.* This is, in effect, how close you have to place your eye to the eyepiece to see the entire field of view. Good eye relief means that you do not have to cram your eye right up against the eyepiece, so those who need to wear glasses while observing should choose eyepieces with good eye relief. Incidentally, if you are short- or long-sighted you do not need to wear glasses for observing because you can adjust the focus of the telescope or binoculars to compensate.

Light grasp and resolution

Where a telescope's aperture becomes crucial is in the matters of light grasp and resolution of detail. All other things being equal, a larger aperture shows fainter stars and finer detail than a smaller aperture, but exactly how faint the stars are and how fine the detail is depends on atmospheric conditions, optical quality and the observer's eyesight. Practical experience shows that the faintest stars likely to be visible through amateur telescopes of various apertures are as given in the table below.

TELESCOPIC LIMITING MAGNITUDES	
Aperture	**Limiting magnitude**
50 mm (2 inches)	11.2
60 mm (2.4 inches)	11.6
75 mm (3 inches)	12.1
100 mm (4 inches)	12.7
150 mm (6 inches)	13.6
220 mm (8.5 inches)	14.4

TELESCOPIC LIMITS OF RESOLUTION

Aperture		Limit of resolution (seconds of arc)
50 mm	(2 inches)	2.3
60 mm	(2.4 inches)	1.9
75 mm	(3 inches)	1.5
100 mm	(4 inches)	1.1
150 mm	(6 inches)	0.8
220 mm	(8.5 inches)	0.5

The resolution, or resolving power, of a telescope is expressed in seconds of arc (″). One second of arc is a very small quantity, equivalent to the size a coin would appear when several kilometres away. A telescope's resolution governs the detail that one can see on the Moon or planets, and the closeness of double stars that can be separated. The theoretical limits of resolution for telescopes of various apertures are shown in the table above. Under exceptional conditions, the limits of magnitude and resolution tabulated here may be surpassed; but in many cases, particularly when observing from cities, they will not be achieved.

Transparency and seeing

Finally, we come to the atmosphere itself. Professional observatories are sited on high mountain tops to get above as much of the atmosphere as possible, but most amateurs are stuck with conditions in their own back yard, all too often including urban haze and glare from streetlights. There are two aspects of the atmosphere to take into account: its clarity and its steadiness.

Atmospheric clarity is known to astronomers as the *transparency*. A good index of atmospheric transparency is the magnitude of the faintest stars that can be seen overhead by the naked eye. The zenithal limiting magnitude should always be noted when observing meteors, because it affects the

hourly rates. Transparency needs to be at its best when you are looking for faint objects, particularly nebulae, galaxies and comets.

On the other hand, the steadiness of the atmosphere, termed the *seeing*, is paramount for planetary and double star observation. Turbulence in the atmosphere produces a boiling effect of the image, drastically reducing the resolution. Hot air rising from neighbouring buildings can be a particularly annoying localized form of turbulence. Good transparency and good seeing do not always go hand in hand. On a crystal-clear night after a rainstorm the seeing can be particularly atrocious, whereas slightly misty nights, when the clarity is low, can be the steadiest.

Seeing is estimated on a five-point scale: 1, perfect; 2, good; 3, moderate; 4, poor; 5, very bad. This is sometimes known as the Antoniadi scale after the astronomer who devised it, Eugène Antoniadi. If a close double star cannot be clearly split on a night of indifferent seeing, examine it again on a better night. Transparency and seeing share one characteristic: they are always poorest closest to the horizon, so to get the brightest, steadiest image objects should be observed when they are as high in the sky as possible.

A useful addition to the modern astronomer's armoury is one of the special filters designed to enhance the visibility of faint objects such as nebulae, galaxies and comets. Such filters screw into the barrel of the eyepiece and come in two main types: one type, called Light Pollution Reduction or LPR filters, blocks out light pollution, thereby increasing the contrast of objects against the night sky; more extreme types, called nebula filters, block out all wavelengths except those which nebulae and comets emit most strongly.

Finally, one important but often overlooked part of an astronomer's equipment is warm clothing. Even on a seemingly mild night a tracksuit and training shoes make a practical outfit, while in the depths of winter you will need to dress like a mountaineer, with a quilted jacket, thick trousers, a hat, thermal socks and boots.

INDEX

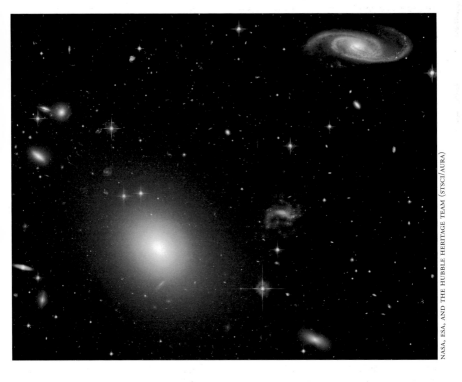

NASA, ESA, AND THE HUBBLE HERITAGE TEAM (STSCI/AURA)

Entries in this Index are listed in strict alphabetical order. All named stars of mag. 2.0 and brighter are included. Star designations such as α (alpha) Centauri and η (eta) Carinae will be found under the relevant constellations. Page numbers in **bold type** refer to photographs or illustrations.